Describing Construction

This third book from editors Rick Best and Jim Meikle brings together and presents insights into a number of key concepts in the study of construction firms, projects and the group of activities that loosely define the construction industry. The value for readers comes from the collection of a variety of topics in a single volume, which provide a basic understanding of the complexities of construction as more than a set of practical concerns such as labour management and materials handling. Instead, the focus is on analysis of the industry and its component parts from the viewpoints of construction economists and others seeking to understand the drivers and challenges that shape an area of economic activity that is a major contributor in all economies.

The aim of this book is to provide an overview and discussion of several aspects of what makes construction tick. It is unlike other industry sectors in many ways, being project-based with often intense competition for work. Where the first book, *Measuring Construction*, focused on particular areas associated with quantifying various aspects of construction activity and the second, *Accounting for Construction*, looked more at how we record and report on construction activity, *Describing Construction* gives readers the views of experts in the field of how the construction industry is described, what its make-up is, and it even asks the question: is construction a single industry? This book will change the way most readers understand the 'construction industry', whatever that may be, not from the point of view of visible on-site activities, but through a scientific approach to analysis and understanding of how projects, firms and various sectors of the industry work and how things are changing and may continue to change in future. It is essential reading for students and researchers in construction management, quantity surveying, architecture and engineering.

Rick Best is an Honorary Adjunct Associate Professor of Construction Management and Economics at Bond University. He has produced numerous book chapters and papers over a 25-year career as an academic as well as co-editing five books and co-authoring one quantity surveying textbook. His research in recent years focused on the problems associated with making valid comparisons of construction industries across countries, and he contributed to the development of a recent construction data collection project within the International Comparison Program.

Jim Meikle retired as a partner of Davis Langdon LLP in 2005 and, since then, has worked as an independent consultant with UK government departments and agencies, foreign governments, international organisations including the EU, the World Bank and the African Development Bank, and private clients. His main interests are international comparisons, construction industry policy and construction data.

Describing Construction

Industries, Projects and Firms

Edited by
Rick Best and Jim Meikle

Routledge
Taylor & Francis Group
LONDON AND NEW YORK

First published 2023
by Routledge
4 Park Square, Milton Park, Abingdon, Oxon OX14 4RN

and by Routledge
605 Third Avenue, New York, NY 10158

Routledge is an imprint of the Taylor & Francis Group, an informa business

British Library Cataloguing-in-Publication Data
A catalogue record for this book is available from the British Library

Library of Congress Cataloging-in-Publication Data
Names: Best, Rick, editor. | Meikle, Jim, editor.
Title: Describing construction: industries, projects and firms/edited by Rick Best and Jim Meikle.
Description: Milton Park, Abingdon, Oxon; New York, NY: Routledge, 2023. | Includes bibliographical references and index. | Summary: "Describing Construction gives readers the views of experts in the field of how the construction industry is described, what its make-up is, it even asks the question: is construction a single industry? This book will change the way most readers understand the 'construction industry', whatever that may be. It is essential reading for students and researchers in construction management, quantity surveying, architecture and engineering"- Provided by publisher.
Identifiers: LCCN 2022012749 (print) | LCCN 2022012750 (ebook) | ISBN 9780367608903 (hardback) | ISBN 9780367608910 (paperback) | ISBN 9781003102403 (ebook)
Subjects: LCSH: Construction industry. | Construction projects.
Classification: LCC HD9715.A2 D34 2022 (print) | LCC HD9715.A2 (ebook) | DDC 624.068 – dc23/eng/20220321
LC record available at https://lccn.loc.gov/2022012749
LC ebook record available at https://lccn.loc.gov/2022012750

ISBN: 978-0-367-60890-3 (hbk)
ISBN: 978-0-367-60891-0 (pbk)
ISBN: 978-1-003-10240-3 (ebk)

DOI: 10.1201/9781003102403

Typeset in Goudy
by Apex CoVantage, LLC

Contents

Contributors

Rick Best is an Honorary Adjunct Associate Professor of Construction Management and Economics at Bond University. He has produced numerous book chapters and papers over a 25-year career as an academic as well as co-editing five books and co-authoring one quantity surveying textbook. His research in recent years focused on the problems associated with making valid comparisons of construction industries across countries, and he contributed to the development of the most recent construction data collection project within the International Comparison Program.

John Connaughton is Professor of Sustainable Construction in the School of the Built Environment at the University of Reading. John has more than 40 years of experience in the construction sector and has particular expertise in sustainability in construction and the built environment, and in construction management and economics. Prior to joining the University in 2012, John was a partner in Davis Langdon, one of the world's largest construction cost and project management businesses. He was Chair of the Executive Board of the UK Construction Industry Research and Information Association (CIRIA) from 2014 to 2021.

Gerard de Valence taught construction economics in the School of the Built Environment at the University of Technology Sydney (UTS) from 1992 to 2020, and at the Bartlett School of Construction and Project Management at University College London from 2014 to 2020. At UTS he was the founding Course Director of the Master of Real Estate Investment (2015–17), Course Director of the Master of Property Development (2007–14) and the Master of Facilities Management (2005–07). Between 2002 and 2011 he was Coordinator of the research group on Construction Economics CIB W55. With Rick Best, he was co-editor of the three-volume *Building in Value* series of books published between 1999 and 2003.

Roger Flanagan is Professor at the School of Construction Management and Engineering, University of Reading, UK. He is Honorary Professor at Tsinghua University, Beijing. Roger's industrial experience includes previously being a member of the Board of Directors of Skanska AB, Sweden, and a director of consulting and construction companies in the UK, US, Canada and Switzerland. He is a past president of the Chartered Institute of Building.

Mark Hackett is a chartered quantity surveyor specialising in dispute resolution. He has been involved in a wide variety of construction schemes and has given evidence in numerous UK trials and international arbitrations. Mark is also the co-author of *The Presentation and Settlement of Contractors' Claims*, published by Routledge in 2001, and the co-editor of The Aqua Group Guide to Procurement, Tendering & Contract Administration, published by Wiley-Blackwell in 2016.

Inna Kolyshkina has a doctorate in mathematics and extensive experience in academia and consulting in the field of data science for many prominent organisations, including PwC and Westpac. She has authored many peer-reviewed publications and regularly presents at international conferences. She is Adjunct Professor at Western Sydney University, working on cutting-edge data science research. Previously, as Adjunct Professor at the University of South Australia, she helped develop a masters course in data science. Inna was the founding chair for the Institute of Analytics Professionals Australia and one of the pioneers of data science in Australia.

Alexandr Kulakov is Visiting Researcher at the Bartlett School of Sustainable Construction, from where he recorded a first-class BSc in project management in construction. His PhD research within the school builds on his undergraduate dissertation on the business models of contractors and developers and prediction of their insolvency risks. His ongoing doctoral thesis focuses on innovating methods of productivity measurement, linking micro, meso and macro datasets on the wider built environment sector and its industries. His use of extensive data on company financial accounts contributes to the empirical body of knowledge of industrial research.

Jim Meikle retired as a partner of Davis Langdon LLP in 2005 and, since then, he has worked as an independent consultant with UK government departments and agencies, foreign governments, international organisations including the EU, the World Bank and the African Development Bank, and private clients. His main interests are international comparisons construction industry policy, construction data and international comparisons.

Alex Murray is a built environment economist specialising in private financing of public infrastructure. Starting his career in KPMG, Corporate Finance, his research into operational infrastructure led to his PhD in whole life cost and value. As a researcher and lecturer at UCL's Bartlett School of Sustainable Construction, he produced the analysis for a 2013 government commission on supply chain finance in construction, published with the 2013 UK Government construction strategy. He was later a Senior Analyst at the Infrastructure and Projects Authority. Alex joined Preqin as Vice President in September 2020, where he specialises in coverage of unlisted infrastructure.

Pat O'Donnell has many peer-reviewed publications and regularly presents at industry and academic conferences. He has extensive experience in the construction industry as a project director and general manager working for a tier 1 construction contractor, with a strong track record of successful project delivery across multiple sectors. Now, as a director in a professional consulting firm, he leads a team of young professionals providing a full range of quantity surveying and project management services as well as preparing expert reports for disputes and litigation matters. He was a building contractor before becoming an academic, teaching building and construction at the (then) University of Western Sydney (now WSU).

George Ofori is a graduate of the University of Science and Technology, Ghana, and University College London, where he obtained his PhD degree and was also awarded a DSc degree. He is a Fellow of the Ghana Academy of Arts and Sciences and a Fellow of the Chartered Institute of Building, Royal Institution of Chartered Surveyors and Society of Project Managers (Singapore). His research interests include construction industry development, sustainable construction, leadership and professionalism and ethics. He has undertaken assignments advising many governments and international agencies. He is Dean of the School of the Built Environment and Architecture at London South Bank University.

Steve Rowlinson is Emeritus Professor in the Department of Real Estate and Construction, Faculty of Architecture, the University of Hong Kong. He also serves as Visiting Professor at Chongqing University, China, and at Bond University, Australia. Steve has over 40 years of international experience in teaching and consultancy in construction project management. His research areas include procurement systems and integrated project delivery, digital construction, construction innovation and occupational health, safety and well-being in construction. With over 250 publications and 50 PhD graduates to date, he is recognised worldwide as a leading academic in construction project management.

Hedley Smyth is Emeritus Professor in the Bartlett School of Sustainable Construction at UCL in London, UK. He has a long-standing teaching and research interest in the strategies of construction firms and organisational behaviour in the firms and at the interface with projects. Particular areas of recent research interest cover business models, decision making and relationship and interaction management for the co-creation of value. Substantive areas have covered business development and the sales process, knowledge management and health safety and well-being.

Preface

This book is the third in a series that presents a total of 36 chapters written by a range of academics and construction professionals with the aim of providing readers with a sound basis for further exploration of construction issues such as productivity measurement, the collection and dissemination of construction industry data and statistics and industry comparisons. The first in the series, *Measuring Construction*, explored matters related to measuring factors such as output, productivity and costs and prices; the second, *Accounting for Construction*, focused more on matters related to the ways that construction activity is recorded and analysed, with some emphasis on the bigger picture presented in national accounts and issues related to the broader industry such as the impact of the so-called shadow economy on measurements of construction activity.

In this book, as the title suggests, the emphasis is more on describing the industry (or industries), entities and activities that are typically grouped together, often in a very arbitrary way, and referred to as 'the construction industry'. A key topic, which has been addressed to some extent in the earlier books, is explored further here, and that is the fundamental question of exactly what is meant by the terms 'construction' and 'construction industry'. Other chapters provide food for thought on topics such as the education of construction professionals, the nature of construction in developing economies and the future of construction in a changing world.

We hope that this series of books will help students and practitioners to develop a better understanding of the construction sector beyond the practicalities of materials, methods and management which are so often the focus of the construction curriculum. Those practicalities are obviously important, but we believe it is good to complement that detail with a broader, more comprehensive view of the industry, or set of industries, that we know as 'construction'.

Note: In this book, as in its companion volumes, the word 'data' is treated as a collective noun and thus it takes the singular form as in '. . . the data is reliable . . .' rather than the somewhat archaic '. . . the data are reliable . . .'. While in Latin the singular is 'datum' and the plural 'data' this book is written in English, not Latin, and the editors are not aware of any common usage of the word 'datum' to describe or identify a single piece of data.

Rick Best – Gold Coast
Jim Meikle – Geelong
February 2022

1 Defining construction

Rick Best and Jim Meikle

Introduction

A typical dictionary definition of *construction* includes 'the act or a mode of constructing', with *construct* meaning to 'make by fitting parts together, build, form (something physical or abstract)' (OERD 1995: 309). Here, of course, the focus is purely on the construction of physical products, more specifically, buildings and other types of facilities and structures. The dictionary definition, however, establishes some boundaries on what 'construction' encompasses as it limits the discussion to the act (or, more precisely, the very large number of discrete acts, or activities, or work items, that contribute to the act) of physically putting materials and components together on building sites. In many areas of research it is necessary to break down complex systems into smaller, more manageable parts, and defining the limits of these parts is important as those limits must also limit the generality of results obtained.

Setting the boundaries

> *The construction sector contributes £117 billion to the UK economy, 6% of total economic output . . . there are 2.4 million jobs in the sector, 7% of UK total.*
>
> (Rhodes 2019)

> *The construction sector is one of the largest in the world economy, with about $10 trillion spent on construction-related goods and services every year.*
>
> (MGI 2017)

Statements such as these, the first taken from a UK government briefing paper and the second from a McKinsey report, are often quoted, but it is not always clear exactly what it is that is being described. Similar data for China suggests that construction represents 29 percent of gross domestic product (GDP) and 63 percent of Gross Fixed Capital Formation (GFCF), yet any direct comparison of the figures for the two economies must be done with caution. The reasons for adopting a cautious approach are discussed in this chapter; they relate mostly to the lack of consistency in the way that construction data is collected and reported around the world.

DOI: 10.1201/9781003102403-1

Rhodes (2019) defines the construction sector as including 'the development and construction of residential and non-residential buildings; construction work on civil engineering projects; and specialist construction activities (such as plumbing and electrical installation)'. While Rhodes talks of the construction 'sector', it is more commonly referred to as the construction 'industry' and that is the term that will be used here. Whatever its title, it is necessary to clarify what is included in the sector/industry, and what is not. Rhodes goes on to say that 'the sector does not include activities such as architectural services or project management which often accompany construction projects', and while that is helpful, in that it sets some boundaries on the industry under discussion, it is only a partial explanation of something that is much more complex.

Even the term 'construction' requires explanation as it is often used interchangeably with 'building', yet sometimes there is a distinction between the two, typically with construction being used to refer to engineering construction (e.g. infrastructure such as bridges and highways) while building refers to the erection of buildings: houses, office blocks, warehouses and so on. The picture becomes less clear when the term 'social infrastructure' appears as that is used as a descriptor for physical assets (including buildings such as schools and hospitals) that serve as infrastructure for society. In the following discussion the term 'construction' is used to include all types of building and construction, and the construction industry will be taken to include all sorts of building work and projects as well as the renovation, refurbishment, maintenance and demolition of existing structures.

The definition provided by Rhodes serves as a useful starting point for this discussion, and a couple of points are worth noting at the outset. He includes the development, and not only the construction, of residential and non-residential buildings and engineering projects but excludes key components of the development and construction phases, namely architectural and project management services. These exclusions raise further questions, for example, are other design services such as engineering design included or not? If architectural design is not in the picture, then it would be logical to also exclude structural and civil engineering. Similarly, if project management is ignored, does this include specialist cost management such as quantity surveying and cost engineering?

The complexities become more pronounced when we consider multidisciplinary firms that not only build but also provide design and/or project management services for construction projects. It is difficult to separate the value of design services provided in-house by a contractor who is running a design and build project, and even more difficult to get reliable data of that sort, so Rhodes' exclusion of architectural services presents some difficulties.

Rhodes' definition provides no real information on the breadth of the industry, particularly regarding the supply chains that sustain construction activity. There are countless businesses that supply materials, equipment and components to construction firms, and it is arguable that they should be included as part of the broader industry. de Valence (2019) discusses the concept of the broad and narrow construction industries, where the broad includes suppliers and designers as well as other contributors such as legal practices and urban planners, and the narrow is restricted to on-site construction activity. Estimates vary and percentages differ between countries and from year to year, but as an example, de Valence cites Ive and Gruneberg (2000), who suggested that of the total number of people in employment in the UK, those who were 'engaged in the production of the built environment' represented more than 11 percent when the broad industry was considered compared to less than 5 percent when the narrower view was taken. Obviously construction's contribution to GDP would also be much greater when the broad industry is compared to the narrow.

While it may not be impossible to analyse supply chains and thus to measure, or atleast make a reasonable estimate of the extent to which various supply chains contribute directly to construction output, the amount of effort required is likely to far outweigh any benefit obtained; while de Valence (2019) provides a more extensive discussion of the problem, a single example here will serve to illustrate the complexities.

The supply chain dilemma – an example

Aluminium is widely used in construction, but it is also used extensively in many other industries including the manufacture of, among other things, truck bodies and cookware. Aluminium extrusions are the basis of window frames and curtain walling used in buildings all over the world. Their production starts with the mining of bauxite and progresses through smelting, powered by large amounts of electricity, to the production of standard extruded sections. A significant proportion of aluminium production also includes recycled material, which adds more complexity to the supply chain question. Some, but far from all, of that production goes to fabricators who manufacture window frames and similar products which are then delivered to, and installed in, buildings of various types. Inputs to the total process include transport (in several stages), energy to drive the mining process as well as the smelting and so on – the question is, how much of the value of all these parts of the process should be considered to be part of the value of construction?

If we consider that the price paid for a product such as a window (or a hamburger, for that matter) represents an aggregation of input costs that includes rent of premises, transport, raw materials, labour, utilities and so on, plus a factor for overheads and profit, then the portion of total aluminium production and manufacturing that is attributable to construction may be covered in national accounts where the value of purchases by contractors is recorded. In many countries, input-output tables, based on national accounts, provide an analysis of industries' purchases from other industries. If it is not, or it is not covered in fine enough detail, then the task of making such a calculation is a challenging one, and it would need to be repeated for a very large number of construction materials, components and assemblies. The scale of the problem and the problems associated with collecting reliable and relevant data are likely to be too great to make such an exercise viable as the cost-benefit ratio is not likely to be favourable.

Firms and projects

There are other ways of looking at construction apart from trying to consider whole industries, as the construction industry (or sector) is made up of many firms of various types and sizes and they produce, collectively, a wide range of individual projects of differing type and scale. In developed economies it is usual to see contractors of varying sizes who contract with clients to carry out the construction of projects. Such contractors have different names in different places: head contractor, general contractor, prime contractor and main contractor are a few. They are generally categorised by size (and thus by their capacity to deliver projects of varying scale) from the very large international firms, e.g. Bechtel, Bouygues, down to small sub-contracting firms that have only a handful of employees, and in some cases may be just a single person who performs a very specific task, such as installing glass bricks in the cottage building sector, or gilding in heritage projects.

Almost all the on-site construction work is generally done by smaller specialist firms that are appointed by the head contractor. They contract with the head contractor and do not have a direct legal relationship with the client. The head contractor provides very few operatives on site apart from some supervisory and managerial personnel such as site managers and cost controllers plus foremen and some general labourers. Project and construction managers may be involved with more than one project at the same time, or they may be responsible for only some portion of the overall project. There is no standard framework or hierarchy in project teams, nor even standard nomenclature for the participants; people designated as project managers in one firm may be known as construction managers in another. In addition, clients can, and do, appoint their own project managers to control their projects on their behalf. The variations are endless with different project teams formed for different projects.

A *project-based industry*

Regardless of where we set the boundaries of a construction industry, there are several key characteristics of construction that make it different to other industries. Some parallels with manufacturing can be identified, but in many respects the two are quite different, and a key factor is the nature of the products that they produce. Unlike manufacturing, where the emphasis is often on the production of large numbers of similar, or even identical units, and production is an ongoing process, most construction output comprises an array of individual projects, each of which is more or less unique.[1] Most construction projects are individually designed to suit a specific purpose and to fit on a specific site, and the team of managers, designers and contractors (of all types) that realises the client's vision dissolves once the project is complete.

A *competitive environment*

Contractors obtain work either by competing against each other by tendering or bidding for projects or by negotiation or direct order. Whatever the process is, contractors exist in a very competitive environment and firms' survival depends on them being successful in their efforts to secure a regular flow of work. Maintaining a steady flow of work is a challenge given that every project is a separate undertaking and as projects are completed fresh projects are needed to keep the business going. On the other hand, firms that take on more work than they can handle can end up in trouble when their workload exceeds their capacity to carry out the necessary work, so it is something of a balancing act, particularly in view of the tendency for contracts to run over time and thus for forward time planning to be disrupted by delayed completion of projects.

Construction activity tends to be cyclical in developed economies, with periods of boom and bust as demand ebbs and flows over time. Competition for work, which is keen at the best of times, becomes very strong when activity is down, as contractors vie for the chance to build the smaller number of projects that are going ahead. During industry downturns, contractors cut their margins in the hope of winning jobs so they can maintain some cashflow and thus retain key staff and keep their firms afloat while they wait for better times. This is one factor that contributes to the adversarial nature of the industry which is evident in many countries. Having bid low in order to win jobs, contractors will look for ways to generate additional income on projects by looking for opportunities to claim for variations during construction. Even when the industry is buoyant, there are disputes about variations, extensions of time and the like, and the industry has a reputation for being a tough environment with disputes that require some form of organised dispute resolution (e.g. arbitration and experts conclave) and even litigation being unfortunately common. There have been many attempts to find better ways to run projects, such as partnering and joint venturing, but disputes between the participants, particularly between client and contractor, as well as contractor and sub-contractors, are still common.

Type of output

The World Bank and many other bodies (e.g. the Australian Bureau of Statistics) divide output into three *basic headings*: residential, non-residential and engineering construction. While this seems self-explanatory there is a catch, and that is in how high-rise apartment buildings are treated; a 50-storey apartment block has little in common with a one- or two-storey housing development and much more in common with high-rise office buildings, at least from a structural point of view. The major difference lies in the degree of internal fit-out: office blocks are typically *shell and core*, with little internal fit-out except in common areas such as bathrooms and lobbies, with office fit-out left to incoming tenants. In contrast, high-rise residential buildings will almost always be fully fitted out with floor coverings, complete kitchens and bathrooms, built-in cupboards and similar. Thus time and cost to complete may be similar for the structure and envelope of these buildings, but total cost and time will be greater for residential than for shell and core offices. High-rise hotel buildings present further differences in level and type of fit-out and in the amenities included, such as swimming pools and banquet/conference facilities.

Renovation, refurbishment, maintenance and repair

There is much construction work carried out that is not related to the creation of new buildings or infrastructure; indeed, in some countries (Italy is a good example) where there is an extensive stock of buildings that have been in use for many years, centuries and even millennia in a few cases, a substantial proportion of construction activity is devoted to work done on existing buildings. It is not just 'historic' buildings such as ancient churches and monuments that are the focus of such work, but also millions of dwellings that have been occupied for a long time, often for hundreds of years, with constant, if not regular, repair and renovation. For example, there are 28.5 million dwellings in the UK; 55 percent were built before 1964, 35 percent before 1944 and almost 20 percent before 1919 (Piddington *et al.* 2020). That means that there are around six million dwellings more than 100 years old. Most UK dwellings will have had windows, whole kitchens and bathrooms, heating systems, roof coverings and other elements replaced, some many times; and redecoration and minor repairs go on all the time. Much of this work is done by small firms and even individual tradespeople; some is done by the building owners themselves. Even in richer countries that have national statistical offices, data on the true value of such work is unlikely to be complete or comprehensive if it is collected at all. This adds yet another dimension to the problem of defining a construction industry or sector.

Such work is not restricted to places with large numbers of very old buildings; routine maintenance and repair of built assets is ongoing, and it may start quite soon after a building is completed. Shopping malls, which are now a feature of many cities around the world, include large numbers of individual tenancies, and the leases on these tenancies routinely require tenants to refit their shops every few years, regardless of their condition. The aim is to keep the malls looking immaculate at all times, as well as continually changing the look and feel, all with a view to maintaining public interest and engagement and thus to maintain maximum rental returns. Major refurbishment of malls occurs far less frequently but when it does occur, contract value can be on a par with newbuild. Construction research tends to be focused on new construction yet that represents only part of the totality of the work undertaken by the industry.

Informal construction

Much of the construction work, particularly the building of basic shelters (huts, cottages, barns, sheds and the like) carried out around the world, is done without planning permission or any sort of building contract, formal or informal. From the *favelas* of Brazil to the slums of Mumbai, makeshift dwellings are constructed of salvaged and scrap materials, with the building work being done by the occupants themselves, often with help from friends and neighbours. Everything about the process is informal, including the occupation of the land as such areas often begin with squatters taking possession of vacant land and simply moving in. Construction activity like this is seldom, if ever, recorded, and if it is included in national accounts it is on the basis of estimates of value which may or may not reflect reality. In truth, it is difficult to measure the value of a shanty that is built using largely unskilled, unpaid labour from materials that have been collected, salvaged or even stolen. There is no meaningful value of inputs nor any way to assess the value added in the construction process.

Not all informal construction, however, is makeshift or temporary. In many cities, housing is built that does not comply with planning or building regulations, but it is indistinguishable from formal housing in terms of design, materials and construction. Informal construction is undertaken in all kinds of locations, rich and poor; do-it-yourself (DIY) work is done by owners and occupants of dwellings even in the most affluent societies, and there is always the potential for work to be done with discounts for payments in cash which will not be reported as income by those doing the work. Attempts are made to estimate the value of such work, which forms a significant part of what is known as the 'shadow' or 'underground' economy; Chancellor *et al.* (2019a) discuss a variety of methods that have been used to estimate the value of construction within the shadow economy from data that exists in national accounts, but there is no method available that can accurately capture the value of informal building work in poorer countries where much of the population lives either on the land (e.g. rural India) or in depressed urban areas such as Soweto.

Construction productivity

The OECD (2020) gives this broad definition of productivity:

> *Productivity is commonly defined as a ratio between the output volume and the volume of inputs. In other words, it measures how efficiently production inputs, such as labour and capital, are being used in an economy to produce a given level of output.*

Productivity is of interest not only to governments, but to firms, as improved productivity can make them more competitive. While capital productivity is taken into account, it is principally labour productivity that is the focus of research, and attempts at measurement, in the construction context. The OECD (2019) defines labour productivity as 'gross value added per hour worked'. In truth, neither of these definitions is particularly helpful when trying to understand construction productivity at any level, especially above that of discrete on-site activities such as laying bricks. The purpose of productivity measurement is to establish benchmarks for comparisons in order to identify best practice and then to look for ways to match, or exceed, those benchmarks. In practice, however, measuring and comparing the time taken (which is the typical measure of labour input) to carry out basic construction activities such as placing concrete, laying tiles or hanging doors, is not as straightforward as it may first appear.

Once again, the heterogeneity of construction output is a complicating factor. Projects differ at all levels, which makes valid comparisons of like with like challenging. At the micro level of individual on-site tasks, differences in building size, design, access, layout, material choice, and surrounding environment can, and do, affect how efficiently work is done. For example, the most likely measure of productivity when laying floor tiles will be the area covered per unit of time, and that will depend on a number of factors, apart from the skill and/or speed of the person(s) doing the work. Even if it is only the amount of cutting of tiles that is considered, which is going to affect the time taken to do the work, then the type and size of the tiles used, the scale of the area to be covered, the regularity of the plan area of the work (e.g. are there non-right angles or circular work), and the incidence of cutting around other building elements such as columns, floor drains, stair wells and similar will have an impact on the time required to do the job.

Productivity measurement at this level, while it may be of concern to smaller firms such as specialist sub-contractors, is of minimal interest to larger firms and governments. Construction firms may compare productivity between projects that they complete, and try to identify best practice within their own organisation as a means of improving their performance on future projects, but it is not easy to make comparisons with the performance of their competitors. In any case, differences between projects can even make intra-firm comparisons problematic; on rare occasions it is possible to compare like with like, however, and the construction of the Petronas Towers in Malaysia is an example (Merchant 2020). In that case different contractors were appointed to construct each of the two essentially identical towers and work on the two structures progressed in parallel so there was an opportunity to compare the performance of the two construction teams, however, such opportunities are few and far between.

The OECD definition quoted earlier mentions 'output volume'; quantifying output is fundamental to the notion of productivity but in the context of construction it is not easy to define clearly. For a firm manufacturing toasters, the volume of output is reasonably easy to measure – counting toasters (or televisions, or cars or any other manufactured article) is largely a matter of recording how many are made, and typically productivity will be calculated on the basis of the number produced in a given time. When the output comprises buildings that take anywhere from three months to five years to complete, and which are more or less unique in terms of size, materials, purpose and so on, there are relatively few options for measuring the volume of output except where the product comprises roughly similar units such as houses of a standard design that has few variations from the base design. Early estimates of construction project cost are usually based on floor area (e.g. $/m^2) but when that is done different rates are required for different types and quality of output as, for example, the cost per unit area for a warehouse will be quite different to the rate for an office block or luxury hotel. Best and Meikle (2015) outline the challenges of measuring the volume of construction output and look at a variety of ways that it may be done. All have some shortcomings and often there is little alternative but to measure output in terms of value (i.e. cost).

There has been, for some time, concern that improvement in construction productivity in general lags well behind that of other industries. MGI (2017) states that labour productivity growth in the global construction sector has averaged as little as 1 percent per year over the past 20 years compared with 3.6 percent for manufacturing and 2.8 percent for the total world economy. In the USA, while productivity in some sectors (e.g. agriculture and manufacturing) has increased by as many as 15 times since 1945, construction productivity is said to have barely changed (MGI 2017). Some challenge this view of the stagnation of construction productivity, given that it is hard to imagine that in spite of significant developments in materials, tools and equipment, as well as widespread adoption of digital technologies in the past 20–25 years, that construction productivity has hardly improved.

Bernstein (2003) suggested that buildings had become more complex and therefore there was an unfounded perception that productivity had decreased when in fact the apparent decrease was the result of increased complexity and quality of the product; if that was true early in the century then it is probably even more relevant today. More recently, Chancellor *et al.* (2019b) put forward the idea that it may be the measures used to determine productivity levels that are the problem and that the industry could be experiencing productivity growth that is just not being detected. Again, this could be a function of increased complexity in building design. Another contributing factor is the impact of ever more stringent workplace health and safety regulations, particularly in industrialised countries; the days of steel workers running around on narrow beams high above the street are largely gone and the extra time required to work at heights, while attached by a safety harness, inevitably slows workers down. There is also additional cost and time associated with the supply, installation and removal of temporary safety fencing, enclosed scaffolding and a host of other measures that are now mandated on construction sites that were not required in earlier times. In countries (e.g. the UK) where industrial manslaughter is now a crime that carries lengthy gaol sentences, management can no longer turn a blind eye to non-compliance in regard to workplace safety, and there is extra cost to contractors with every increase in the demands of workplace safety law.

Industrialised building

There have been many attempts to make construction more like manufacturing through an increased focus on off-site manufacture (or prefabrication) of building parts, with spectacular results in some cases (see, for example, *The Guardian* 2015)[2].

Some building parts, such as structural steel frames and precast concrete components (e.g. wall panels, column sections, stair flights), have been manufactured in factories and fabrication yards and delivered to site ready for installation in the works, following the example provided by Paxton in the mid-19th century, whose Crystal Palace, erected in London's Hyde Park in just nine months in 1851, measured an impressive 560 × 125 m on plan by 22 m high. Its wrought and cast iron frame was manufactured in pieces, many of which were identical and could thus be produced quickly and in large quantities, away from the site, and then fitted together on site (Beaver 1970).

Early in the 20th century the Swiss-French architect, Le Corbusier, proposed mass production of housing based on his 'Dom-ino House' concept. His proposal involved the creation of houses using manufacturing methods to produce standard modules that could be utilised to create standard slab and column structures which allowed highly flexible interior layouts and a variety of eternal cladding types and variable floor plans. The model was a reflection of his vision for construction to be more like the manufacturing that had emerged from the Industrial Revolution rather than the craft-based industry that it had always been. Buckminster Fuller pursued similar ideas in his Dymaxion (1927) and Wichita House (1941) designs, although instead of reinforced concrete, Fuller applied the materials and manufacturing techniques of aircraft making to house construction.

Offsite fabrication of the structural steel framework was also a key factor that enabled completion of the 103-storey (380 m) Empire State Building in New York in a little over one year (410 days) in 1930–31 (Tauranac 1995). Windows and spandrels were similarly mass produced and delivered ready for installation.

Undoubtedly, there are potential advantages in the application of such methods, with on-site activity largely reduced to installing and connecting pre-made units ranging from framing members to complete bathrooms and pre-finished room units that come to site complete with all services installed and ready for connection. Fabricating units in a controlled factory environment improves quality control (e.g. closer tolerances, better finishes), no interruptions due to bad weather, and reduced instances of injury and death on sites due to shorter construction times and fewer people on site. The counter argument is that buildings created using industrialised methods are limited to repetitive modular units, which restricts what designers can do and thus the buildings produced are less interesting. Large-scale off-site manufacturing also requires up-front capital expenditure and guaranteed long-term demand to make it viable.

Building activity in developed economies now typically displays a combination of traditional methods (e.g. the on-site fabrication of made to measure formwork and the placement of wet concrete into those forms), use of modular systems (e.g. suspended ceiling systems based on standard pre-fabricated components that are cut to size during installation to suit specific layouts in individual projects), and the installation of bespoke components that are routinely fabricated to order off-site and delivered ready for installation (e.g. roof trusses and steel framing members) as well as standard manufactured components (e.g. pre-glazed metal-framed window units that are selected from catalogues of standard designs and sizes). There are, however, buildings with unique bespoke designs that are virtually 'hand-made' (e.g. some churches, high-cost houses) as well as simpler buildings (e.g. accommodation for mining camps and disaster recovery) that are mass-produced in a factory-like environment and transported in units that can be bolted together and connected to services, with little other on-site work required other than site preparation and some form of substructure (footings and so on) to support the buildings.

Construction in the digital age

The rapid developments in digital technology in the last 40 years have had major impacts on most businesses and industries and construction is no exception, although a visit to a construction project in 2020 shows that much of the on-site work has not changed a great deal apart from the use of power tools in place of traditional hand tools and greater reliance on equipment such as scissor lifts. Where the changes are more evident is in the design and management of construction projects through the adoption of technologies such as CAD (which originally stood for computer-aided design) in the latter decades of the 20th century, and more recently, BIM (building information modelling).

With the advent of the personal computer in the early 1980s, the move away from paper-based work (e.g. the traditional architect's drawing board) to computer-based work became available to most firms. Quantity surveyors were early adopters of technology that greatly improved workflow in routine activities such as the preparation of bills of quantities[3] while architects and engineers moved progressively to CAD-based systems for design and documentation of projects.

Computer-driven fabrication has followed from digital design, with machines that are controlled by computers rather than humans now able to mass produce complex building components comprising shapes that would once have been nearly impossible to make at all. Advanced CAD software that allows designers to visualise complex shapes can also instruct laser cutters and 3D printers to produce designs of great complexity with compound curves and even structures inside other structures that would be virtually impossible to make using traditional methods. There are already examples of complete houses that have been 'printed' using large 3D printing rigs.

Computer modelling of projects is now the norm, and research continues in such areas as virtual reality, robotics and artificial intelligence that could see construction robots increasingly replacing human workers on sites, particularly where designs incorporate large proportions of components and assemblies made off site, which will require minimal human input when they are assembled.

Conclusion

This chapter outlines some basic characteristics of, and issues related to, the broader construction industry/sector and sets the scene for other ways of looking at construction industries, projects and firms. The essential point is that the activities and products of construction are so varied as to defy simple description.

Notes

1 There are standard building designs, particularly for housing units, but the finished products are often not completely identical due to modifications that are made in response to differing site conditions (e.g. slope, aspect, access), client preferences, and availability of particular materials or components, to name a few.

2 In this case a 57-storey block was erected in just 19 days; this was only possible due to a combination of factors that included a very high level of offsite manufacture of components and modules that could be hoisted and fixed in position, a flat site with good all-round access and a simple, repetitive floor plan.
3 Bills of quantities were individually typed which was not only time-consuming but provided very limited opportunities for making amendments without re-typing whole pages, sections or complete documents. Digital documents could be amended easily, just as CAD drawings could be updated much more easily than traditional paper drawings.

References and further reading

Beaver, P. (1970) *The Crystal Palace* (London: Hugh Evelyn).

Bernstein, H. (2003) Measuring Productivity: An Industry Challenge. *Civil Engineering*, 46–53.

Best, R. and Meikle, J. (eds.) (2015) *Measuring Construction: Prices, Output and Productivity* (Abingdon: Routledge).

Chancellor, W., Abbott, M. and Carson, C. (2019a) Measuring Construction Industry Activity and Productivity: The Impact of the Shadow Economy. In: Best, R. and Meikle, J. (eds.) *Accounting for Construction* (Abingdon: Routledge).

Chancellor, W., Abbott, M. and Carson, C. (2019b) Productivity and Levels of Output in the Construction Industry. In: Best, R. and Meikle, J. (eds.) *Accounting for Construction* (Abingdon: Routledge).

de Valence, G. (2019) Accounting for the Built Environment. In: Best, R. and Meikle, J. (eds.) *Accounting for Construction* (Abingdon: Routledge).

The Guardian (2015) Chinese Construction Firm Erects 57-Storey Skyscraper in 19 Days. *The Guardian*, 30 April. www.theguardian.com/world/2015/apr/30/chinese-construction-firm-erects-57-storey-skyscraper-in-19-days.

Ive, G.J. and Gruneberg, S.L. (2000) *The Economics of the Modern Construction Sector* (London: Macmillan).

Langston, C. (2015) Performance Measures for Construction. In: Best, R. and Meikle, J. (eds.) *Measuring Construction: Prices, Output and Productivity* (Abingdon: Routledge).

Merchant, C. (2020) *The History and Construction of the Petronas Twin Towers* (mm2h Malaysia Is My Second Home). www.mm2h.com/the-history-and-construction-of-the-petronas-twin-towers/

MGI (2017) *Reinventing Construction: A Route to Higher Productivity* (McKinsey Global Institute). www.mckinsey.com/industries/capital-projects-and-infrastructure/our-insights/reinventing-construction-through-a-productivity-revolution.

OECD (2019) *OECD Compendium of Productivity Indicators 2019* (Paris: OECD Publishing). https://doi.org/10.1787/b2774f97-en.

OECD (2020) *Productivity* (Paris: OECDiLibrary). https://doi.org/10.1787/0bb009ec-en.

OERD (1995) *The Oxford English Reference Dictionary* (Oxford: Oxford University Press).

Piddington, J., Nicol, S., Garrett, H. and Custard, M. (2020) *The Housing Stock of the United Kingdom* (Watford: BRE Trust).

Rhodes, C. (2019) *Construction Industry: Statistics and Policy*. Briefing Paper Number 01432 (London: House of Commons Library), 16 December.

Tauranac, J. (1995) *The Empire State Building: The Making of a Landmark* (New York: Scribner).

Tilley, P.A., Wyatt, A. and Mohamed, S. (1997) Indicators of Design and Documentation Deficiency. In: Tucker, S.N. (ed.) *5th Annual Conference of the International Group for Lean Construction* (pp. 137–148) (Gold Coast, Australia: International Group), 16–17 July.

Editorial Comment

Anything that can go wrong probably will.

This well-known observation, in various forms, is known as Murphy's (or Sod's) Law; it has many corollaries, such as the Law of Selective Gravitation:

When something falls, it will fall in such a way as to do maximum damage to both itself and whatever it lands on.

In a similar vein, although probably less well-known, is Kilpatrick's Law:

Interchangeable parts don't.

This 'law' is relevant here because, in the following chapter, the authors examine further the concept of construction as a sector, comprising a set of inter-related and interdependent parts that are often treated as if they are interchangeable when in fact they are not. The authors add weight to Kilpatrick's proposition by developing the theme of the industry being composed of a number of industries and sub-industries, populated with firms of different sizes, which together make up a diverse sector that ranges from sole practitioners such as tradespeople who operate alone (e.g. a handyman or plumber who does only small domestic jobs, often for cash) all the way up to international contracting firms with many offices and tens of thousands of employees spread across multiple countries. The sector includes not only firms that actually perform the construction work but also the wide range of firms and activities that supply and support construction: material and component manufacturers and suppliers, for example.

The relevance of Kilpatrick's Law can be extended to include the differences between buildings that may be generally described in the same way – airport terminals and hospitals are good example – which may differ enormously in terms of characteristics such as their scale, complexity, engineering services and so on. Even a 'typical' free-standing house that accommodates a family of four in, say, Norway, will be a very different building to buildings of the same category built in Italy or Malaysia or New Zealand. While they share many of the same attributes: bedrooms, bathrooms, kitchens and the like, the shape, size, materials and finishes, among other things, will be quite different in different places. In short, they are all single-family houses but they are not interchangeable and it makes as little sense to lump them all together as it does to talk of a construction industry that includes all firms from the local handyman to integrated multi-disciplinary, multinational firms that operate around the world.

The authors provide some new insights on how we might better describe and analyse construction as a sector rather than as a single industry.

2 Construction products and producers

One industry, or several?

Jim Meikle and Gerard de Valence

Introduction

Construction is referred to in statistical datasets and national accounts as an industry, and it is often perceived as a single monolithic industry. It has been labelled as backward (Woudhuysen and Abley 2004), inefficient and unproductive (McKinsey 2017; WEF 2016) and described as fragmented, and adversarial (Latham 1994; Egan 1998); and has been compared – badly – with manufacturing (Gann 1996). This chapter, however, argues that construction is not a single industry, it is rather a large and extremely varied set of economic activities, and that it is not possible or sensible to generalise about the characteristics of all of these activities or to compare them, without adjustment, with other industrial sectors.

Construction ranges from minor domestic repairs to major infrastructure, and it is part of something larger, the built environment, that consists of on-site construction work and a number of related activities, including material production, property development, design and consultancy (Pearce 2003). Most construction works are made to order, are specific to a location and take a long time to produce; with a few exceptions they are not made in factories and there are few standard off-the-shelf final products. Most structures also last for a long time although many will have elements repaired or replaced, often many times (Brand 1994). Construction also has a relatively high proportion of informal activity – work that is undertaken outside the regulated economy (Blades and Roberts 2002). This is a particular issue in developing countries, where the majority of construction work can be classified as 'informal', but it is also common enough in developed economies (Blades *et al.* 2011; Schneider and Enste 2000).

The purpose of this chapter is to review the various types of construction, the main producers of construction and the different processes used to commission construction works in the UK, and to demonstrate the diversity of each. The chapter suggests that at least some of the combinations of products, producers and processes are more or less separate industries and, while it is often not possible to clearly delineate these, it is important to recognise that the differences exist and need to be taken into account by government, industry, researchers and others. Considering construction as a single industry leads to analyses and prescriptions that may be appropriate to some parts of the sector but are certainly not applicable

DOI: 10.1201/9781003102403-2

to all. The chapter also compares data on construction with data on manufacturing in general and motor vehicles in particular.

The chapter is in seven parts including this introduction. The next two parts present definitions of, and statistical concepts for, construction in the UK and discuss what is included in each. A fourth section reviews the products, the producers, and the processes of construction, a fifth compares construction with other industrial sectors and a sixth reviews government and industry reports on the UK construction industry. A final section draws together conclusions from the different parts and makes some recommendations.

Definitions

The UN System of National Accounts (SNA) provides a framework for the preparation of each country's national accounts and, within that, the International Standard Industrial Classification (ISIC) provides a structure for the recording of production data in national accounts (UN 2008). The Nomenclature Generale des Activities Economiques dans les Communautes Europeenes (NACE) produced by the Statistical Office of the European Community (Eurostat) represents another framework that is compatible with the SNA and is the immediate statistical basis for the UK national accounts. The statistical classification of economic activities in the European Community (Eurostat 2008) states that NACE Rev 2 is 'based on ISIC Rev 4 and adapted to European circumstances'. The UK SIC is published by the UK Office of National Statistics (ONS). The first version of the UK SIC was published in 1948 and there have been seven versions since; the current version was published in 2007(ONS 2007). It describes Section F: Construction as follows:

> *'This section includes general construction and specialised construction activities for building and civil engineering works. It includes new work, repair, additions and alterations, the erection of prefabricated buildings or structures on the site and also construction of a temporary nature.'*
>
> *'. . . . This section also includes the development of building projects for building or civil engineering works by bringing together financial, technical and physical means to realise the construction projects for later sale.'*

That wording indicates the range of activities covered by 'construction'.

Although construction is included as an industry in the ISIC, Eurostat and ONS systems, it is unlike many other manufacturing industries with which it is typically grouped as a secondary economic activity. Generally, it is not produced in a factory and it includes not only construction produced by construction contractors but also work produced by households and others. And, unlike most manufacturing industries, it includes both new production and repair, maintenance and improvement of existing buildings and works (Ive *et al.* 2004). In official statistical systems most production industries only include the manufacture of new products; repair and maintenance of products appear as separate industries. Most manufacturing industries, however, include design services while construction design services are typically included in professional services, a separate SIC category.

Another unique characteristic is that construction appears in both the production and the expenditure versions of the national accounts, the only 'industry' to appear in both versions (de Valence and Meikle 2019), although what is included in each version is different. Descriptions of the content of different construction concepts are set out in the next section.

There are other important categorisations of construction, including geographic or locational, and particular types of work. The latter could be government buildings where there are security issues, scientific buildings where there are special requirements or historic buildings where particular skills and expertise are needed. Location can be an issue, for example, both on remote sites with transport and employment issues, and on congested urban sites with access and public nuisance issues. ONS provides a geographical breakdown of output and new orders but only by UK countries and English regions.

A significant proportion of construction work in the UK is on existing buildings and structures – routine repair and maintenance or more major renovation, conversion or extension works. It is undertaken on all types and sizes of projects for all types and sizes of customers, from households to large industrial enterprises. Some of it will be undertaken when users are occupying and operating their facilities; this typically involves major inconvenience to contractors, including out-of-hours working, in order to ensure minimum inconvenience to users. It carries a premium price but not all contractors are willing or able to carry it out.

Concepts

The ONS uses a number of concepts to represent UK construction activity: contractors' output, construction turnover, construction value-added, gross construction output and construction investment. Similar but not necessarily identical concepts are used in other countries. Table 2.1 sets out what the UK concepts comprise. These are simplified descriptions rather than formal definitions and, besides each is a 2017 value, the latest year for which data is available for all concepts. Differences in the value of the five concepts are due to inclusions and exclusions that affect the size of the sample used.

Contractors' output data in Great Britain from Construction Statistics (CS) is produced and published primarily for construction industry use; it represents the value of construction activity by enterprises registered to construction SICs and is provided by type of work in some detail and whether the customers are classified as public or private. Construction turnover from the Annual Business Survey (ABS) sets out financial data on construction businesses by SIC category; it is provided in SIC categories by types of expenditure and within each is broken down in some detail by intermediate consumption (purchases from other industries) and value added (principally depreciation, remuneration of employees and gross surplus). Construction value-added is construction's contribution to Gross Domestic Product (GDP) in the production version of the national accounts; it consists of the

Table 2.1 Statistical concepts for construction activity in the UK[1]

Concept	*Source*	*Inclusions*	*Exclusions*	*Value 2017 GBP billion*
Contractors' construction output	Construction Statistics, Great Britain	Construction output by contractors in Great Britain classified to construction SICs.	Non-construction activity by construction contractors and construction activity by non-contractors. Construction activity in Northern Ireland	164
Construction turnover	Annual Business Survey	Construction turnover by contractors classified to construction SICs, including double counting of sub-contractors, plus the informal sector and self-build by households in the UK.	Own-account construction activity by public and private entities not classified to construction SICs.	269
Construction value-added	National accounts	Value-added (crudely: labour, profit and depreciation) of all activities by all producers of construction classified to construction SICs, informal construction activity and self-build construction by households in the UK.	Purchases from other industries and own-account construction activity by public and private entities not classified to construction SICs.	117+
Gross construction output	Input-output/supply and use tables	Construction turnover by construction contractors, including double counting of sub-contractors, plus the informal sector and self-build by households in the UK.	Own-account construction activity by public and private entities not classified to construction SICs.	301 (182)*
Construction investment	Gross Fixed Capital Formation (GFCF)	Construction expenditure on capital works by all public and private entities and households in the UK, including related professional services.	Expenditure by entities and households on repair and maintenance.	208

* This figure represents gross construction output less the construction industry's purchases from industries registered to construction (sub-contracting).
+ 2018 data.

total value of construction less purchases by the producers of that construction. Gross construction output in input-output tables sets out construction's purchases from other industries plus its value-added. Construction investment, construction in Gross Fixed Capital Formation (GFCF), is part of the expenditure version of the national accounts and represents the purchases by all customers of capital construction work (new work and major improvements); breakdowns are provided for residential buildings and other types of construction and by type of investor, private and public.

Construction in the national accounts includes an estimate of informal construction activity (work undertaken by small contractors or individual tradesmen that is not officially reported). It does not, however, include maintenance work undertaken by households themselves. Redecoration of a house by tradesmen, formally or informally, is part of national economic activity; redecoration by an individual in their own house is not. This treatment of domestic repair and maintenance is consistent with national accounts treatment of other housework.

None of the concepts of construction listed and described earlier represent the total value of all construction activity in the UK. An estimate of total – official – construction output in the UK in 2017 would be construction investment (GBP 208 billion, from GFCF) plus repair and maintenance (R&M) by construction contractors and other public and private organisations (from CS) but not by households themselves (because the last is not included in national accounts). Contractors' R&M output on non-residential and infrastructure work is known (GBP 26 billion, from CS) and contractors' residential R&M can be crudely estimated as 50 percent of what is recorded as contractors' housing R&M (GBP 13 billion, because housing R&M in GB includes major refurbishment, again from CS). UK total construction output is, therefore, of the order of GBP 247 billion (GBP 208 billion plus GBP 26 billion and GBP 13 billion), 51 percent more than contractors' output and 36 percent more than gross output (less double counting of sub-contracting).

Strictly speaking, further adjustments are necessary, including for construction professional services (included in investment but excluded from contractors' output), repair and maintenance work by organisations not registered to construction and the fact that contractors' output is for GB not UK (that is, output in Northern Ireland is excluded). The UK is not necessarily representative of the situation in other countries, but the figures indicate the kinds of differences that can obtain. The point is that care must be taken when referring to the different concepts for construction activity.

Green (2019) provides a comprehensive description of the history, development and composition of contractors' output in CS. Mahajan (no date), in a note for the ONS Consultative Committee on Construction Industry Statistics, sets out a reconciliation of contractors' output from CS and construction turnover from the ABS. The content of UK construction investment in GFCF and comparisons with data from a number of other countries is in de Valence and Meikle (2019).

The scope of the different concepts emphasises the range of different activities and actors included as construction (de Valence 2019). It can include work undertaken by construction firms, other enterprises, the informal sector and households. It can include design work (construction in GFCF) or exclude it (all other concepts). And it can include all construction activity – except DIY by households – or only construction investment (new work and major improvements). There has been ongoing interest in issues around the definitions of these different concepts (Cannon 1994; UN 1998; Briscoe 2006) and the modelling used to produce estimates of output (de Valence and Meikle 2019). The problems Briscoe identified are still relevant:

> *Problems with reliable and accurate data collection and statistical analysis include defining the scope and coverage of the industry; measuring industry outputs and their allocation across different types of activity; identifying construction firms; and measuring capital formation and capital stock; inconsistencies in employment statistics and labour market variables . . . . and the resolution of the discrepancies between different registers.*
>
> (Briscoe 2006: 220)

Characteristics of UK construction

This section reviews the main characteristics of construction, some of which are common to other economic activities but some of which are not. It includes discussion of the types of work and shares of activity in UK construction and the types of producers of construction. The intention is to show how different some construction activity is from others.

UK construction data has been subject to a number of changes over time. The data used here is largely as published by ONS and is the best available source. Some of the data presented comprises averages, for example, of employment or turnover per enterprise. These have some analytical value but give no sense of the composition of variables that make up the averages. In some cases, however, for example, the size structure of firms is available, and this presents a more holistic impression of construction activity.

The products of construction

There are two main measures of the products of construction contracting in UK data: construction output and construction new orders. Both can be subdivided into various types of residential and non-residential buildings and civil engineering work (infrastructure). It is also possible to identify other characteristics, including shares of type of work and the sizes of firms producing that work. The most detailed breakdown of all contracting output comes from Construction Statistics (CS). Table 2.2 presents contractors' output by customer (public and private), by different combinations of new work, major renovation and repair and

Table 2.2 Contractors' output in current prices, Great Britain, 2020

<table>
<tr><th rowspan="2"></th><th colspan="4">2020, GBP million</th></tr>
<tr><th>New work</th><th>Major renovation</th><th>Repair and maintenance</th><th>All work</th></tr>
<tr><td>Public residential building</td><td>4,763
3.2%</td><td colspan="2">6,365
4.2%</td><td>11,128
7.4%</td></tr>
<tr><td>Private residential building</td><td>32,992
21.8%</td><td colspan="2">19,146
12.7%</td><td>52,138
34.5%</td></tr>
<tr><td>Public non-residential building</td><td colspan="2">9,901
6.6%</td><td>4,883
3.2%</td><td>14,784
9.8%</td></tr>
<tr><td>Private non-residential building</td><td colspan="2">29,477
19.5%</td><td>11,585
7.7%</td><td>41,062
27.2%</td></tr>
<tr><td>Civil engineering/infrastructure</td><td colspan="2">22,517
14.9%</td><td>9,483
6.3%</td><td>32,000
21.2%</td></tr>
<tr><td colspan="4">All contractors' output</td><td>151,114
100.0%</td></tr>
</table>

Source: Output in the construction industry (ONS 2021a)

maintenance and by type of work (housing, non-residential building and infrastructure). Summary data is shown for 2020.

The SIC system recognises the distinct characteristics of the three main construction sectors – residential and non-residential building and civil engineering work (infrastructure) – and ONS identifies differences between public and private work and between new work and work to existing structures.

All public work, including infrastructure – whether it is publicly owned or not – represents 38.4 percent of all contractors' output; private sector residential and non-residential building represents 61.6 percent. All housing work represents 41.7 percent; non-residential building work, 37.3 percent; and infrastructure, 21.0 percent. It is important, however, to note again some features of this data. It includes only output produced by contractors registered to construction SICs; it excludes output by households, informal contractors and individuals and enterprises not registered to construction. It also includes only contractors' output in Great Britain (UK less Northern Ireland). In the published data, new private non-residential output is broken down into commercial and industrial work; and non-residential building repair and maintenance (R&M) is given as one total. As a result, it is not possible to produce accurate totals for any of new work, major renovations or repair and maintenance.

The distinctions between public and private also need comment. In the case of residential building, 'public' includes work for Registered Social Landlords (RSLs) although these are typically private not-for-profit enterprises, partly funded by government; and some RSL units are built for shared ownership tenure that may over time become fully privately owned. Private non-residential construction includes work produced under public private partnerships (PPPs) for public sector use, except Defence PPPs which are included as public building. Infrastructure output is not distinguished between public and private; historically most of this work would have been public but, with utility privatisation, much is now privately commissioned and owned.

ONS uses individual price deflators to convert values of output from current to constant prices. There are deflators for work types, whether work is public or private and for new work and repair and maintenance. The different measured price trends for each are indicative of different market dynamics for different types of work. There are, however, no deflators for different sizes of projects although these may in fact indicate further and more significant differences in price trends.

New orders comprise all work included as 'new construction' and exclude work described as 'repair and maintenance'. New orders tend to indicate construction market conditions at a point in time rather better than construction output: they represent the value of orders for new work placed in a particular period; output, on the other hand, represents the total value of work completed in a period (including work starting, work in progress and work ending) – contractors' turnover.

Extensive data on the size of new construction projects used to be published regularly for orders for new work by detailed project value bands and durations but that is now much reduced. Table 2.3 presents size data on new orders for non-residential construction projects for 2012, the latest year for which this data is available. The data in Table 2.3 comprises new construction, including major renovation on projects over GBP 100,000 in value; the data excludes construction work by non-contractors and construction work in Northern Ireland.

Table 2.3 indicates that the top 20 projects were very large, particularly the top ten and the top 50 projects represented almost 30 percent of all new orders over GBP 100 thousand. The same proportion, however, was on projects of less than GBP 2 million in value and 41 percent was on projects of more than GBP 10 million, each with an average value of around GBP 29 million; the balance of just less than 30 percent was in projects of between GBP 2 million and 10 million. And, although there may have been some large repair and maintenance projects with a value of over GBP 10 million in value, there will have been relatively few, so the total number of construction projects more than GBP 10 million was likely to be in the high hundreds at the most. This data, of course, is rather dated but current trends are likely to be similar.

When construction projects are viewed as products, the differences between the three sectors are more important than the similarities. The characteristics of the work and how it is organised, the scale of projects and their value cannot be

Table 2.3 Breakdown of non-residential new orders, 2012

New orders	*GBP million current prices*	*Percent of total*	*Number of jobs*	*Average value GBP million*
Top 10 jobs	6,660	15	10	666
Top 20 jobs	8,974	20	20	449
Top 50 jobs	12,630	29	50	253
Jobs over GBP 10 million	17,864	41	625	29
Jobs over GBP 2 million	30,530	70	3,753	8

Source: ONS: New orders, Top 50 jobs (ONS 2013)

easily compared between residential building, non-residential building and civil engineering construction. As the data aforementioned shows, these three sectors are grouped together in the SIC definition of construction on the basis of shared inputs of materials (brick, steel, concrete, etc.), despite the differences between the houses, factories and airports that are produced. Within each of these sectors, there is a wide variety of products (e.g. industrial, commercial and institutional non-residential building) but the differences between the sectors is larger. So large are these differences that the three sectors should be regarded as separate industries

Producers of construction

Industry observers define industries differently from national statisticians. The management expert, Michael Porter, suggests the following:

> *Any definition of an industry is essentially a choice of where to draw the line between established competitors and substitute products.*
>
> (Porter 1980: 32)

Porter's definition focuses on construction enterprises rather than construction output and suggests that small-scale domestic repair and maintenance is a different industry to large-scale construction of roads and bridges or the refurbishment and conservation of historic buildings. This definition is self-evidently sensible in industrial analysis terms but it is clearly very different from the definitions used in statistical systems. It suggests that construction is in fact a collection of industries that are grouped together because they use broadly similar resources (concrete, bricks, timber, steel, etc.) and produce broadly similar types of products (completed buildings and civil engineering works), although the latter can be very different in terms of their scale, complexity and purpose.

In Porter's terms, the broad categories of housing, non-residential building and infrastructure are distinct industrial groupings or industry sectors and it is partially possible to match construction contractors and construction customers with types of work. CS divides residential and non-residential building into public and private work although, as noted previously, these categories are not precise and they give no further information on type of customer. Non-residential work is divided into commercial and industrial building but, again, there is no further information on type of customer. Infrastructure is given as a single category. New orders provide more detail on non-residential construction by type of work but not by type of customer although some can be implied, particularly in infrastructure.

Most new private housing is produced by speculative house builders; most non-residential building (new and major renovation) is produced by general building contractors for a wide range of customers; and most infrastructure work (new and work to existing) is undertaken by civil engineering contractors for utility companies and other customers of built infrastructure. There is also data on contractors by size of enterprise. And there are specialist contractors in particular market sectors, for example repair and restoration of historic buildings, that will rarely, if ever, undertake other types of work. Demand for these broad types of work is driven by different economic and social factors – they are different but related market sectors or industries.

The 'most' that prefixes many of the earlier statements is important. Some new housing is produced by general building contractors and some is self-built by households and some larger contractors undertake both building and civil engineering work. And larger contractors will undertake large-scale refurbishment and repair and maintenance work while smaller projects of that type are generally undertaken by smaller firms. Size – of firms and projects – is important and size of construction firms is looked at in more detail in the following text.

The ABS contains data on construction enterprises by type of work, type of firm and size of firm. There are firms that undertake building and civil engineering work and operate as main and specialist contractors. Most but not all specialist contracting firms and some general contracting firms work as subcontractors but some specialist firms, painters and decorators and mechanical and electrical services contractors, for example, work directly for building customers, particularly in repair or renovation works. A firm's SIC classification is determined by its main activity; however, firms typically self-select the SIC category they belong to although they can work across different categories on different projects.

Table 2.4 sets out key data for all construction and the three main SIC subdivisions of construction in the UK. The largest grouping by the number of enterprises and employment is specialised construction, typically single trade contractors; there are 17 individual industries or trades under SIC 43. The largest group by turnover is building contractors, including residential and non-residential building with only two SIC sub-categories. Civil engineering contractors have the smallest number of enterprises and employment but the highest average number of employees and highest average turnover per enterprise. Civil engineering work is typically relatively large scale compared to building work.

Data on construction turnover by size of firm is available from the ABS. That turnover, of course, includes the value of subcontracting and construction work

Table 2.4 Construction turnover in the UK, 2019

SIC	*Type of enterprise*	*Total turnover GBP million*	*Number of enterprises*	*Turnover per enterprise GBP thousand*	*Total employment thousand*	*Employees per enterprise number*	*Turnover per employee GBP*
F	All construction	309,123	348,598	886,760	1,562	4.48	197,902
41	Building	144,035	104,674	1,376,034	498	4.76	289,227
42	Civil engineering	54,885	25,711	2,134,689	212	8.25	258,891
43	All specialised	110,203	218,213	505,025	853	3.91	129,195

Source: Annual Business Survey (ONS 2021b)

by non-contractors. The distribution of construction turnover by number and size of firm and average turnover per firm is set out in Table 2.5.

The data indicates that, broadly, 99 percent of construction firms have fewer than 50 employees and are responsible for just over 50 percent of turnover and 94 percent of firms have fewer than ten employees and are responsible for around 35 percent of turnover. At the other end of the size scale, fewer than 1 percent of firms, those with 50 or more employees, are responsible for the other almost 50 percent of turnover. Around 0.1 percent, a few hundred, are responsible for around 30 percent of turnover and each of these has an annual turnover averaging around GBP 275 million.

Slightly different data on construction contractors by turnover size band is also available from CS (in Table 3.6: Construction firms: Number of firms by turnover size band). In 2019, there were 290,374 enterprises classified to construction in Great Britain. A total of 265,542 (91.4 percent) had turnovers of less than GBP 1 million; 26,492 (8.2 percent) had turnovers of between GBP 1 million and 20 million; 847 (0.3 percent) had turnovers of between GBP 20 million and 100 million; 237 (0.1 percent) had turnovers of between GBP 100 million and 500 million; the remainder, 35 firms, (0.01 percent) had turnovers of more than GBP 500 million. There are only 272 firms in 2019 with turnover of more than GBP 100 million.

This size data on construction enterprises, employment, output and turnover confirms the substantial differences between the small number of very large firms and the large number of very small firms. Clearly, enterprise strategies for these businesses, industry policies and the approaches of their customers need to recognise these differences when developing responses. There are also middle-sized categories of firms which segue into the large and small categories at both extremes but, themselves, have particular characteristics.

The processes of construction

Construction typically involves a client and a constructor as the main contracting parties although there are other essential project participants and other services that are sometimes included in construction arrangements. Table 2.6 outlines the

Table 2.5 Construction enterprises and turnover by employment size band, 2019

Employee size band	*No. of firms*	%	*Total turnover GBP million*	%	*Average turnover per firm GBP thousand*
1–9	328,936	94.4	107,076	34.6	325
10–49	17,197	4.9	57,991	18.8	3,372
50–249	2,132	0.6	52,284	16.9	24,523
Over 249	333	0.1	91,771	29.7	275,588
All	348,598	100.0	309,123	100.0	887

Source: Annual Business Survey (ONS 2021c)

broad categories of project processes in use in the UK nowadays. The descriptions are not comprehensive, and the examples are only illustrative. Table 2.6 illustrates a progression from providing a simple product to providing a comprehensive combination of product and services. Not all of the models will include a constructor as a partner in the service delivery team; in some cases the constructor will only be contracted to produce construction works. But, increasingly, particularly on major works, construction contractors are members of consortia or Special Purpose Vehicles (SPVs) set up to undertake large projects. All new projects require design services, construction work and finance and some can also involve operation of completed facilities. Work to existing buildings and infrastructure works can also involve complex management issues, particularly if the facility is to remain in operation while construction is undertaken.

It is not possible to get reliable data on the use of different processes and how these are changing over time. There are surveys of procurement methods and contracts used, some regular over time and some one-off, but all suffer from relatively low response rates and, as a result, questionable representativity and validity. According to the NBS[2] National Construction Contracts and Law Survey (with almost 1,000 respondents), the indications are that traditional procurement, broadly defined, is the most common method, followed by design and build, although the latter has been catching up in recent years (Hughes *et al.* 2006). Other methods, including management forms, cost plus, public private partnerships (PPPs) and partnering/alliancing, are much less commonly used although when they are used they tend to be on larger projects.

The different processes used by construction clients broadly reflect the differences between the three sectors, although there are overlaps (Drew and Skitmore 1997). Small residential projects are typically delivered with a traditional design-bid-build process. Some larger residential and many non-residential building projects use D&B contracts, and some of the latter have finance and operation included. The processes used for large engineering construction projects like alliances and BOOT are rarely found in building projects.

Contractors and their customers cannot be matched precisely to each other or to project types or processes and there will be overlaps in many cases. In terms of size, generally, larger customers and contractors undertake large projects

Table 2.6 Generic construction process models

Content	*Brief description*	*Comments and examples*
Build only	A direct contract for only the construction work between a construction client and a contractor.	Typically, finance will be provided by the client and any design and other consultancy services will be separately engaged by the client.
Design and build (D&B)	A contract with a client where the contractor takes responsibility for both design and construction.	This has become increasingly common in recent years. Design will often be by a separate design firm employed by the contractor.
Finance, design, build and sell	A project undertaken by a developer trader that is sold on completion. There will typically be a contract between the developer and a contractor but they can be the same organisation.	Examples are the speculative housing developer model or the commercial developer trader. Some or all aspects of design and construction may be sub-contracted.
Finance, design, build and own	A project undertaken by a developer investor that is retained and managed by the developer on completion.	This is the commercial developer owner model used, for example, for office, warehouse or retail investments.
Finance, design, build, own and operate	A project undertaken by an organisation that will own and operate the completed structure.	Examples include infrastructure projects provided and managed by private sector operators including constructors, e.g. toll roads and bridges.
Finance, design, build, own, operate and transfer	A project that provides and operates a piece of 'public' infrastructure for an agreed period of time after which it reverts to public ownership.	Examples include public private partnerships (PPPs), Private Finance Initiative (PFI) projects and Build Own Operate Transfer (BOOT) projects.

and smaller customers and contractors undertake smaller projects (Skitmore 1991). But most contractors will operate to some extent at least on either side of their principal process method. Many large contractors will operate under most, if not all, processes at some time; smaller contractors will however be more restricted in what they can take on. A small design and build contractor may also undertake build only projects and occasionally finance their own small project, a single house for example. The processes indicated earlier represent more or less distinct market sectors and, at the extremes, very different industrial settings.

Market conditions can influence the types of work contractors take on. When demand is high, contractors will tend to stay in the markets they are comfortable in or, may, perhaps attempt to move up-market. When demand is low, large contractors may move down market and small new-work contractors may take on very small work or maintenance projects that they would reject in busier times.

Comparing construction with other industries

Construction is often compared, usually negatively, with other economic activities. It is particularly compared with manufacturing and the comparisons are typically between all of construction and all of manufacturing (Egan 1998; McKinsey 2017, for example). The problem with that is that both measures are averages of extremely varied activities. This section attempts to break down both construction and manufacturing and make more useful comparisons between the two.

The shortcomings of averages as analytical units have been commented on earlier; they do not represent how activity is distributed among firms and employees. The dilemma is what measures we should use. This chapter uses SIC data from the Annual Business Survey (ABS) where both construction and manufacturing can be broken down. The breakdowns are not ideal but they are better than the larger portfolio categories. Table 2.7 sets out higher level data for the main components of the non-financial business part of the UK economy: Agriculture (Section A), Mining and quarrying (B), Manufacturing (C), Electricity, gas, steam and air conditioning (D), Water supply, sewerage, waste management and remediation (E), Construction (F) and Wholesale and retail trade, repair of motor vehicles and motorcycles (G).

Table 2.7 indicates that construction has the third largest number of enterprises, is the third largest category by turnover but has the second smallest average number of employees per enterprise. Mining, manufacturing and utilities are relatively large-scale activities by average numbers of employees per enterprise while agriculture and construction are relatively small-scale by that metric; trade and repairs lie in-between.

Manufacturing comprises 24 two-digit industrial groups (SIC 10 to SIC 33), for example, food products (SIC 10), manufacture of paper and paper products

Table 2.7 Key data for selected industrial groups in the UK, 2019

SIC code	*Activity*	*Turnover GBP millions*	*Number of enterprises number*	*Turnover/ enterprise GBP thousand*	*Employment thousands*	*Turnover/ employee GBP*	*Employees/ enterprise number*
A	Agriculture	5,983	15,160	394.66	54	110,796	3.56
B	Mining	36,380	1,297	28,049.34	57	638,246	43.95
C	Manufacturing	570,611	136,585	4,177.70	2,484	229,715	18.19
D/E	Utilities	145,911	13,632	10,703.57	324	450,343	23.77
F	Construction	348,598	309,123	1,127.70	1,562	223,174	5.05
G	Trade and repairs	1,543,378	395,995	3,897,468	4,864	317,306	12.28

Source: Annual Business Survey (ONS 2021b)

(SIC 17) and manufacture of motor vehicles, trailers and semi-trailers (SIC 29); and 325 individual industries. Manufacturing of fabricated metal products except machinery and equipment (SIC 25) is the largest two-digit group with 22 individual industries, 26,301 total group enterprises and total group turnover of GBP 23.6 billion; the smallest is the single industry group of manufacture of tobacco products (SIC 12) with nine enterprises and a turnover of GBP 12 million. Manufacturing is not only relatively large but extremely diverse and industry policies have reflected that by targeting specific industries such as IT and automobiles for example.

Construction and manufacturing are similar in terms of turnover per employee and are the most common subjects of comparison. Table 2.8 compares construction and manufacturing by size of enterprise. Manufacturing enterprises are a little more spread across the size categories in terms of numbers of enterprises but turnover is much more concentrated in the larger manufacturing firms.

It is possible to look more closely at and compare individual industries in the two industrial groups. Table 2.9 shows construction and its main components and manufacturing and the manufacture and repair and maintenance of motor vehicles. Construction includes work on existing structures (improvements and repair); manufacture of motor vehicles does not, and maintenance of vehicles is taken from SIC category G to make the groups more comparable.

The table shows that construction turnover is 50 percent of all manufacturing turnover and almost 60 percent of all manufacturing employment and much larger than any of the individual manufacturing industries. The largest manufacturing industry in 2018 was, in fact, motor vehicles with 22 percent of construction turnover and 5.5 percent of construction employment. And it is manufacture of motor vehicles that is often compared with construction (Gann 1996; Egan 1998). 'Why can't construction be more like the car industry?' is a frequent complaint of construction's critics. There are undoubtedly lessons to be learned from manufacturing in general and the car industry in particular but there are also major differences that need to be recognised (Green *et al.* 2005).

Table 2.8 Construction and manufacturing by size of firm

Employee size band	*Construction*					*Manufacturing*				
	No. of firms	*%*	*Total turnover GBP m*	*%*	*Average turnover per firm GBP k*	*No. of firms*	*%*	*Total turnover GBP m*	*% %*	*Average turnover per firm GBP k*
1–9	328,936	94.4	107,076	34.6	325	107,177	78.5	29,003	5.1	271
10–49	17,197	4.9	57,991	18.8	3,372	21,523	15.8	56,798	10.0	2,639
50–249	2,132	0.6	52,284	16.9	24,523	6,466	4.7	118,166	20.7	18,275
Over 249	333	0.1	91,771	29.7	275,588	1,419	1.0	366,645	64.2	258,383
All	348,598	100.0	309,123	100.0	887	136,585	100.0	570,611	100.0	4,178

Source: Annual Business Survey (ONS 2021c)

Table 2.9 Turnover, employment and turnover per worker, selected industries, 2018

SIC code	*Activity*	*Turnover GBP millions*	*Number of enterprises*	*Turnover per enterprise GBP*	*Employment thousands*	*Turnover per worker GBP*	*Employees per enterprise*
F	All construction, of which	309,123	348,598	886,761	1,562	197,902	4.5
41	• Building construction	144,035	104,674	1,376,034	498	289,227	4.8
42	• Civil engineering construction	54,885	25,711	2,134,689	212	258,892	8.2
43	• Specialist construction	110,203	218,213	505,025	853	129,195	3.9
C	All Manufacturing, of which	570,611	136,585	4,177,699	2,484	229,715	18.2
29.1	• Motor vehicles	60,885	10,660	5,705,910	86	707,965	8.1
G	All Trade and repair, of which	1,543,378	395,995	3,897,468	4,864	317,306	12.3
45.2	• Maintenance of vehicles	28,807	45,047	639,488	256	112,527	5.7
29.1 and 45.2	Motor vehicles and their maintenance	89,692	55,707	1,610,067	342	262,257	6.1

Source: Annual Business Survey (ONS 2021b)

As noted, all construction includes not only new construction but also the repair and maintenance of existing buildings and works. In order to adjust for this, maintenance of vehicles (SIC 45.2) is added to manufacture of vehicles (SIC 29.1) in Table 2.9. When maintenance is added to manufacture, turnover increases by 58 percent but employment increases by almost 180 percent. This allows a more realistic comparison and reveals that motor vehicles and their maintenance (SIC 29.1 plus SIC 45.2) has almost the same turnover per worker as building construction (SIC 41). Turnover per worker is a common metric of productivity and, on this basis, all construction is less productive than all manufacturing and much less productive than motor vehicle production but when repair is added to manufacture, the car industry is on a par with building, the largest part of construction.

Broad comparisons between the available statistical data on construction and manufacturing are not particularly useful as an aid to industrial policy and more detailed comparisons between, say, construction and vehicle manufacture are also unhelpful. The minimum requirements of a robust comparison are not fulfilled. Table 2.10 summarises the key features of the construction and motor vehicle comparison and illustrates how different they are.

It is interesting to see the relative scales of the different groups or industries in terms of enterprises, turnover and employment but it is difficult to draw clear conclusions from comparison of their economic performance or productivity. Vehicle manufacture and, to a lesser extent, vehicle repair and maintenance are capital intensive businesses; construction, generally, is not, although a few activities like tunnelling and prefabricated housing are.

Both McKinsey and the World Economic Forum (WEF) make much of the change over time in the productivity of construction and manufacturing (McKinsey 2017; WEF 2016). Construction comes badly out of their comparison but it is a poor one for the reasons outlined earlier. A better comparison – of construction

Table 2.10 Comparing construction and vehicle manufacture

	Construction	*Vehicle manufacture*
Total turnover	GBP 309 billion	GBP 61 billion
Number of enterprises	348,598	10,660
Total employment	1,562,000	86,000
Turnover per worker	GBP 197,902	GBP 707,965
Capital intensity	low	high
Labour intensity	high	low
location	mobile site based	fixed factory based
Product design	mostly in professional services	mostly by vehicle manufacturers
Standard products	very small proportion	vast majority of production
Work on existing products	substantial	excluded
Users in occupation	significant	not applicable

against a combined new vehicle and vehicle repair and maintenance – shows markedly different results.

Although SIC codes – and their contents – have changed and there have, no doubt, been different price trends over the years, it is possible to make a crude comparison of the performance of the construction and car industries over the past 20 years. Construction value added per worker in current price terms has increased by a factor of 2.7 between 1998 and 2019, while the same measure for vehicle manufacture has increased by a factor of 3.3 but for vehicle repair and maintenance has only increased by a factor of 1.4. Over the same period, the turnover of new car production has increased by 1.5 while the turnover of vehicle repair and maintenance has increased by a factor of more than three; either the new car industry has reduced in size or repair and maintenance has increased over the period, or, possibly, a bit of both. As a result, the value added per worker of the combined vehicle sector, again, in current price terms increased by a factor of only 1.7. By that measure, construction outperformed the combined vehicle sector over the period.

At least two things are happening to produce this result: the lower value added per worker of vehicle repair and maintenance and the increased share of repair and maintenance in the combined total. There may also be differences in price trends over time that influence the result.

The reports on construction

There are regular reports on the UK construction industry, its performance and its shortcomings, sponsored by government or the industry. In most cases construction is treated as a single industry with a common set of characteristics and failings across the wide range of projects, project participants and processes that exist. The roles of public clients, the particular needs of large engineering works and some other features of the sector are sometimes identified and there is a separate set of reports on housing (for example, Barker 2004; Callcutt 2007) but these do not suggest that there are distinct construction markets that might call for very different analyses and prescriptions.

A dozen construction reports between 1944 and 1998 are summarised and discussed in Murray and Langford (2003), who concluded that those reports agreed on the poor performance of construction with minor differences between their explanations for that and recommendations for improvement. The last two of those reports by Latham (1994) and Egan (1998) became particularly influential as the UK government became the leading advocate of reform.

That this series of reports (and many others not included in Murray and Langford 2003) was required, averaging over two a decade for 50 years, shows how ineffective they were in developing policies to address the issues raised. The explanation for this policy ineffectiveness offered by Latham and Egan is industry culture, broadly seen as the custom and practices underlying the business model in UK construction. Latham focused on procurement and contractual relations with recommendations to change an adversarial culture, calling for more collaboration between clients, contractors, subcontractors and consultants, and more cooperative practices. He recommended 'Partnering' between clients and contractors to realise this.

Egan began his report arguing industry improvement required changing the industry culture, recommending lean production methods using examples from car manufacturing, steel-making, grocery retailing and offshore engineering, and setting ambitious performance targets for the industry. The government followed Egan by promoting offsite manufacturing in the *Modernising Construction* (National Audit Office 2001) and *Accelerating Change* (Strategic Forum 2002) reports, and supported the reform movement with legislation and by establishing Rethinking Construction, the Construction Best Practice Programme and the Movement for Innovation, which were brought together in 2004 as Constructing Excellence 'to achieve a step change in construction productivity by tackling the market failures in the sector and selling the business case for continuous improvement' (Strategic Forum 2002: 2).

Prior to Egan the reform movement relied on industry participation, with little effect on how projects were procured and delivered. Contractual relationships were the focus of much of the reform agenda to improve industry performance. Egan introduced benchmarking against best practice to improve productivity, and Constructing Excellence documented demonstration projects. Murray and Langford thought the 'demands on the industry cannot be met and so lead to an industry that cannot attract staff to deliver buildings on time, with increased costs and questionable quality' (2003: 7). Other critics attacked the reform movement for its technocratic and managerial approach (Green and May 2003; Green 2011) and the language used (Fernie *et al.* 2006). More relevant was a review of progress since Egan by Wolstenholme (2009), which found there had been little change in the industry: clients still awarded projects to the lowest bidder while contractors offloaded risks and maximised profits.

Sixty-five years after the Simon Committee report on building contracts (the first in Murray and Langford) Wolstenholme again called for cultural change 'to integrate and embrace the complex picture of how clients and contractors interact' (2009: 8). Industry culture is clearly important, but it is also clear that culture is not malleable and does not change easily or quickly. A better explanation for the lack of impact of these reports, their recommendations, and the ineffectiveness of public policy in reforming construction is required. Simmons (2015) blames the policy making process as resistant to evidence and subject to ministerial whims and churn, with issues becoming politicised once they enter public debate. Carroll (2010) suggests that regulatory proposals typically don't have a convincing evidence base and there is poor integration of impact assessment with policy development processes. Wond and Macaulay (2010) argue that generic 'problem-inspired' strategies developed by central policy-makers have to be interpreted by the 'problem-solving' implementers responding to nuances of local context and capability.

This chapter argues that construction is better viewed as three industries when the differences between residential building, non-residential building and engineering construction are taken into account and that there are also differences by size of firm or size of project and by the processes adopted. If the culture in each of the three industries is different, recommendations and policy directed at construction as a single industry are unlikely to be relevant across the different industrial groupings, and will thus be disregarded by many firms and clients. Clients are also different and can be generalised as households, businesses and the public sector, and their relationships with contractors varies accordingly. Another example is design, where house builders have pattern books, commercial building uses architects, and infrastructure is designed by engineers. These structural differences between the three industries affect the way clients, contractors, designers and suppliers will interact, thus each industry has developed individual characteristics over time within the broader culture of construction that become that particular industry's culture (de Valence 2018). The specific nature of these industry cultures makes recommendations and policy directed at construction as a single industry ineffective.

Conclusion

Although the SIC groups all construction firms into a single category, that is for statistical convenience and is based on conventions that were developed primarily for classifying manufacturing. The exclusion of design from construction output while included in other industries is one result. Another is the view of construction as a single industry, producing and maintaining buildings and structures, despite their many different types and differences in the producers and processes used in their delivery.

The distribution of projects, firms and output all support the idea that construction is a collection of industries, not one single industry, albeit with overlaps among them. There is probably no need for precise definitions of construction markets or industries or of strict boundaries between them, following the case-by-case approach used by competition policy. It is sufficient to know and understand that there are different groupings that need different analyses and prescriptions. For example, the broad subdivisions of new construction, major improvement and repair and maintenance call for different approaches and the break down below that of work with users in occupation and work with no users in place call for different levels of expertise and management.

The broad categories of residential and non-residential building and civil engineering construction have wide ranges of customers and projects and are different enough to call for different types of contractors and implementation processes. Therefore, this chapter has argued these should be regarded as separate industries and government policy needs to recognise their differences.

Size matters. The generalisation of a large number of small contractors undertaking a large number of small projects that represents a significant but not the majority of construction projects and a small number of large contractors undertaking a relatively small number of large projects that represents a similar proportion of construction output is borne out by the data. There is also a similar proportion of output undertaken by medium-sized contractors on medium-sized projects.

The world of big firms and major projects is important but does not represent the whole of construction. The analysis presented here suggests that there are fewer than 1 percent of firms (around 2,500 firms) that represent around 45 percent of construction output in the UK and the majority of that output is undertaken by a few hundred very large firms. The vast majority of firms and projects, however, are small and medium sized; firms with fewer than ten workers (around 330,000 firms) represent 35 percent of construction output; medium-sized firms (around 17,000) and projects represent the balance of output. The groupings will, of course, change over time, partly as project values are affected by price changes over time and partly as industry structures change.

Within these size groupings there are subsets of parties and processes. The needs and practices of frequent or experienced customers of customers are necessarily different from those of occasional and one-off customers. The first group will have, or will have ready access to, professional advice that they have confidence in; the second group generally will not. Speculative developers, developer investors, utility undertakings and many public agencies typically fall into the category of expert customers. On the simple construction-only to service provision continuum, there will be a large number of small, medium and large construction contractors at the construction and design and build end but, at the service provision end, the number of providers will be much smaller and will overlap with the small number of very large contractors.

Construction markets consist of customers and suppliers ordering and delivering construction works using a variety of processes and mediated by market conditions. Construction demand tends to be more volatile than some other economic sectors and the economy as a whole but some construction markets are more volatile than others. Repair and maintenance and smaller work generally tends to be more stable as does public work. The most dramatic fluctuations in demand tend to be in speculative housing and commercial building work.

All construction activity data, however presented, is an average of an extremely wide variety of types and sizes of projects. To be useful, discussion or analysis of construction productivity, industry shortcomings or market conditions needs to be of distinct construction settings. And prescriptions for improvement or recommendations for different processes also need to be matched to these settings. Prescriptions and recommendations for all construction are unlikely to be appropriate to all types of work, all parties or all processes.

It is important to appreciate the different needs of different groups when industry or government policies are being developed. Prequalification, registration or bidding procedures, for example, that are appropriate for large firms and large projects may present insurmountable barriers to smaller firms when applied on smaller projects. Including design and construction may prove problematic for smaller firms. Alliancing and partnering are only appropriate for frequent customers of construction. Different types of turnkey projects call for consortia of experienced and usually large and well financed participants.

Although distinct construction industries or markets are difficult to define there is a case for developing typologies of construction – within each of residential, non-residential and civil engineering – and characterising construction work by combinations of key variables, for example:

- Process, with a range of models from construction-only to full-service delivery;
- Type of contractor, from micro and local to mega and global;
- Type of customer, from one-off or occasional to frequent and expert;
- Size, measured by both volume and value;
- Complexity, from simple to complex;
- Technology, from traditional to offsite fabrication; and
- From new work on vacant sites to major refurbishment with users in occupation.

There will be scales of 'less demanding' to 'more demanding' for each of these variables and there will be more variables.

Within the different types and sizes of project there will be common issues – selection of contractors, resolution of disputes, reconciliation of the cost-time-quality-triangle discussed by Hackett in Chapter 5 in this book, for example – but their resolution will be context dependent. There are generic solutions or options for very large or very small projects and their variants in the three main sectors but each case is subtly different and construction policies and construction reports should allow for that variation and eschew simple 'one size fits all' recommendations.

A number of reports and researchers have called for construction to adopt similar production practices to manufacturing (Egan 1998), particularly car manufacturing (Gann 1996). There are some parts of construction where this could be applicable – factory made standard structures and components, for example. The opportunities for standard buildings are, however, limited although they should be investigated. The opportunities for standard construction products and assemblies are much more widespread and are currently pursued by specialist construction contractors and product manufacturers although the latter are typically not classified to construction. Lessons from other industries and their production methods and processes can be useful and informative but comparing performance between industries is very difficult without adjustments to make the subjects comparable.

With separate industries and separate cultures, separate policies are required. A broad industry policy of the sort that targets construction as a single industry will be challenged by three deeply entrenched cultures with limited similarities. Research and reports that treat construction as a single industry have the same problem, although there is an economic activity called construction in the SIC the characteristics of the three sectors make them different industries. The manufacturing SIC includes glass, wood products, steel, plastics and concrete, but they are regarded as separate industries and are not grouped together under a construction products SIC. An industry policy for the steel industry is not thought to apply to plastics or concrete because it is not relevant to those industries. The same applies to the differences between residential building, non-residential building and civil engineering construction.

Notes

1 Data sources are noted in the body of the table.
2 The origins of NBS are in the UK's National Building Specification, originally developed by the Royal Institute of British Architects (RIBA). In 2020 the NBS was purchased by Byggfakta Group, a leading construction data and software provider, based in Sweden.

References and further reading

Barker, K. (2004) *Review of Housing Supply* (London: HMSO).
Blades, D., Ferreira, F. and Lugo, M. (2011) The Informal Economy in Developing Countries: An Introduction. *Review of Income and Wealth*, **57** (5), 1–7.
Blades, D. and Roberts, D. (2002) *Measuring the Non-Observed Economy* (Vol. 5) (Paris: OECD).
Brand, S. (1994) *How Buildings Learn* (New York: Viking Press).
Briscoe, G. (2006) How Useful and Reliable Are Construction Statistics? *Building Research & Information*, **34** (3), 220–229.
Callcutt, J. (2007) *Review of Housing Delivery* (London: HMSO).
Cannon, J. (1994) Lies and Construction Statistics. *Construction Management and Economics*, **12** (4), 307–313.
Carroll, P. (2010) Does Regulatory Impact Assessment Lead to Better Policy? *Policy & Society*, **29** (2), 113–122.
Chancellor, W., Abbott, M. and Carson, C. (2019) Measuring Construction Industry Activity and Productivity: The Impact of the Shadow Economy. In: Best, R. and Meikle, J. (eds.) *Accounting for Construction: Frameworks, Productivity, Cost and Performance* (Abingdon: Routledge).
de Valence, G. (2018) *Construction as a Mature Technological System* (Wolverhampton: First International Conference on Construction Futures).
de Valence, G. (2019) Accounting for the Built Environment. In: Best, R. and Meikle, J. (eds.) *Accounting for Construction: Frameworks, Productivity, Cost and Performance* (Abingdon: Routledge).
de Valence, G. and Meikle, J. (2019) Construction Output as Gross Fixed Capital Formation. In: Gruneberg, S. (ed.) *Global Construction Data* (New York: Routledge).
Drew, D. and Skitmore, M. (1997) The Effect of Contractor Type and Size on Competitiveness in Bidding. *Construction Management & Economics*, **15** (5), 468–489.
Egan, J. (1998) *Rethinking Construction* (London: Department of the Environment, Transport and the Regions).
Eurostat (2008) *Statistical Classification of Economic Activities in the European Communities* (Luxembourg: Office for Official Publications of the European Communities).
Fernie, S., Leitinger, R. and Thorpe, T. (2006) Change in Construction: A Critical Perspective. *Building Research & Information*, **34** (2), 91–103.
Gann, D. (1996) Construction as a Manufacturing Process? Similarities and Differences Between Industrialized Housing and Car Production in Japan. *Construction Management and Economics*, **14** (5), 437–450.

Green, B. (2019) The Challenges of Measuring British Construction Output. In: Best, R. and Meikle, J. (eds.) *Accounting for Construction: Frameworks, Productivity, Cost and Performance* (Abingdon: Routledge).

Green, S.D. (2011) *Making Sense of Construction Improvement* (Oxford: Wiley-Blackwell).

Green, S.D., Fernie, S. and Weller, S. (2005) Making Sense of Supply Chain Management: A Comparative Study of Aerospace and Construction. *Construction Management & Economics*, **23** (6), 579–593.

Green, S.D. and May, S.C. (2003) Re-Engineering Construction: Going Against the Grain. *Building Research & Information*, **31** (2), 97–106.

Hughes, W., Hillebrandt, P., Greenwood, D. and Kwawu, W. (2006) *Procurement in the Construction Industry: The Impact and Cost of Alternative Market and Supply Processes* (London: Taylor & Francis).

Ive, G., Gruneberg, S., Meikle, J. and Crosthwaite, D. (2004) *Measuring the Competitiveness of the UK Construction Industry* (Vol. 1) (London: Department of Trade and Industry).

Latham, M. (1994) *Constructing the Team*. Final Report of the Government/ Industry Review of Procurement and Contractual Arrangements in the UK Construction Industry (London: HMSO).

Mahajan, S. (no date) *Brief Background to ONS Construction Industry Estimates of Output and Gross Value Added*. Note for the ONS Consultative Committee on Construction Industry Statistics (CCCIS), London.

McKinsey (2017) *Reinventing Construction Through a Productivity Revolution* (London: McKinsey Global Institute).

Murray, M. and Langford, D. (2003) *Construction Reports 1944–98* (Oxford: Wiley-Blackwell).

National Audit Office (2001) *Modernising Construction*. National Audit Office (London: The Stationery Office).

ONS (2007) *Standard Industrial Classification of Economic Activities*. Office of National Statistics (London: The Stationery Office).

ONS (2013) *New Orders in the Construction Industry* (Office of National Statistics). https://webarchive.nationalarchives.gov.uk/ukgwa/20160131093003/http:/www.ons.gov.uk/ons/publications/re-reference-tables.html?edition=tcm%3A77-311472.

ONS (2021a) *Output in the Construction Industry* (Office of National Statistics). www.ons.gov.uk/businessindustryandtrade/constructionindustry/datasets/outputintheconstructionindustry.

ONS (2021b) *Non-Financial Business Economy, UK: Sections A to S* (Office of National Statistics). www.ons.gov.uk/businessindustryandtrade/business/businessservices/datasets/uknonfinancialbusinesseconomyannualbusinesssurveysectionsas.

ONS (2021c) *Non-Financial Business Economy, UK: Employment Size-Band* (Office of National Statistics). www.ons.gov.uk/businessindustryandtrade/business/businessservices/datasets/uknonfinancialbusinesseconomyannualbusinesssurveyemploymentsizeband.

Pearce, D. (2003) *The Social and Economic Value of Construction: The Construction Industry's Contribution to Sustainable Development* (London: nCRISP).

Porter, M. (1980) *Competitive Strategy* (New York: Free Press).

Schneider, F., Buehn, A. and Montenegro, C.E. (2010) *Shadow Economies All Over the World: New Estimates for 162 Countries From 1999 to 2007* (World Bank). https://openknowledge.worldbank.org/handle/10986/3928.

Schneider, F. and Enste, D. (2000) *Shadow Economies Around the World*. IMF Working Paper WP/00/26 (Washington, DC: IMF).

Simmons, R. (2015) Constraints on Evidence-Based Policy: Insights From Government Practices. *Building Research & Information*, **43** (4), 407–419. https://doi.org/10.1080/09613218.2015.1002355.

Skitmore, M. (1991) The Construction Contract Bidder Homogeneity Assumption: An Empirical Test. *Construction Management & Economics*, **9** (5), 403–429.

Strategic Forum (2002) *Accelerating Change, Rethinking Construction* (London: Strategic Forum).

UN (1998) *International Recommendations for Construction Statistics* (New York: United Nations Publications).

UN (2008) *International Standard Industrial Classification of All Economic Activities (ISIC)* (New York: United Nations Publications).

UN (2009) *System of National Accounts 2008* (New York: United Nations).

WEF (2016) *Shaping the Future of Construction: A Breakthrough in Mindset and Technology* (Geneva: World Economic Forum and the Boston Consulting Group).

Wolstenholme, A. (2009) *Never Waste a Good Crisis: A Review of Progress Since Rethinking Construction and Thoughts for Our Future* (London: Constructing Excellence).

Wond, T. and Macaulay, M. (2010) Evaluating Local Implementation: An Evidence-Based Approach. *Policy & Society*, **29** (2), 161–169.

Woudhuysen, J. and Abley, I. (2004) *Why Is Construction So Backward?* (Chichester: Wiley-Academy).

Editorial Comment

In the preceding chapter the authors expand on the notion that construction should be considered as a sector comprising numerous sub-sectors, rather than as a single industry; this is a theme that has also been explored from different angles in several chapters in the previous books in this series (Best and Meikle 2015, 2019). There are two other recurring themes in these books that are linked to this idea; one is to do with how built environment sector (BES) statistical data is recorded, aggregated (and, perhaps, disaggregated) and presented in national accounts and by organisations such as the OECD and the World Bank, and the other concerns the measurement of construction productivity.

There are important linkages between the different components of the sector and between these themes, and understanding those linkages is necessary if valid observations and measurements of construction activity, output and productivity are to be made. For example, 'value added' is often reported in the statistics and used as a basis for statements about productivity in the construction industry, often as part of unflattering comparisons with productivity, particularly productivity improvement, achieved in other industries. Such observations require a great deal of unpacking because virtually every part of these analyses depends on how some basic questions have been answered; they include, but are not limited to:

- What industry or sector or sub-sector is being examined?
- Is 'construction' limited to on-site construction or does it include other sub-sectors such as material supply and professional services (e.g. supervision)?
- Is repair and maintenance of existing structures included?
- What definition of 'value added' is being applied?

Valid comparisons can only be made where, as the saying goes, we compare apples with apples, and it is all too often the case that apples are compared with watermelons or bananas or potatoes. They may all be food items with varying nutritional attributes, which we grow, but they are quite different in many other ways. The same can be said, for example, of structures; is it, for example, logical (or useful) to compare the productivity of bridge builders by comparing their performance in the construction of a reinforced concrete arch road bridge that spans a narrow creek bed with that of a steel truss type railway bridge that spans a wide river? Even with appropriate allowances and adjustments for the obvious difference in scale, the different characteristics of materials and methods, as well as locational (climate, access, remoteness and so on) and other effects make such comparisons of dubious value.

In the following chapter, the author looks in more detail at the composition of the sector using the BES in Australia as a case study through analysis of various relevant datasets published by the Australian Bureau of Statistics (ABS) with the aim of furthering our understanding of what constitutes the so-called construction industry and whether it should be seen as a broader sector within economies, as well as providing further insights into how the industry/sector works in an economic sense.

3 Output and employment in Australian built environment industries 2007–2019

Gerard de Valence

Introduction

The fact that the construction industry, as measured in the national accounts or the industrial classification system, is only one part of the creation and maintenance of the built environment was emphasised by Turin (1969) and Strassman (1970), one an architect and the other an economist who both focused on the role of construction in economic development. Because the construction industry links to other industries in a wide variety of ways, those links to other sectors of the economy give the industry an important macroeconomic role. Using data on inter-industry flows of products and services such as input-output tables produced by national statistical agencies, the structure and pattern of construction's relationship with other industries have been extensively studied (see Bon 2000). However, because this data only shows flows between industries and is released infrequently, it does not provide useful data such as the current number of people employed or the value of output (Gretton 2013).

Output and employment are the fundamental economic characteristics of an industry, and changes in these over time reveal the dynamics of an industry as it develops. That data is typically found across a number of statistical collections, for example employment from a workforce survey and output from a business activity survey, and released in different publications and formats (Gruneberg 2019). There are also industry statistics produced using administrative data such as tax and corporate returns required by regulators. How to identify the individual contributions of industries that participate in the production, management and maintenance of the built environment is discussed in the next section of the chapter. That is followed in part three by a survey of previous research on how to use the available data to estimate the total economic contribution of those industries, which are collectively called the built environment sector (BES).

Part four applies the method developed in previous research to measure the size and extent of the built environment sector in Australia. The size and structure of the Australian BES is based on data provided in the Australian Bureau of Statistics annual publication *Australian Industry* (ABS 8155), which collects data for output and employment for 16 relevant industries and sub-industries between 2007–08 and 2018–19. These are industries with a direct, physical relationship

DOI: 10.1201/9781003102403-3

with the built environment that are included in *Australian Industry*. Part five is on the economic role of the BES and has two examples. The first discusses the effect of the increase in public building work done during the global financial crisis in 2007–10 on the built environment industries and flow-on effects of these large fiscal expenditures on output and employment. The second looks at the role of the BES and residential building during the macroeconomic transition required at the end of Australia's mining boom in 2014. Part six concludes.

Production, management and maintenance of the built environment

Data on economic and industry activity is so diverse some system of classification and categorisation is necessary (UN Statistical Division 2008). With the development of the Standard Industrial Classification (SIC), industries have come to be defined by the profile created for them by the data collected by national statistical agencies. Thus, the industry called 'Construction' is made up of the three sub-industries of residential building, non-residential building and engineering construction in the SIC. These statistics also divide construction firms between building contractors, engineering contractors and subcontractors (often called special trades). Because statistical data on work done and employment is presented following these systems of classification, most of the discussion and reporting of the industry also follows this pattern. But is this data truly reflective of an industry as diverse and wide-ranging as Construction? For example, Bygballe *et al.* (2013) discuss four different models of construction as an industry.

Industries are groups of firms with common characteristics in products, services, production processes and logistics, subdivided by the SIC into a four-level structure. The highest level is alphabetically coded divisions such as Agriculture, forestry and fishing (A), Manufacturing (C) and Information and communication (J). The classification is then organised into two-digit subdivisions, three-digit groups, and four-digit classes. SIC codes are therefore two, three and four-digit numbers representing industries, defined as firms with shared characteristics (UN Statistical Division 2008).

The SIC definition of the construction industry captures the on-site activities of contractors and subcontractors, and this data on building and construction work is taken to represent the industry. However, on-site work brings together suppliers of materials, machinery and equipment, products and components, and all the other inputs required to deliver the buildings and structures that make up the built environment. Consultants provide professional services such as design, engineering, cost planning and project management services as inputs into building and construction (Connaughton and Meikle 2013). There are also inputs from urban planning, transport, finance and legal services. Many industries are structured around such value chains and production networks (Hughes 1983), and when enough firms share sufficient characteristics they are often described as an industry cluster (Cruz and Teixeira 2010) or sector (Porter 1985).

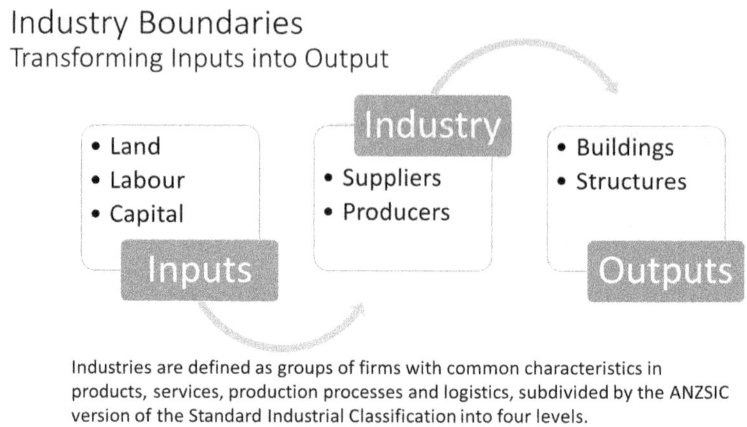

Figure 3.1 Industry interdependence

The boundaries around an industry are tightly defined by the SIC, to allow identification of individual industries as producers of goods and services and measurement of their contribution to output and employment in the economy. However, where the boundaries around a cluster or sector actually are is an open question. Turin (1969), Strassman (1970) and the Bartlett International Summer School series in the 1980s (Groak 1992) advocated looking at the sector that produces the built environment in broad and integrative terms, and Ofori (1990) proposed a broad definition of construction:

> *as that sector of the economy which plans, designs, constructs, alters, maintains, repairs and eventually demolishes buildings of all kinds, civil engineering works, mechanical and electrical engineering structures and other similar works.*
>
> (Ofori 1990: 23)

Measuring the broad construction industry

The idea that the construction industry is only one part of the creation and maintenance of the built environment has been developed by researchers into a method for measuring the broad construction industry. Ive and Gruneberg used UK data and the SIC classifications to identify the industries involved in production of the built environment at the four-digit SIC level (Ive and Gruneberg 2000: 25–28). From this data they estimated the output and employment of firms directly and indirectly involved, finding 'over 11.2 per cent of employees in employment are engaged in production of the built environment. This compares with the figure of 4.7 per cent taking the construction category alone.' (Ive and Gruneberg 2000: 11). This approach was also the basis of industry estimates given in Pearce (2003), which found, for the UK in 2001, contractors accounted for around half of both the total number of firms, employment and total gross value added.

For Australia, de Valence (2001) compared the size of construction, using data from the construction industry survey done by the Australian Bureau of Statistics (ABS 1998), from an industry cluster perspective. As percentages, with the On-site services and Project firms segments added together for the site-based measure of construction, the results look similar to the UK data given earlier, with the narrow industry 51 percent of income and 48 percent of employment of the broad industry. Carassus *et al.* (2006) compared the size of the 'construction sector system' in seven countries, again using a wide definition to include property management, repair and maintenance, and the institutional actors involved. Ruddock and Ruddock (2009) also followed this approach and used SIC data on it to get estimates for the construction sector in 20 European countries, which ranged between 12 and 22 percent of GDP, with an average of about 17 percent. Another example of the use of SIC categories is Taylor (2010), on the definition and valuation of offsite construction in the UK.

Squicciarini and Asikainen (2011: 690) compared the European NACE and North American NASIC classes that can be included in what they call a value chain approach: 'The definition includes the core NACE section F codes and supplements this with selected four-digit NACE classes that relate to manufacturing and services activities in the pre-production, support and post-production construction segments.' They include a wide range of industries with varying degrees of connectedness to production and maintenance of the built environment.

If the built environment encompasses the entirety of the human built world, then the Built Environment Sector (BES) is the collection of industries responsible for producing, managing and maintaining the buildings and structures that humans build (de Valence 2019). To be included in the BES an Industry needs a direct physical relationship with buildings and structures. Those industries are divided into ones on the demand side and those on the supply side, like materials or specialised trades, Demand side industries like property developers and facility managers pull output from the supply side, both for new output and for servicing and managing existing assets. Therefore, the BES is a sector more like defence than tourism, because it also produces long-lived assets for clients outside the sector (governments and owners respectively) that require repair and maintenance, which generates significant ongoing revenue for firms across the broad industry sector that produces those assets.[1]

A final requirement is that data on the industries included in the BES needs to be available at a level of detail that separates out BES components of industries like manufacturing and professional services. The data used here excludes industries such as retail and wholesale trade, transport, legal and financial services. These industries clearly play a role in the BES, but that role is hard to identify in the ABS industry statistics because of availability or the level of aggregation in the data does not allow a built environment share to be estimated. The concept of the BES is broad and extensive, so cannot be precise and exact. While the boundaries of industries and markets are important, in practice the data and SIC definitions are the starting point for the data used. The industries included are selected because they clearly have a relationship with the construction, management and maintenance of the built environment. This may not capture every single contribution to the BES, but it does allow the development of a profile of the sector.

Measuring the Australian Built Environment Sector

The method used to measure the size and extent of the Australian BES is to collect the SIC data for output and employment for relevant industries and sub-industries, following the method used in previous research. These are industries with a direct, physical relationship with the built environment. The data is provided in the Australian Bureau of Statistics annual publication *Australian Industry* (ABS 8155), produced annually using a combination of directly collected data from the annual Economic Activity Survey conducted by the ABS, and Business Activity Statement data provided by businesses to the Australian Taxation Office. The data includes all operating business entities and Government owned or controlled Public Non-Financial Corporations. *Australian Industry* excludes the finance industry and public sector,[2] but includes non-profits in industries like health and education and government businesses providing water, sewerage and drainage services. The selected industries included account for around two-thirds of GDP.

The analysis is based on Industry value added (IVA) and Industry employment. IVA is the estimate of an industry's output and its contribution to gross domestic product (GDP), and is broadly the difference between the industry's total income and total expenses. IVA is given in current dollars in *Australian Industry*. The data is presented at varying levels for industry divisions, subdivisions and classes. The most recent issue is for 2018–19.

Data and method

First issued for 2007, there is data at the two-digit subdivision level for the industries included in *Australian Industry*, including industries wholly within the BES like Construction services and Property operators and real estate services. This is not the case for all industries, for example Manufacturing has many subdivisions like textiles, chemicals and petroleum, not directly linked to or part of the BES, so the tables here collect the relevant subdivisions as BES Manufacturing.

However, for the subdivisions Professional, scientific and technical services and Building cleaning, pest control and other services, the data at the subdivision level includes contributions from other classes outside the BES. Therefore, for these industries the two-digit subdivision estimates have to be weighted using the four-digit class data for the BES component, and this can be done because the ABS over the last two years has collected detailed information for a number of specific industries in the annual Economic Activity Survey, one of the data sources for *Australian Industry*. These are released as supplementary tables and typically provide data at the subdivision and class level. Professional, scientific and technical services were included in 2015–16, and in 2016–17 this data was provided for two divisions: Rental, hiring and real estate services, with subdivisions Rental and hiring services (except real estate), and Property operators and real estate services;

and Administrative and support services, with subdivisions Administrative services and Building cleaning, pest control and other support services.

The data from the survey years for these subdivisions is used to weight the non-survey years. For the other years, employment and IVA are weighted by applying the BES proportion in the subdivision's survey year to the other year estimates. The tables combine the identifiable relevant industries that contribute to production, maintenance and management of the built environment in Australia, from the ABS data provided for industries and their subdivisions. Note that the current year and previous year estimates are revised with each new release, usually these are minor but some are significant.

The data is not complete because some industries cannot be separated into the relevant classes from *Australian Industry*. For example, rental of heavy machinery and scaffolding (class 6631) is in subdivision 66 but the data is not available. Also, services such as marketing, legal and financial are important but again not identifiable. Government spending on infrastructure and portfolio investment in departments like health and education is included through the BES supply industries, although any maintenance and work done internally will generally not be included. That also applies in industries like retailing and transport where some (unknown) proportion of work is done in-house.

There is also leakage around the boundaries of industry statistics: some glass is used in mirrors, some in car windscreens; textiles are used in buildings; architects design furniture; engineers repair machines as well as structures, and so on. Because *Australian Industry* uses tax and business register data, it is the classification of firms to SIC industry classes that fundamentally determines the structure and scope of that data. Needless to say, such classifications are not perfect, particularly in regard to large multi-unit or multi-divisional organisations.

The concept of the BES is to include industries where they clearly have a direct physical relationship with buildings, structures, and the built environment. The extent of that relationship can be debated, likewise contributions from outside the data available from *Australian Industry* such as financial services. The Australian

Table 3.1 Industries included in the Australian BES

Supply industries	*Demand industries*	*Maintenance industries*
Non-metallic mining and quarrying	Residential property	Water, sewerage and drainage
Building construction	Non-residential property	Waste collection, and disposal
Heavy and civil engineering	Real estate services	Building and industrial cleaning
Construction services		Building pest control services
Architectural services		Gardening services
Surveying and mapping services		
Engineering design and consulting		
Manufacturing industries		

BES combines data for 16 industries included in *Australian Industry* that together form one of the largest and most important industrial sectors in the economy.

When economic activities are spread across a wide range of individual industries the contribution of the whole is not obvious. Therefore the ABS and the Australian tourism industry developed an annual tourism satellite account (ABS 5249), to bring together the contributions of relevant industries like accommodation, tour operators and entertainment to estimate their total output and employment (ABS 2009). The contribution of tourism to GDP was 3.1 percent in 2017–18, and the share of employment was 5.2 percent. A satellite account uses the ABS Input-Output Tables (ABS 5209), a much broader but less timely data set than *Australian Industry*.

Australian BES output and employment

In an industrial sector like the BES changes in output, structure and dynamics will typically be slow and gradual. The Australian data confirms that is the case. Both output and employment have grown consistently since 2007. The general trend of an increasing share of services in output holds, with Professional, technical and scientific services and Property and real estate services both increasing their shares of BES output, by one and six percent, respectively. The industry that has grown fastest is Property operators and real estate services, which doubled IVA between 2007 and 2016 and increased its share of total BES output from 20 to 26 percent (see Figures 3.2 and 3.3).

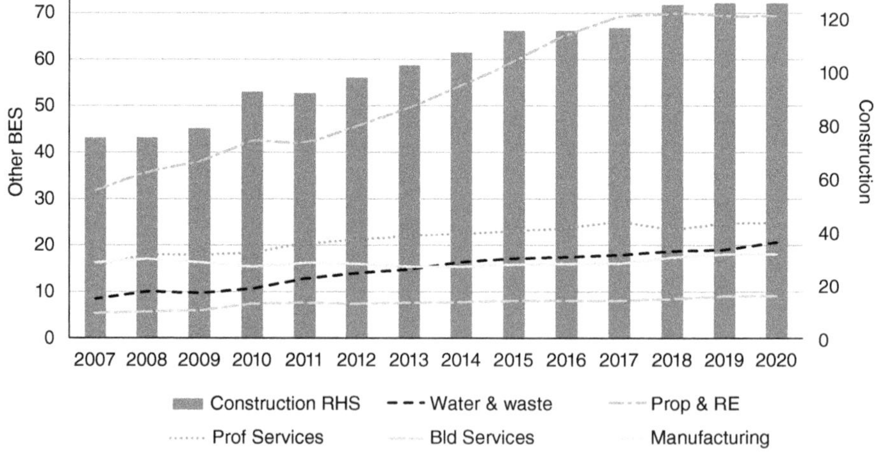

Figure 3.2 Output of Australian built environment industries

Source: ABS 8155

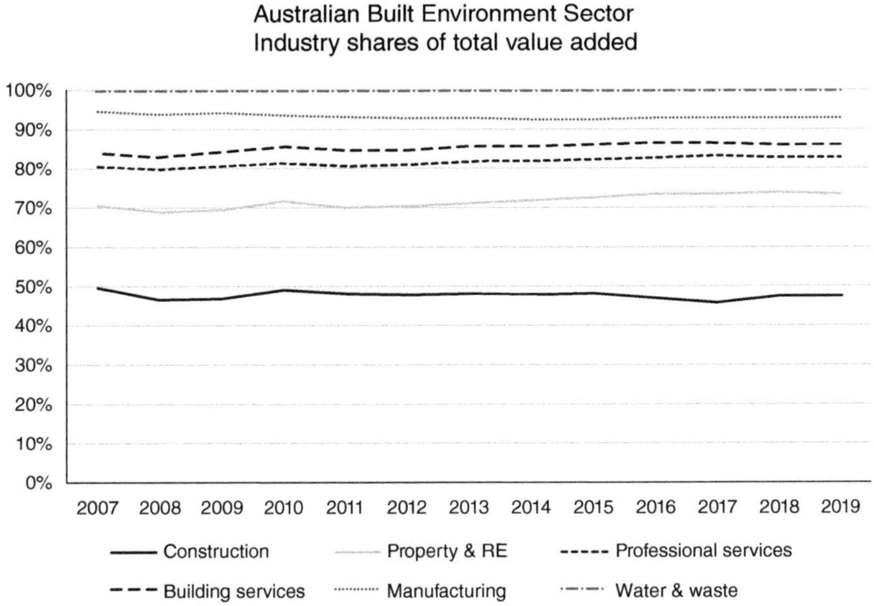

Figure 3.3 Output shares of Australian built environment industries

Source: ABS 8155

Table 3.2 Output of the Australian BES (ABS 8155, various years)

ABS 8155 ANZSIC Industry	*Industry value added AUD billion*			
	2015-16	*2016-17*	*2017-18*	*2018-19*
09 Non-metallic mineral mining and quarrying	2.6	2.7	2.9	3.3
28 Water sewerage and drainage	13.2	13.1	14	13.9
29 Waste collection, treatment and disposal	4.2	4.5	5	5.3
30 Building construction	26.2	29.3	33.3	33.6
31 Heavy and civil engineering construction	21.2	18.6	20	19.8
32 Construction services	68.1	68.6	72.5	72.6
Construction	115.5	116.6	125.8	126
6711 Residential property operators		11.1		
6712 Non-residential property operators		42.6		
6720 Real estate services		15.5		
67 Property operators and real estate services	65.6	69.2	69.8	69.2
6921 Architectural services	3.9			
6922 Surveying and mapping services	1.9			

(Continued)

Table 3.2 (Continued)

ABS 8155 ANZSIC Industry	*Industry value added AUD billion*			
	2015-16	*2016-17*	*2017-18*	*2018-19*
6923 Engineering design and consulting services	17.9			
BES Professional, scientific and technical services	23.7	25	23.3	24.7
7311 Building and cleaning services		6		
7312 Building pest control services		0.7		
7313 Gardening services		1.4		
BES Building cleaning, pest control, other services	7.9	8.1	8.5	9.6
BES Manufacturing	16	16	17.2	18
Total BES IVA	248.8	255.2	266.5	269.5

As found in the previous studies on the broad construction industry discussed earlier, total construction accounts for nearly half of the BES total. That share, however, has fallen from 53 percent of BES IVA to 47 percent between 2007 and 2019, at the same time construction fell from 54 to 49 percent of total BES employment. Note the share of employment is one or two percent higher than the share of IVA, due to the labour-intensive nature of Construction services.

Because BES IVA is measured in current dollars, changes in its value reflect changes in both prices (paid for goods and services) and the quantity of output. To find changes in the quantity of output and the real rate of growth, an implicit price deflator is used to adjust for the effect of price changes. Thus, changes in quantities are found by deflation, calculated by dividing the current-dollar value of GDP by an appropriate price index with the reference-year value set to 100. A price index is appropriate if its definition and coverage closely match those of the series being deflated. The ABS produces chain volume measures of industry gross value added, which are constant price estimates of industry output. Unfortunately, these are at the two-digit SIC level and do not map closely to the 16 BES industries.

Employment

As a sector, the BES accounts for 16.5 percent of total employment in Australia. The total number of people employed in the BES has increased steadily but not dramatically, reaching 2,126,000 in 2018–19. Over that time, the significant changes in the composition of employment across the BES have been the rise in Professional, technical and scientific services from 10 to almost 12 percent, and the fall in Manufacturing BES from 9 to under 7 percent, reflecting the fall in the manufacturing share in the overall economy. Employment in Professional, technical and scientific services increased from 189,000 to 259,000 between 2015–16

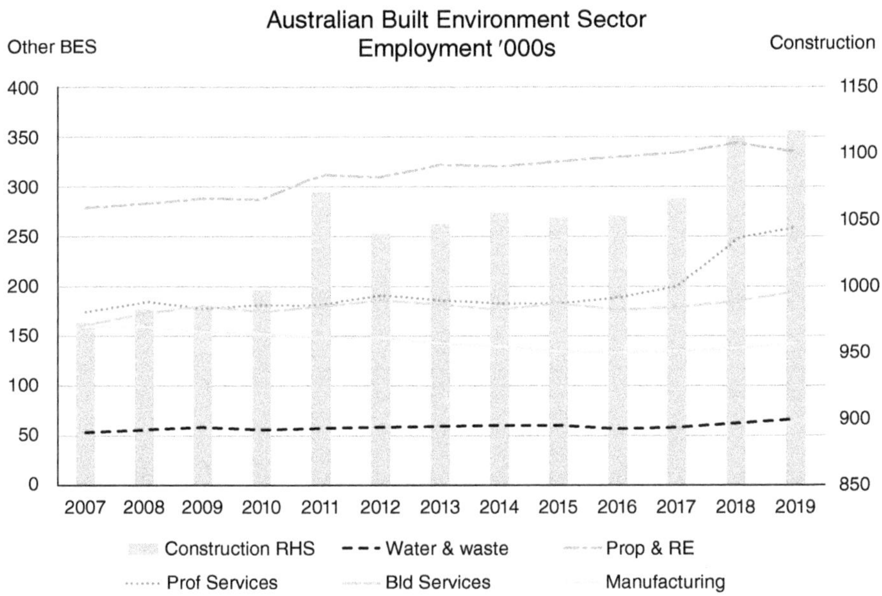

Figure 3.4 Employment in Australian built environment sector

Source: ABS 8155

Table 3.3 Employment in the Australian BES

ABS 8155 ANZSIC industry	*Employment '000s*			
	2015-16	*2016-17*	*2017-18*	*2018-19*
09 Non-metallic mineral mining and quarrying	12	12	13	13
28 Water sewerage and drainage	27	27	28	29
29 Waste collection, treatment and disposal	29	31	34	36
30 Building construction	208	212	224	225
31 Heavy and civil engineering construction	122	110	119	115
32 Construction services	723	748	771	777
Construction	1,053	1,070	1,115	1,117
6711 Residential property operators		50		
6712 Non-residential property operators		146		
6720 Real estate services		137		
67 Property operators and real estate services	331	334	343	336
6921 Architectural services	40			
6922 Surveying and mapping services	16			
6923 Engineering design and consulting services	133			
BES Professional, scientific and technical services	189	199	247	259

(Continued)

Table 3.3 (Continued)

ABS 8155 ANZSIC industry	*Employment '000s*			
	2015-16	*2016-17*	*2017-18*	*2018-19*
7311 Building and cleaning services		150		
7312 Building pest control services		10		
7313 Gardening services		21		
BES Building cleaning, pest control, other services	175	179	186	194
BES Manufacturing	133	134	138	142
Total BES Employment	1,949	1,986	2,101	2,126

Source: ABS 8155

and 2018–19, both the largest and fastest increase across the BES. Table 3.5 shows percentage changes in people employed and annual growth rates of each industry.

Although the industry shares in total employment look stable, the growth in employment in Professional services is notable, which increased from 9.6 to over 12 percent of BES employment between 2007 and 2019. After Construction the Property operators and real estate services industry is the largest employer, followed by Professional services.

The industry shares of BES output and employment have in fact changed over time, as shown in Table 3.4. The decline of manufacturing and decreasing share of Construction has seen the Property and real estate services industry increase its share of IVA from 16 to 26 percent, and share of employment from 15 to 21 percent between 2007 and 2019. The Water and waste and Property and real estate services industries have positive differences between the shares of IVA and employment, reflecting their higher capital requirements and investment in buildings, structures, plant and equipment.

In the input-output research cited earlier by Bon (2000) and others, 'stability' is a low rate of change of the technical coefficients in the input-output models, and 'maturity' means that the technical coefficients of a sector show stability. Using these models, these researchers studied construction's backward and forward linkages from the manufacturing and service sectors. Special attention was given to changes in construction technology that led to changes in the relative shares of the manufacturing and service sectors as inputs to construction, and over time these shares do increase as the wider economy develops. In an industrial economy the input shares generally do not change much over time, because construction is already using current technology. Gregori's (2008) review of construction input-output research found a steady increase in the share of services into construction is widespread, which reflects the wider post-industrial trend of increasing the intangibles and services percentage of GDP (Bostic *et al.* 2016). Unlike services, changes in manufacturing inputs have varied greatly internationally.

With construction as a technologically mature industry (de Valence 2018) at the core of the collection of industries that make up the BES, the gradual changes in

Table 3.4 Output and employment industry shares of BES total 2007 and 2019

		Property & RE	*Professional services*	*Building services*	*Manufacturing*	*Water & waste*	*Construction*
Employment	2007	15.4%	9.6%	8.9%	8.8%	2.9%	53.7%
	2019	15.8%	12.2%	9.1%	6.7%	3.1%	52.5%
IVA	2007	20.47%	10.20%	3.46%	10.33%	5.49%	48.69%
	2019	25.61%	9.15%	3.63%	6.66%	7.22%	46.63%

Table 3.5 Average annual growth rates 2006–07 to 2018–19 and 2014–15 to 2018–19, percent

Industry	*Employment 2006-07 to 2018-19*	*5 year avg 2014-15 to 2018-19*	*IVA 2006-07 to 2018-19*	*5 year avg 2014-15 to 2018-19*
Water and waste	1.7	1.7	9.7	3.6
Construction	1.1	1.2	5.2	3.4
Property and real estate	1.6	0.9	9.1	5.4
Professional services	3.8	8.2	4.3	2.1
Building services	1.6	2.0	5.3	3.6
Manufacturing	–0.8	0.4	1	3.5
Total BES	1.3	1.8	5.7	3.8

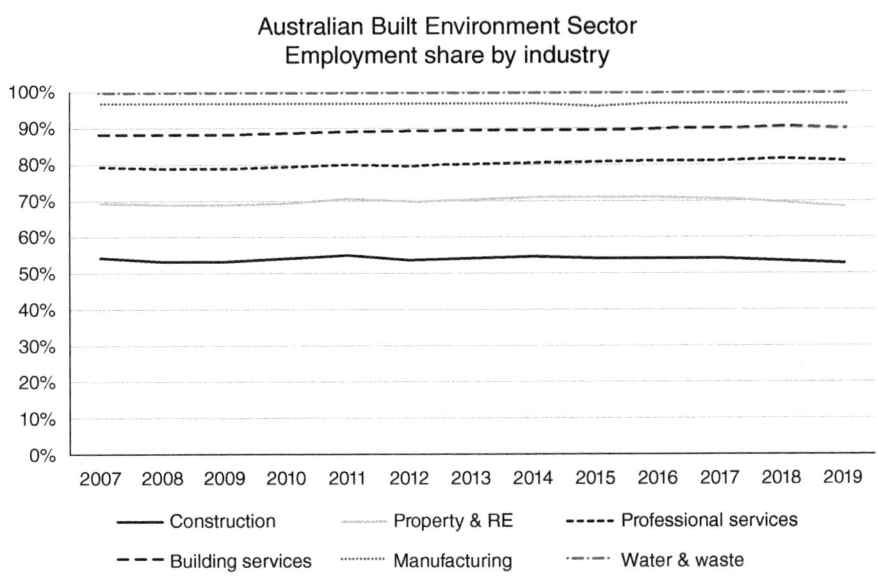

Figure 3.5 Employment shares in Australian built environment sector

Source: ABS 8155

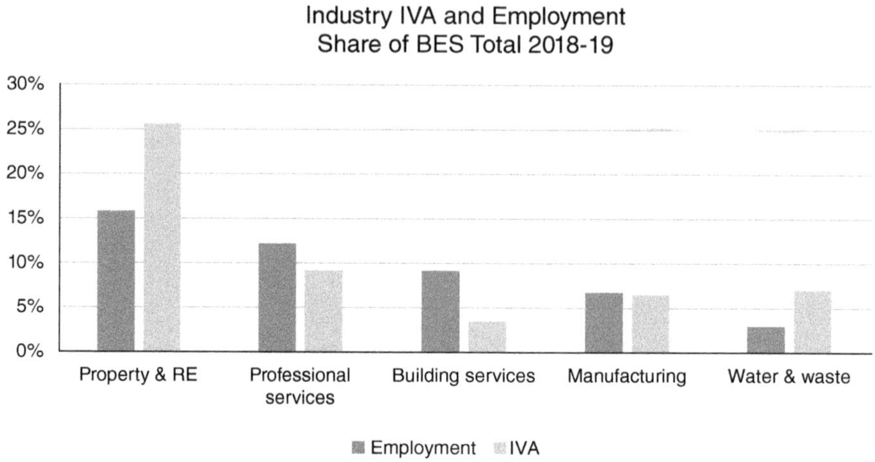

Figure 3.6 BES industry shares of output and employment

Source: ABS 8155

shares of IVA and employment found here are consistent with input-output studies of Australian construction (Zhu *et al.* 2020). The BES is a mature technological system, with a diverse cluster of industries in deep layers of specialised firms creating a dense network of producers, suppliers and materials. Basic construction materials like cement, concrete, glass, and components like building management systems, interior walls, plumbing fixtures, lifts and elevators are well-established industries in a mature supply chain. Hughes (1987) argues that over time such a mature technological system, one that is a long way into its life cycle, produces a specific culture of technology, embodied in the firms and social institutions of the system of production. This creates a tendency for an industry to develop along defined technological trajectories unless or until deflected or disrupted by a powerful external force. Therefore, a mature technological system will have considerable stability in the long-run structure of output and employment. However, that does not mean there are no changes in the shares of individual industries within the sector, because it is becoming more services intensive over time or because relative prices are changing.

Industry comparisons

Across the 16 industries included in the Australian BES there are significant differences between them in their growth rates of employment and output. Since 2007 employment in Professional services has grown an average of 3.8 percent a year, twice as much as other industries, and since 2014 by over 8 percent a year. Because IVA is in current dollars and not adjusted for price changes, unlike the national accounts data for example, growth rates are in nominal values and appear high. Water and waste and Property and real estate have both increased by over 9 percent a year.

The BES IVA and employment data give the contribution of the individual industries and the structure of the BES, as shown earlier. This data can also be used to reveal other characteristics of these industries and to make comparisons between them, starting with IVA per person employed. Broadly, the stability of IVA per person employed across the BES implies both employment and output have increased at about the same rate, and there is little evidence of substitution of capital for labour through increasing mechanisation and automation (Chan 2019).

IVA per person employed reflects the capital structure of an industry, which is the investment required for physical capital like machinery and buildings and intellectual capital like patents and processes. The higher the capital requirements, or capital intensity, of an industry the higher the level of IVA per person employed is expected to be because workers with more capital are more productive. Both excavators and shovels require one operator but the former shifts more soil. That effect is seen across the BES, where services like cleaning and construction trades have the lowest level of IVA per person employed, but also have lower capital requirements than the higher IVA per person employed industries of Water and waste, Property and real estate, Professional services, Engineering and Mining. These industries in Figure 3.7 are also the high gross fixed capital formation (GFCF) industries in Figures 3.8 and 3.9.

As a combination of output and employment IVA per person employed looks like a measure of productivity, but while it is indicative that is not the case. Output is not adjusted for price changes and employment is not given in hours worked

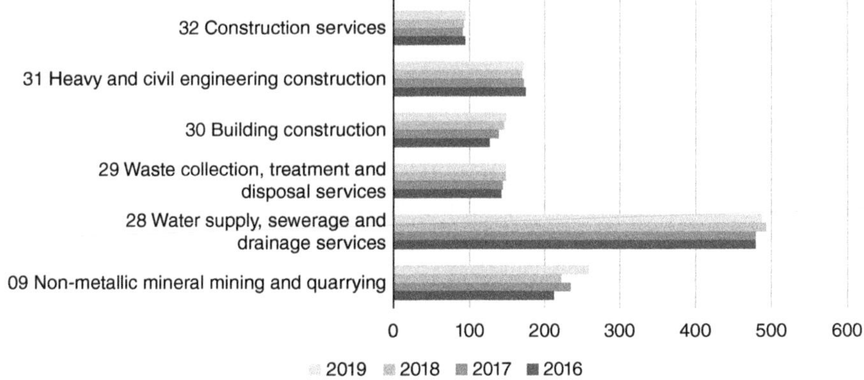

Figure 3.7 BES output per person employed. Nine industries

Source: ABS 8155

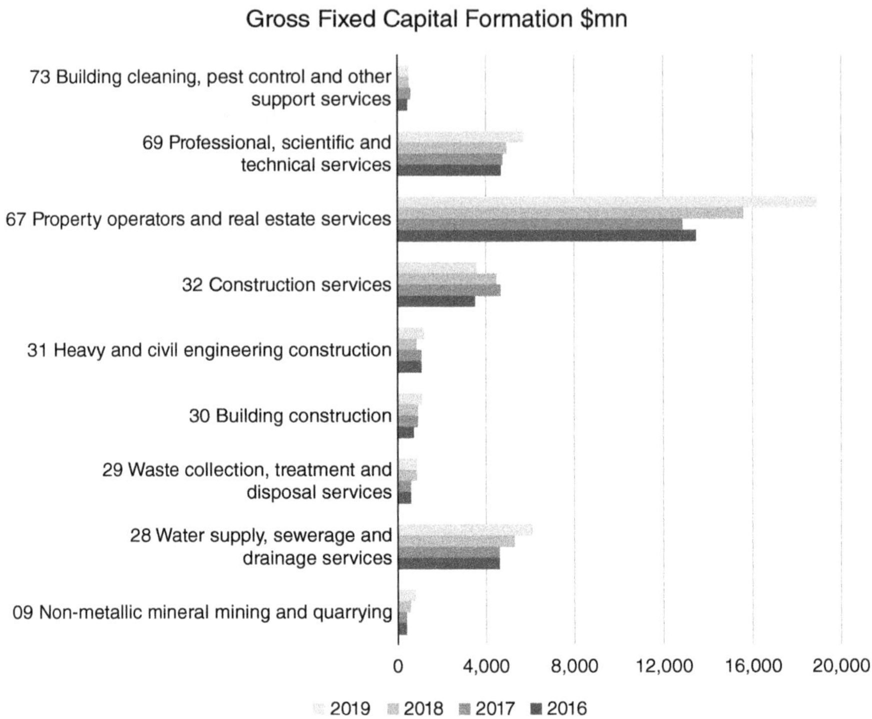

Figure 3.8 BES gross fixed capital formation. Nine industries.

Source: ABS 8155

in *Australian Industry*, therefore the usefulness of IVA per person employed as a proxy for productivity per person is limited. Although these appear to be similar to the output and input data needed to calculate productivity, indexes of output and input are used for productivity analysis. Nevertheless, the small changes seen in Figure 3.7 reflect the lack of productivity growth many studies have found in construction and related industries (see de Valence and Abbott 2015 for a review).

IVA per employee highlights differences in the capital requirements of industries. Capital expenditure by firms is their purchases of buildings, structures, software and machinery, known as gross fixed capital formation (GFCF). 'Gross' means the expenditure is measured without deducting the consumption of fixed capital through wear and tear caused by its use in production. GFCF has two types of assets, material and intellectual, the latter includes mineral exploration; computer software and databases; and entertainment, literary and artistic originals. In the long run, investment in GFCF determines industry growth rates and their level of labour productivity.

Over time, annual GFCF becomes the capital stock of an industry, the quantity of assets used in production, and industries range from labour intensive to capital intensive. Capital intensity is typically measured as the ratio of fixed capital to

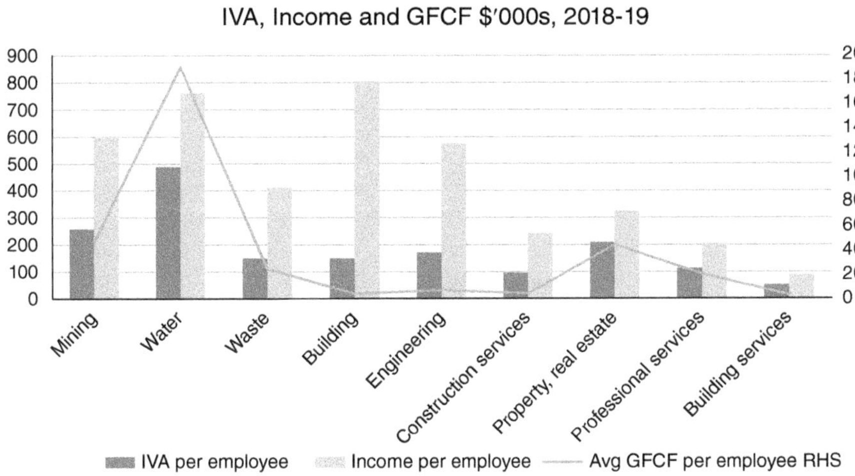

Figure 3.9 Industry comparison of IVA, income and GFCF per person employed

Source: ABS 8155

labour, or of assets to revenue in a company's accounts. Industries that are capital intensive like cement, water and sewerage, and real estate require large amounts of capital, and therefore high levels of GFCF. In the absence of capital stock data at this level, GFCF is an alternative measure of capital intensity across the BES. However, annual GFCF figures are highly variable, so here a four-year average is used. In Figure 3.9, when this average GFCF per employee between 2015–16 and 2018–19 is compared to IVA per employee there is a close match, industries with high IVA per employee also have high expenditure on GFCF per employee. Water, sewerage and drainage, Property and real estate, and Professional and technical services are the industries with relatively high levels of GFCF.

The three construction industries all have a low level of GFCF, primarily because they typically lease or rent large machinery and equipment as required. Many Construction services are labour intensive and average GFCF per employee is low because it is spread over many employees. Also, when the value of IVA per employee is compared to income per employee, the Building and Engineering industries convert relatively low shares of their revenue into value added, reflecting a contracting business model based on managing cash flow. At the other extreme, Water and waste and Property and real estate convert around 65 percent of income per employee into IVA.

Economic role of the Australian BES

Taking a broad view of an industrial sector provides perspective on its role and significance in economic and technological development. The IVA of the 16 built environment industries contributed 14.2 percent to Australian GDP in 2018–19, within their long-run range between 14 and 15 percent of GDP since 2006–07. The BES share of total employment fluctuated between 16.5 and 17.5 percent of

Table 3.6 Australian BES: Summary statistics

2018-19	*Employment (persons)*	*IVA ($ billion)*
Total Australian Built Environment Sector	2,126,000	270
Total Australia Employment and GDP	12,867,000	1,801
BES Percent of Australia total	16.5%	14.2%

Sources: ABS 8155, ABS 5206, ABS 6202

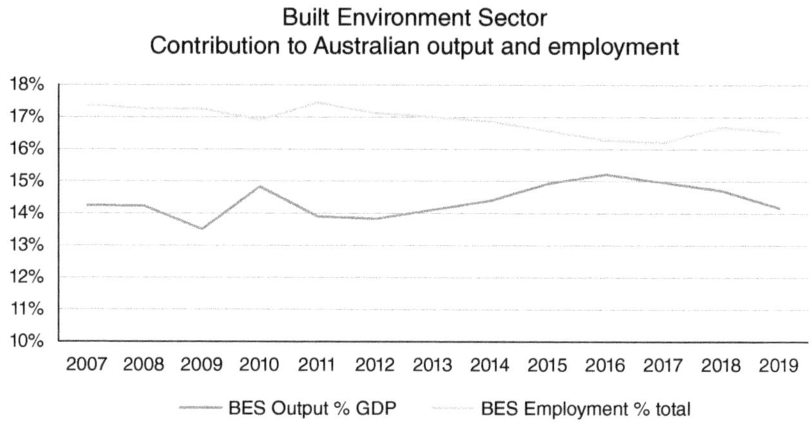

Figure 3.10 Economic role of the Australian built environment sector

Sources: ABS 8155, ABS 5206, ABS 6202

total employment, reaching a high in 2011 after a major fiscal stimulus during a period of exceptionally high mining investment expenditure.

Research shows construction has strong demand-pull linkages from other industries but relatively weak supply-push linkages, which means the contribution to total output through the industry's domestic supply chain has an important macroeconomic role (Lewis 2008; Meikle and Gruneberg 2015). Using the input-output tables produced by national statistical agencies, this characteristic has been found in the Italian construction industry (Pietroforte *et al.* 2000), the Australian construction industry (Song and Liu 2006; Liu and He 2016; Zhu *et al.* 2020), the Turkish construction industry (Gundes 2011), and South Asian construction (Ali *et al.* 2019). Through those linkages the impact of construction activities on other parts of the economy is much greater than their direct contribution (Pietroforte and Gregori 2003). These linkages with other sectors can also be estimated through the industry's multiplier effects (Lean 2001; Gregori and Pietroforte 2015).

The scale and continuity of the BES contribution to output and employment is a significant component in macroeconomic stability, directly through household incomes and indirectly through aggregate demand. Recognising this, policies to increase building and construction work are a common response to slowdowns in growth during the contraction phase of the business cycle (Briscoe 2009), and increased infrastructure spending is seen as an effective policy to support demand in the short run while developing assets for the future (Gruneberg and Francis 2019). The Australian experience managing the global financial crisis after 2007 and the end of the mining boom in 2014 are given in the following text as examples of the importance of the BES to macroeconomic policy.

Fiscal policy and the global financial crisis

The economic stimulus used by many governments during the global financial crisis that started in 2007 followed the Keynesian macroeconomic fiscal policy framework of increased government expenditure. The Australian Government's response to the crisis included a major school building program and a home insulation scheme, as well as large and small infrastructure projects. As the following Figures show, the counter-cyclical timing and impact of the Rudd Government's fiscal stimulus and building program is an example of the effectiveness of policies that engage the entire BES network of firms and organisations.

Separating the BES from the other industries in the ABS data shows that the increase in Commonwealth Government spending in 2009–10 on schools, buildings and infrastructure flowed through to the wider economy over the following years. The big increase in spending was concentrated on building work in the 2009–10 and 2010–11 budgets, in Figure 3.11. With the increase in public building there was a very large increase in the IVA of the built environment industries. The BES IVA increased by 12 percent over 2009–10, at a time when nominal GDP growth was under 2 percent, in Figure 3.12.

Australia was the only G20 country to avoid a recession during the global financial crisis, and one reason was the increase in Commonwealth construction expenditure. After starting to increase in 2008, in 2009–10 public sector building work done more than doubled to over 1.5 percent of GDP. Australian Government net debt had previously peaked in 1996 at 18 percent of GDP, then fell to zero by 2006, and was back to 12.5 percent of GDP in 2013–14 after the GFC and years of large budget deficits.[3]

Monetary policy and the transition

Rebalancing the economy after the mining boom ended in 2014 was another major macroeconomic challenge. To support aggregate demand the Reserve Bank of Australia lowered interest rates and encouraged banks to lend for mortgages and property development, see Figure 3.14. During the subsequent residential boom in apartment building from 2013 to 2018, the BES supported output across the economy as the mining boom ended and engineering construction and business investment fell from 18 percent of GDP to 8 percent. Over that period residential building rose from around 120,000 to over 200,000 commencements a year, due to an increase in high density apartment developments. In

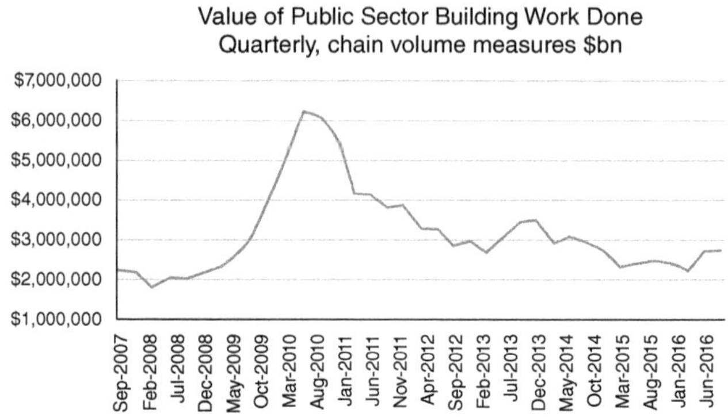

Figure 3.11 Government expenditure on building

Source: ABS 8752

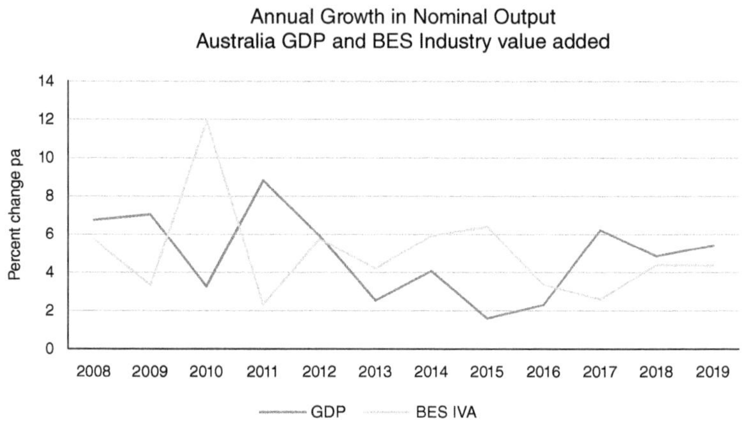

Figure 3.12 Annual percentage change in output

Sources: ABS 8155, ABS 5206

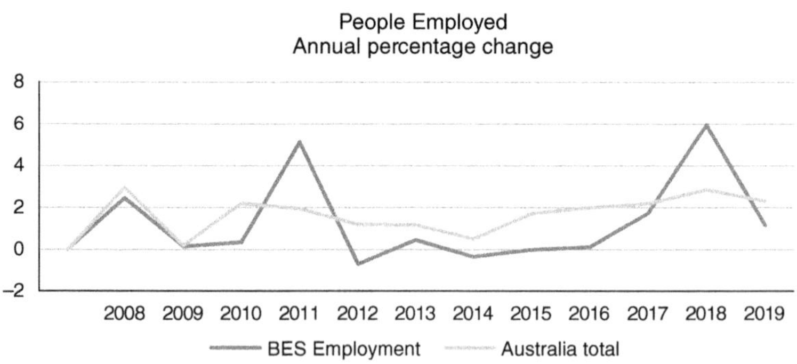

Figure 3.13 Annual percentage change in people employed

Sources: ABS 8155, ABS 6202

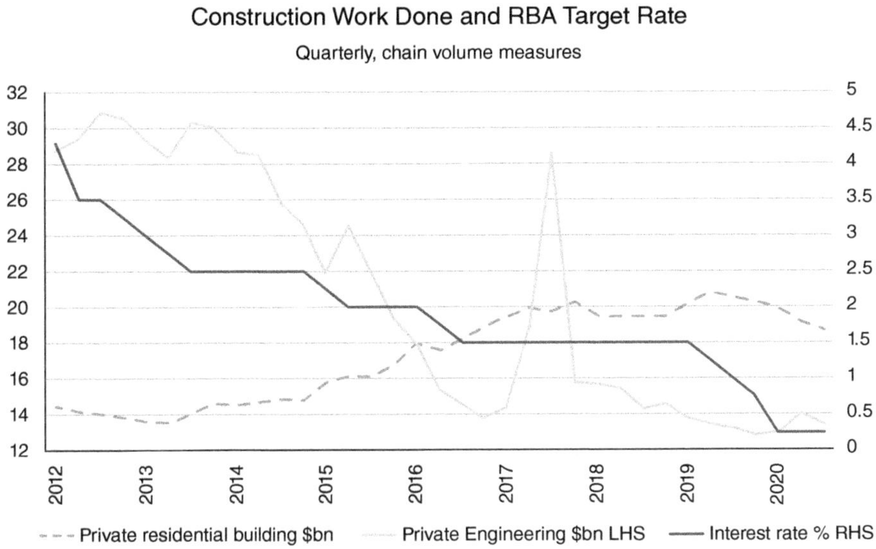

Figure 3.14 Private sector construction and interest rates

Sources: ABS 8752, ABS 8762, Reserve Bank of Australia

2017–18 BES employment growth peaked at around 6 percent, at the top of the residential cycle.

Australia's mining boom started in the early 2000s, and between 2006–07 and 2013–14 Engineering construction more than doubled its share of Construction IVA, increasing from 12 to 24 percent. Over that period the share of Construction services fell from 67 to 55 percent of total Construction IVA, while Building was around 20 percent. By 2016–17 Engineering had fallen to 16 percent of Construction IVA, Building had increased to 25 percent, and Construction services were 59 percent. The shift from Engineering to Building meant BES IVA was growing around twice as much as GDP between 2003 and 2017, and the strong backward linkages between industries meant the effect on the economy was stronger compared to the mining boom because of the large amounts of imported plant, machinery and equipment included in the Engineering work statistics (included are oil and gas platforms for example).

The BES clearly has a significant role in the economy, as the examples of the effects on BES output and employment of fiscal policy in the global financial crisis and monetary policy in the transition after the mining boom show. In the first case, a very large increase in 2009–10 on public sector building work saw an increase in BES IVA of 12 percent and employment of 6 percent. In the second case, after 2013 as engineering construction work fell from the highs of the mining boom, interest rates were lowered and the increase in residential building work supported the economy during a difficult macroeconomic transition.

Within Construction, internal dynamics saw significant changes as the IVA shares of the three component industries rose and fell with changes in the composition of Construction output: Construction services between 67 and 53 percent;

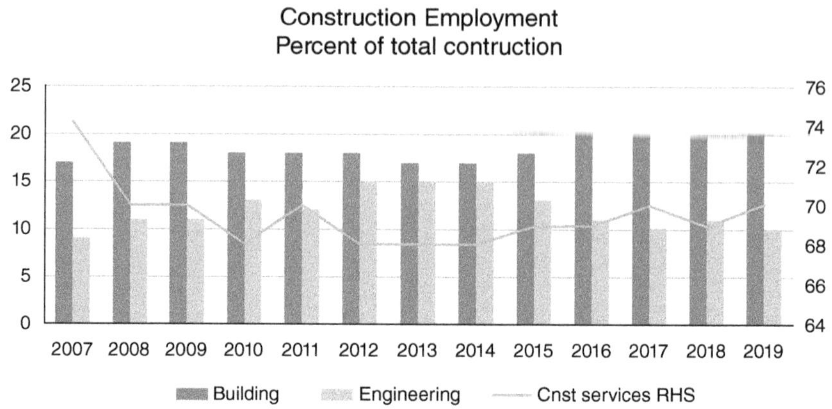

Figure 3.15 Industry shares of construction employment

Source: ABS 8155

Building between 20 and 27 percent; and Engineering between 12 and 24 percent of Construction IVA. While those fluctuations in output were occurring there was a shift in employment away from Construction services as Engineering increased, which was reversed after 2014 as Building work increased and Engineering fell.

Why measure the BES?

Construction is made up of three of the 16 industries involved in the production of the built environment. On-site work links suppliers of materials, machinery and equipment, products and components, and all other inputs required to deliver the buildings and structures that make up the built environment. Consultants provide design, engineering, cost planning and project management services. Once produced, buildings and structures then need to be managed and maintained over their life cycle, work done by another group of related industries. The built environment also needs infrastructure and services like water and waste disposal, provided by yet more industries. As Figure 3.16 shows, construction only accounts for around half of the total output and employment of the BES.

A dense network of many different firms and participants such as this is often called an industrial or economic network (Hughes 1983) or sector, because it is too diverse and distributed to be a cluster in the conventional geographic sense (Porter 1985). There is no specific definition of an industrial sector, as it is a broad collection of firms with one or more common characteristics, like 'manufacturing' or 'the business sector', though firms in these sectors come from many different industries. This is also the case with the diverse collection of industries involved in constructing and maintaining the built environment.

In a time of rapid urbanisation and great social and environmental challenges, infrastructure, the built environment and in recent years city policies have

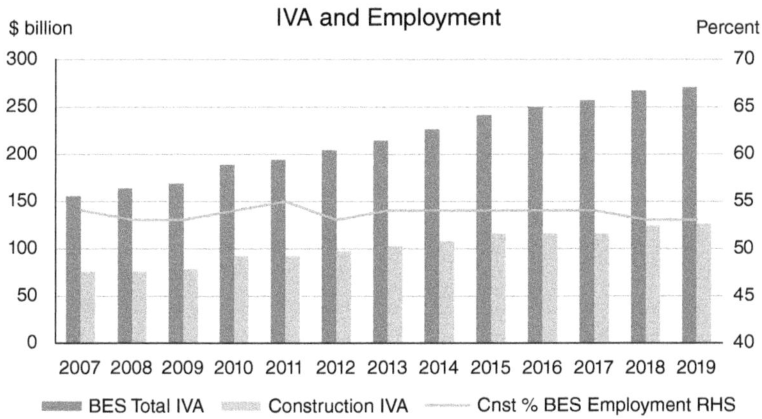

Figure 3.16 BES and construction IVA and employment

Source: ABS 8155

become central issues in public policy. The quality of the built environment the BES delivers is a major determinant of the quality of life. Further, in a fundamental sense, how cities function depends on how well the BES can deliver the projects required, and cities are at the centre of the modern economy. There are many issues affecting the built environment, many of which are wicked problems of great complexity that range widely across industries, institutions and regulatory systems. How measuring the BES helps is by providing an overview of a value chain, from suppliers to end users, and offering a view, or views, of pathways to future policy goals. It does this by allowing possibilities for deeper integration between participants across the sector.

Taking a sector approach to the built environment allows for a wider perspective on industrial development, which is important when work is project-based. Economies grow by upgrading the products they produce and export, but the technology, capital, institutions, and skills needed to make newer products are more easily adapted from related products with common labour and capital requirements. This network of relatedness between products means industries move through their product space by developing goods and services close to those they currently produce, based on their ability to adapt with new products and services. (Hausmann and Hidalgo 2011). Under these conditions, the set of options available to a narrow SIC industry is strongly influenced by its position in the product space of the wider sector it contributes to. With digital twins and the wide range of new production technologies currently emerging, such as 3D printing of concrete, automated machinery and buildings made with new materials like engineered wood, the BES is a laboratory for the fourth industrial revolution. Because it is not possible to know now which of these technologies will work at scale, a role of policy as facilitator can be through providing opportunities for new methods of production, organisation and management to be tested and trialled on demonstration projects throughout the industries that contribute to the construction and maintenance of the built environment. Other important parts of the economy such as tourism and defence are described as industrial sectors, although firms in these sectors come from many different industries there is a framework for policies to support industry development and capabilities and maintain a pipeline of projects.

Conclusion

There are 16 industries with data available from the annual ABS publication *Australian Industry* that can be classified as contributing to the construction, management and maintenance of the built environment. The collective significance of these industries is obscured by their diversity, ranging from architecture to waste disposal, and their geographic distribution. A method to measure the contribution of built environment industries to output and employment using SIC codes as developed by construction economists has been applied to Australia using the data from *Australian Industry*.

Between 2007 and 2019 the Australian BES accounted for 14–15 percent of GDP and 16–17 percent of total employment. It has a significant role in the economy, with examples given of the effects on BES output and employment of fiscal policy in the global financial crisis and monetary policy in the transition after the mining boom. In 2009–10 expenditure on public sector building work led to an increase in BES IVA of 12 percent and employment of 6 percent. In 2013–14, as engineering construction work fell from the highs of the mining boom, interest rates were lowered and increasing residential building work supported the economy during a difficult macroeconomic transition over the next three years.

The three industries that make up total Construction are the largest component of the BES, but their share has fallen gradually from 49 to 47 percent of total BES IVA. Construction employment, however, has remained at 53 percent of total BES employment. Within Construction the internal dynamics saw significant changes as shares of the three component industries rose and fell with changes in the composition of construction output. Adapting to these shifts in demand and changes in the type of work, industries across the BES have demonstrated the characteristic flexibility and responsiveness of project-based production.

There has also been a long-run trend in the BES reflecting the increasing share of services in the wider economy. Within BES total output this can be seen in the increased share of Property and real estate services in IVA from 20 to 26 percent between 2007 and 2019. BES employment shares of the total have been relatively stable except for Professional services, which had the biggest increase in employment, of 20 percent, and fastest growth, of 8 percent a year in the 5 years to 2019. Since 2007 the share of Manufacturing in BES IVA and employment has fallen from 10 to 7 percent, although the nominal value of IVA produced by manufacturing-related industries has had a small increase.

In the same way manufacturing is not itself an industry, but a collection of industries that make up an industrial sector of the economy where firms have similarities in products and processes, industries that contribute to the construction and maintenance of the built environment can also be collected and their contribution to output and employment measured. The economic role of the BES is important and better data can contribute to economic policy decisions that, through the BES and its dense network of linkages between industries, significantly affect macroeconomic outcomes.

Notes

1 Tourism is not an industry in the conventional sense and is defined in an ABS tourism satellite account as an amalgam of other industries including transport, accommodation, food service provision, retail trade, entertainment and education (ABS 2009). Multipliers for a satellite industry such as tourism double count multipliers of component activities.
2 Excluded are ANZSIC Subdivisions 62 Finance, 63 Insurance and superannuation funds, 64 Auxiliary finance and insurance services, 75 Public administration, and 76 Defence.
3 See the Commonwealth Government's Budget Paper No. 1, *Budget Strategy and Outlook* for the years 2006–07 to 2013–14 for details. The fiscal stimulus totalled over 5 percent of GDP over 2009–12.

References and further reading

ABS (1998) *Construction Industry Survey 1996–97*. Cat. No. 8772.0 (Canberra: Australian Bureau of Statistics).

ABS (2009) *Information Paper: Introduction of Revised International Statistical Standards in the Australian Tourism Satellite Account*. Cat. no. 5249.0.55.002 (Canberra: Australian Bureau of Statistics).

AEGIS (1999) *Mapping the Building and Construction Product System in Australia*. Australian Expert Group on Industry Studies (Canberra: Department of Industry, Science and Resources).

Ali, Y., Sabir, M. and Muhammad, N. (2019) A Comparative Input-Output Analysis of the Construction Sector in Three Developing Economies of South Asia. *Construction Management & Economics*, **37** (11), 643–658.

Best, R. and Meikle, J. (2015) *Measuring Construction* (Abingdon: Routledge).

Best, R. and Meikle, J. (2019) *Accounting for Construction* (Abingdon: Routledge).

Bon, R. (ed.) (2000) *Economic Structure and Maturity: Collected Papers in Input-Output Modelling and Applications* (Aldershot: Ashgate).

Bostic, W.S., Jarmin, R. and Moyer, B. (2016) Modernizing Federal Economic Statistics. *American Economic Review*, **106** (5), 161–164.

Briscoe, G. (2009) The Impact of Fiscal, Monetary and Regulatory Policy on the Construction Industry. In: Ruddock, L. (ed.) *Economics for the Modern Built Environment* (London: Taylor & Francis).

Bygballe, L.E., Håkansson, H. and Jahre, M. (2013) A Critical Discussion of Models for Conceptualizing the Economic Logic of Construction. *Construction Management & Economics*, **31** (2), 104–118.

Carassus, J., Andersson, N., Kaklauskas, A., Lopes, J., Manseau, A., Ruddock, L. and de Valence, G. (2006) Moving from Production to Services: A Built Environment Cluster Framework. *International Journal of Strategic Property Management*, **10**, 169–184.

Chan, T.K. (2019) Cost Ratios and Technology Choice. In: Best, R. and Meikle, J. (eds.) *Accounting for Construction: Frameworks, Productivity, Cost and Performance* (pp. 129–154) (London: Taylor & Francis).

Connaughton, J. and Meikle, J. (2013) The Changing Nature of UK Construction Professional Service Firms. *Building Research and Information*, **41** (1), 95–109.

Cruz, S.C. and Teixeira, A.A. (2010) The Evolution of the Cluster Literature: Shedding Light on the Regional Studies – Regional Science Debate. *Regional Studies*, **44** (9), 1263–1288.

de Valence, G. (2001) Defining an Industry: What Is the Size and Scope of the Australian Building and Construction Industry? *The Australian Journal of Construction Economics and Building*, **1** (1), 53–65.

de Valence, G. (2018) *Construction as a Mature Technological System* (Wolverhampton: First International Conference on Construction Futures).

de Valence, G. (2019) Reframing Construction Within the Built Environment Sector. *Engineering, Construction and Architectural Management*, **26** (5), 740–745.

de Valence, G. and Abbott, M. (2015) A Review of the Theory and Measurement Techniques of Productivity in the Construction Industry. In: Best, R. and Meikle, J. (eds.) *Measuring Construction: Prices, Output and Productivity* (pp. 205–223) (London: Taylor & Francis).

Gregori, T. (2008) Input-Output Techniques Applied to Construction. In: Ruddock, L. (ed.) *Economics for the Modern Built Environment* (pp. 60–78) (London: Routledge).

Gregori, T. and Pietroforte, R. (2015) An Input-Output Analysis of the Construction Sector in Emerging Markets. *Construction Management & Economics*, **33** (2), 134–145.

Gretton, P. (2013) *On Input-Output Tables: Uses and Abuses*. Staff Research Note (Canberra: Productivity Commission).

Groak, S. (1992) *The Idea of Building* (London: E. & F.N. Spon).

Gruneberg, S. (ed.) (2019) *Global Construction Data* (London: Taylor & Francis).

Gruneberg, S. and Francis, N. (2019) *The Economics of Construction* (Newcastle upon Tyne: Agenda Publishing).

Gundes, S. (2011) Exploring the Dynamics of the Turkish Construction Industry Using Input-Output Analysis. *Construction Management & Economics*, **29** (1), 59–68.

Hausmann, R. and Hidalgo, C.A. (2011) The Network Structure of Economic Output. *Journal of Economic Growth*, **16**, 309–342.

Hughes, T.P. (1983) *Networks of Power: Electrification in Western Society, 1880–1930* (Baltimore: Johns Hopkins University Press).

Hughes, T.P. (1987) The Evolution of Large Technological Systems. In: Bijker, W.E., Hughes, T.P. and Pinch, T.J. (eds.) *The Social Construction of Technological Systems: New Directions in the Sociology and History of Technology* (Cambridge, MA: MIT Press).

Ive, G.J. and Gruneberg, S.L. (2000) *The Economics of the Modern Construction Sector* (London: Macmillan).

Lean, S.C. (2001) Empirical Tests to Determine Linkages Between Construction and Other Economic Sectors in Singapore. *Construction Management & Economics*, **13**, 253–262.

Lewis, T.M. (2008) Quantifying the GDP Construction Relationship. In: Ruddock, L. (ed.) *Economics for the Modern Built Environment* (London: Routledge).

Liu, C. and He, S. (2016) Input-Output Structures of the Australian Construction Industry. *Construction Economics and Building*, **16** (2), 56–70.

Meikle, J. and Gruneberg, S. (2015) Measuring and Comparing Construction Internationally. In: Best, R. and Meikle, J. (eds.) *Measuring Construction: Prices, Output and Productivity* (Abingdon: Routledge).

Ofori, G. (1990) *The Construction Industry: Aspects of Its Economics and Management* (Singapore: Singapore University Press).

Pearce, D. (2003) *The Social and Economic Value of Construction: The Construction Industry's Contribution to Sustainable Development* (London: nCrisp).

Pietroforte, R., Bon, R. and Gregori, T. (2000) Regional Development and Construction in Italy: An Input-Output Analysis 1959–1992. *Construction Management & Economics*, **18** (2), 151–159.

Pietroforte, R. and Gregori, T. (2003) An Input-Output Analysis of the Construction Sector in Highly Developed Economies. *Construction Management & Economics*, **21**, 319–327.

Porter, M.E. (1985) *Competitive Advantage* (New York: The Free Press).

Ruddock, L. and Ruddock, S. (2009) The Scope of the Construction Sector: Determining Its Value. In: Ruddock, L. (ed.) *Economics for the Modern Built Environment* (London: Routledge).

Song, Y. and Liu, C. (2006) The Australian Construction Linkages in the 1990s. *Architectural Science Review*, **49** (4), 408–417.

Squicciarini, M. and Asikainen, A.-L. (2011) A Value Chain Statistical Definition of Construction and the Performance of the Sector. *Construction Management & Economics*, **29** (7), 671–693.

Strassman, P. (1970) The Construction Sector in Economic Development. *Scottish Journal of Political Economy*, **17** (3), 391–409.

Taylor, M.D. (2010) A Definition and Valuation of the UK Offsite Construction Sector. *Construction Management & Economics*, **28** (8), 885–896.

Turin, D.A. (1969) *The Construction Industry: Its Economic Significance and Its Role in Development* (New York: UNIDO).

UN Statistical Division (2008) *Standard Industrial Classification of All Economic Activities* (New York: United Nations).

Zhu, R., Hu, X. and Liu, C. (2020) Structural Analysis of Inter-Industrial Linkages: An Application to the Australian Construction Industry. *Construction Management & Economics*, **38** (10), 934–946.

Editorial Comment

In 1881 a US Army General, Montgomery C. Meigs, designed a new building in Washington DC to house the Pension Bureau. The Pension Building, which now houses the National Building Museum, is a five-storey brick building in the Italianate Renaissance Revival style which occupies an entire city block with a footprint 400 × 200 feet (approx. 120 × 60m). Meigs was both architect and engineer, and he supervised the construction which ran from 1882 to 1887. He included a number of innovations, most notably a natural ventilation system which achieved up to 30 air changes per hour with no mechanical assistance. This was possible because Meigs, as the only designer, integrated all facets of the design.

Through the 20th century building design became more and more specialised and fragmented, with project teams formed that included architects, engineers of various types (mechanical, electrical, civil and so on) and cost consultants, among others. Traditional design-bid-build procurement meant that those who eventually constructed the projects were not involved until the design and documentation was more or less complete. Disputes were, and still are, commonplace, with confrontation often more likely than co-operation or collaboration.

In the latter part of the century, as interest in 'green' buildings increased, the need for a more integrated design approach became apparent, as features such as passive ventilation required different parts of a building to work together if the system was to function efficiently. Equally, what was designed had to be built correctly or the system would fail. In 1987, 100 years after the completion of Meigs' building, the new NMB Bank building in Amsterdam was completed. It, too, featured natural ventilation, along with a raft of other 'green' innovations. The design was carried out by an integrated, collaborative design team that included physicists, artists, and the construction company that would build the project as well as the usual architects and range of engineers. Collaboration was the key to its success and since that time there have been many instances of this sort of integrated building design around the world. Early contractor involvement (ECI) is a form of collaborative contracting that has evolved as a way of getting input from builders prior to finalisation of the design and the tendering process.

A move to greater cross-disciplinary collaboration in the design and construction of projects raises questions of how built environment professionals – architects, builders, quantity surveyors, engineers and others – are educated so that collaboration and teamwork can be fostered and developed. In the following chapter the author considers a range of issues related to education in these disciplines with training for collaboration as its focus.

4 Education for collaboration in construction

Challenges for universities and institutions

John Connaughton

Introduction

Construction industries around the world are frequently characterised by generally poor performance across key processes of design development, coordination, management and construction leading to projects that are late-running, over-budget and poorly built. Factors implicated in this include the organisational and technical complexity of construction projects, disciplinary fragmentation in design and construction, and commercial barriers to effective joined-up working between the parties involved. Improving collaborative working among construction participants as a way of tackling these problems has attracted the attention of scholars for some time (see, for example, Xue *et al.* 2010; Monson *et al.* 2015). Industry-based practitioners have also joined calls for more collaborative working (Morrell 2015; Farmer 2016). In spite of this, and of further arguments that improving collaboration will contribute to more durable behaviour change that will benefit construction as a whole in the long term (Smyth and Pryke 2008; Sunding and Ekholm 2015), achieving more collaborative design and construction processes is still an elusive goal. Questions about how those involved in construction projects can work more effectively together to improve outcomes (Gottlieb and Haugbolle 2013) remain broadly unanswered.

Why is this? It is possible to suggest a range of potential causes across a variety of different perspectives and levels. Professional institutions, focused primarily on the development of their specific, mono-disciplines of architecture, engineering, construction and management, and their role in safeguarding their members' interests do not naturally see improved collaboration as their central concern (Foxell 2019; Morrell 2015). Construction businesses, drawing on lengthy industrial traditions and working within an extensive apparatus of regulation, contracts, processes and procedures that essentially enshrine these institutional divisions do not generally cope well with innovation and change (Reichstein *et al.* 2005) and have little incentive to do so. The public, including the industry's clients, may have little direct interest in collaboration between professional disciplines either. This is arguably because they rely heavily on the professional knowledge in discipline-specific areas to safeguard their interests, and their scope for redress in cases of failure is formed out of sharp distinctions between them (e.g. Lee *et al.* 2020).

DOI: 10.1201/9781003102403-4

And construction education, which to a large extent mirrors these wider institutional and societal structures and arrangements, is also implicated. A key criticism is that higher education prepares graduate construction professionals for a life of mono-disciplinary practice that effectively ignores both the need for them to work effectively together with other disciplines and, crucially, the development of skills that would enable them to do so (MacDonald and Mills 2013).

A central dilemma then is that while many associated with construction wish to see more, and more effective collaboration between different disciplines, many of the industry's underlying structures and arrangements are at best not designed with collaboration in mind and, worse, may present significant obstacles to achieving it. Within such context, this chapter focuses on understanding the current and potential contribution that construction education at tertiary/professional level can make in supporting collaboration. To help do that, it will be argued that an important and frequently overlooked challenge to improving collaboration lies in the nature of what collaboration actually is, and a lack of clarity over what work it involves. So before turning to an examination of the role of construction education, attention turns first of all to concepts and practices of collaboration.

What is collaboration, and how is it done?

There is now a somewhat extensive and growing literature on collaboration in the construction domain. At a very general level, collaboration may be viewed as a form of collective human action in pursuit of a common goal (Wood and Gray 1991). Beyond that, however, there is less agreement about what is involved, and considerable variety in how it is treated in terms of its foundational concepts, the circumstances or conditions that might influence it, the different forms it takes and its underlying purpose.

Key concepts, principles and conditions

Foundational concepts include trust and mutual respect between participants (Kadefors 2004; Nystrom 2005; Gajendran and Brewer 2012; Hughes *et al.* 2012), openness and effective communication (Gajendran and Brewer 2012), and a willingness of participants to work together allied with 'appropriate' behaviours (Lloyd-Walker *et al.* 2014). Favourable conditions are related to such foundational concepts and include relational attitudes and teamworking quality (Suprapto *et al.* 2016), a mutual understanding of project goals and tasks (Okhuysen and Bechky 2009; Mattessich and Monsey 1992), the absence of a range of inappropriate cultural 'barriers' (Mollaoglu *et al.* 2015), and often the presence of a 'convenor' (Wood and Gray 1991) to steer and manage the collaborative endeavour. In terms of the forms that construction collaboration may take, work tends to focus on a wide range of organisational and contractual project arrangements, including partnering, alliancing, joint ventures and networking that are considered to fall within what Hughes *et al.* (2012: 355) describe as an 'umbrella' term of collaboration. In a meta-analysis, London and Pablo (2017: 555–558) draw elements of work on collaboration together and characterise dominant portrayals in mainstream approaches as involving a human activity, led by a convenor and focused on achieving a common goal, seeking integration, accompanied by formal arrangements and 'structures', and taking place in an external environment.

What is collaboration for?

Such contributions are, of course, important in helping to understand the conditions for collaborative working and how they might be formalised in project contractual arrangements. Indeed, they may be seen as directed at overcoming the kind of institutional and commercial obstacles discussed earlier. But they say little about why construction professionals should collaborate (what it is for) nor, more particularly, what collaboration practice actually involves. These questions matter for those concerned with construction education and the challenge of preparing graduates for a future of collaborative working during their professional lives. And here the question of what it is for seems crucial to understanding what collaboration might entail in the day-to-day work of project life – who works with whom, in what ways, and towards what ends? Yet such questions are not always explicit in much of the collaboration literature. Instead, the core purpose of collaboration in particular is often taken as self-evident: construction is a complex process requiring the input of different disciplines, so improved collaboration between them will improve the process. But is that enough? Is collaboration simply about improving process and efficiency, or might it be more to do with improving outcomes and effectiveness, or indeed both? For Hughes *et al.* (2012), for example, 'desirable aspects' of collaboration may include reducing waste and improving efficiency and profit for those involved. But for Schöttle *et al.* (2014), collaboration is about more than this: it is a way of achieving commonly agreed project goals through a joint enterprise in which participants solve problems together and may even share in the risk and reward of doing so.

What does collaboration involve?

And what do such different notions of what collaboration is for imply for what it involves and how it might be done? Perhaps because the purpose is not always clearly defined, the different ways in which collaboration is conceptualised and described are not so clearly distinguished either. For some, collaboration is a general term for working together, and is used synonymously with other terms including coordination and cooperation, as in Bygballe *et al.* (2016), for example, to help explain the quality and intensity of relationships between project participants. But coordination is arguably different, involving more of a temporal or organisational ordering of the respective inputs of project participants working together (e.g. Bechky 2006; Okhuysen and Bechky 2009; Boudeau 2013 and, indeed, in Bygballe *et al.* 2016). And according to Schöttle *et al.* (2014), such coordinating activity requires less integration of the respective contributions of different disciplines than a more active form of collaboration that is about creating a shared enterprise between them. Similarly, cooperation may cover situations where disciplines contribute their respective inputs but stop short of an active sharing of them, retaining ownership of their contributions alongside their professional independence. Again, Schöttle *et al.* (2014: 1273–1275) distinguish this from what they consider to be collaboration, the latter involving participants surrendering some of their independence in sharing responsibility for problem solving in the joint enterprise they create and in which they share the risk and reward of resultant outcomes.

In a sense, then, these different concepts of working together – of coordination, cooperation and collaboration – could potentially represent different degrees or intensities of collaborative working, reflecting also the extent to which contributions from individual participants are integrated. But, as noted, they are not always clearly delineated in treatments of collaboration. A similar and potentially useful categorisation is available from work in the area of collaborative research that takes a disciplinary perspective. Klaassen (2018), drawing on Menken and Keestra (2016) highlights distinctions between multidisciplinary, interdisciplinary and transdisciplinary working in terms of a simple and gradual scale to help describe how knowledge from different sources is progressively integrated. Their scheme is illustrated in Figure 4.1. At the first level, termed multidisciplinary working (Multi), disciplinary contributions are unintegrated. Each discipline contributes a piece of the jigsaw, so to speak, which is largely unaltered in the final outcome and sits alongside other disciplinary inputs. The result is a series of parallel visions of the 'problem area' (or, in construction project terms, the project aims and potential solutions) from different disciplinary perspectives. By contrast, at a higher level of integration and collaborative working, termed interdisciplinary working (Inter), the different disciplinary contributions become more integrated in a synergistic way, developing shared problem solutions (Schöttle *et al.* 2014)

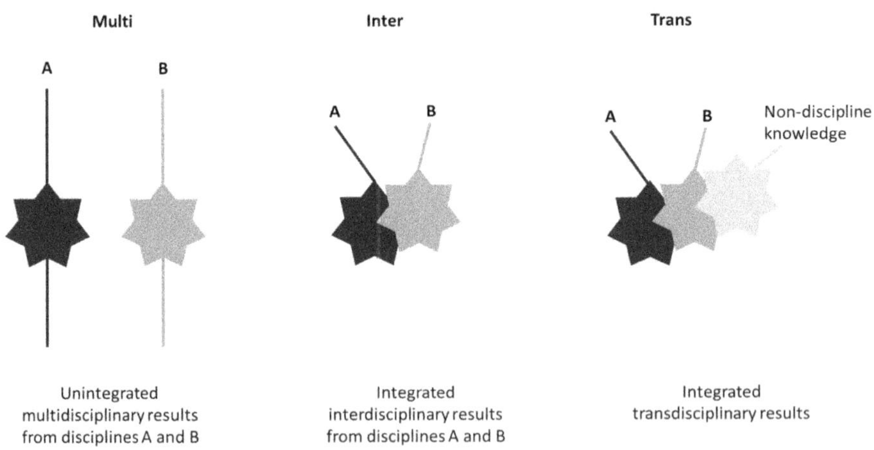

Figure 4.1 Levels of integration in different forms of collaborative working (After Klaassen 2018; Menken and Keestra 2016)

that go beyond the primarily additive outcomes of multidisciplinary working. And finally, transdisciplinary working (Trans) involves stakeholder contributions outside of the normal disciplinary groupings in a search for innovative solutions.

Construction, with its often complex requirements and site-specific contexts that mean solutions are not always known in advance, is particularly amenable to interdisciplinary problem solving. In recent years a focus on the challenges presented by the need for improved sustainability performance requiring novel, inter- and transdisciplinary solutions has brought this into sharp focus (Hill and Lorenz 2011; Pero *et al.* 2017). Yet, in work on collaboration, the distinction is not always clearly made and, as with the interchangeable use of concepts such as coordination, cooperation and collaboration, the implications for what interdisciplinary and transdisciplinary working might mean for construction are not always explicitly recognised or explored.

So, perhaps because of this, little attention is paid to understanding the practices of collaboration and how people work together in different ways, particularly in the kind of interdisciplinary working that involves the sharing and integration of knowledge and decision inputs to create novel, shared solutions. Rather, the focus of a good deal of the literature seems to be more on understanding and developing the *conditions* – contractual and otherwise – favourable to collaboration. And because of this, there is an underlying assumption that really all that is needed to unlock the assumed innate potential of professionals to collaborate are the right conditions to enable them to work together. But what of the actual doing of collaboration, especially in interdisciplinary working which suggests that professionals work across disciplinary boundaries and, indeed, may share responsibility for outcomes? If they already know how to do this, where does this knowledge come from?

Education for collaboration in construction – an overview

Education of construction professionals at undergraduate level tends to be delivered primarily through degree programs in mono-disciplinary schools and departments focusing on architecture, urban planning and a range of engineering, construction management and surveying disciplines (Morrell 2015; MacLaren *et al.* 2017). Such programs generally offer a route to membership of professional institutions and are accredited by them.

Collaboration in degree-level construction education

While MacLaren *et al.* (2017), for example, note how some UK institutions require degree programs to take account of the need for skills in aspects of collaboration, including teamworking and a knowledge of different disciplinary roles, a brief review of the accreditation requirements of the main UK institutions suggests that such requirements are not extensive – see summary in Table 4.1. Indeed, in so far as they acknowledge the need for collaboration at all, they tend to focus mainly on developing an *awareness* of the role and contributions of other disciplines rather

Table 4.1 Accreditation requirements of the major UK institutions for undergraduate degrees: the extent to which they include explicit criteria relating to collaboration

RIBA (RIBA 2014)	Very little on collaboration in the General Criteria and Graduate Attributes at Part 1 and Part 2, apart from General Criteria relating to understanding of 'the role of the architect within the design team' and a knowledge of 'professional inter-relationships in . . . architectural projects' as well as 'contributions of architects and co-professionals to the formulation of the brief' (2014: 59–61).
RICS (RICS 2018)	RICS requirements are somewhat complicated by the range of surveying disciplines covered and the different 'pathways' to membership. Mandatory requirements in the 'Competencies' guide cover 'Diversity inclusion and teamworking' but do not explicitly cover collaboration (2018: 17). Competencies also include leadership and managing people in relation to senior roles (2018: 20–21) but not specifically in relation to collaboration.
CIOB (CIOB 2018)	CIOB acknowledges the need to 'recognize the collaborative linkages and interdisciplinary relationships' and to 'evaluate the challenges of working in a collaborative environment', but emphasis more on teamworking and conflict resolution in terms of skills required for working with others (2018: 13–14, 18).
Engineering (Engineering Council 2014)	Engineering requirements are somewhat complicated by the range of engineering disciplines and institutions involved. In the UK the Engineering Council sets the overall accreditation requirements for higher education engineering programs. In these requirements the emphasis is mainly on technical knowledge, competence and skill development in science and engineering disciplines. There are some requirements in the 'practice' element for an 'Awareness of team roles and the ability to work as a member of an engineering team' and a recognition of the personal responsibility required for teamworking, but no explicit requirement for interdisciplinary collaboration.

than developing particular collaboration skills and knowledge, especially for working at the interdisciplinary level.

And while the relative lack of emphasis on collaboration in institutions' accreditation requirements perhaps underpins the reluctance of universities to focus on it, there are other factors at work. Morrell (2015), for example, speaks about a fragmented and siloed construction education process, observing that many schools and universities lack one or another of the disciplines necessary to establish multi-disciplinary faculties and approaches. Further, for those that have them, single-discipline departments can be isolated from each other by university structures (Morrell 2015: 59). MacLaren *et al.* (2017: 182) highlight higher education funding structures that encourage academic staff to develop deep domain expertise at the expense of competence – and an interest – in cross-discipline collaboration. In earlier work, MacDonald and Mills (2013) also argued that such specialist staff have little incentive to shed their hard-won expertise and associated teaching habits and material developed over many years. In addition, they point to a lack of compensation across faculties for staff involvement in cross-disciplinary working, as well as pressures created by growing class sizes (particularly in Australia, where many academics face minimum class sizes of 80) that present difficult challenges for the development of multi-disciplinary programs (MacDonald and Mills 2013: 97–98).

But this is not to say that tertiary education across construction disciplines has ignored the need for collaboration in construction project teams. Far from it. Indeed, construction degree programs have long recognised that construction professionals need to work together – Selman and Westcott (2005), for example, describe how module design in built environment courses at the University of the West of England has adopted an 'interdisciplinary philosophy' since the early 1990s. And degree programs typically include elements designed to support the development of what Holland *et al.* (2010) describe as 'T-shaped people': professionals that have deep knowledge of their own discipline and a broad understanding of the roles and responsibilities of others in the team. Many ideas have been proposed to address this in which, for example, construction courses could contain common years of study across different disciplines, as well as common subject modules (e.g. Manthe and Smallwood 2007). While this may seem to be a (albeit partial) response to institutional requirements for *knowledge* of related professional disciplines, it does not explicitly address the question of the *ability* required to collaborate actively with them. However, more recent ideas associated with learning in groups (frequently referred to as 'collaborative learning') have sought to shift the emphasis towards the latter.

Education for collaboration – the importance of learning in groups

Ideas of learning in groups are not particularly new in tertiary education. Davidson and Major (2014), for example, chart a growing interest in group-based learning since the 1960s, with theoretical origins in social interdependence theory (see Johnson and Johnson 1994, 2009) and considerable, if at times mixed, empirical evidence supporting its beneficial use (e.g. Springer *et al.* 1999; Smits *et al.* 2002). Indeed, such has been the extent of academic interest that developments in this broad area have led to myriad terms describing a range of approaches to learning in groups, including small-group learning, collaborative learning, cooperative learning, problem-based learning, team-based learning and a variety of other terms (Davidson and Major 2014: 9–10). And while this descriptive terminology varies considerably, so also do ideas about the purpose of group-based learning, ranging from claimed benefits in long-term knowledge retention to the development of skills necessary for future collaboration.

In a helpful general categorisation, Davidson and Major (2014) distinguish three dominant forms of learning in groups: cooperative, problem-based and collaborative learning, each having different theoretical underpinnings and assumptions. In this broad scheme, cooperative learning covers a wide range of approaches concerned with students working in groups in which they communicate with and help each other to complete learning tasks (Davidson 2002; Leite 2016). Problem-based learning puts the task or 'problem' – often presented as a 'real world', messy problem – centre stage to provide both the context and stimulus for the acquisition of knowledge and the development of problem-solving skills (Savery 2006). Collaborative learning departs somewhat from these approaches in its focus on the co-creation of knowledge by students and teachers working together. This essentially interpretivist approach (Bruffee 1993 in Davidson and Major 2014: 20–21) can be task- or problem-focused, the emphasis typically falling on the creation of novel solutions or products, and not simply the teacher's version of them.

A further element may be a requirement for students to take responsibility for their learning and what they create out of the collaborative endeavour (Bruffee 1993). While these broad distinctions become blurred in application and development, they help with understanding both the purpose of learning in groups and what this may involve. As with the concepts of collaborative working illustrated in Figure 4.1, they also suggest a broad (albeit crude) 'scale' for understanding the extent of interdisciplinary collaboration that might be involved. With this in mind, some approaches to learning in groups in construction education will now be briefly discussed. While a comprehensive review is beyond the scope of this chapter, the intention is to illustrate the range of provision, identify some of the more common elements and discuss the extent to which different approaches may address particular concepts of collaborative working and learning discussed earlier.

Some example of how education for construction is done

Construction, with its focus on one-off projects and preoccupation with teamworking, provides considerable opportunity for cooperation and problem-based learning approaches. In the construction domain, Chan and Sher (2014), for example, distinguish a particular form of problem-based learning as project-based learning, with the problem area in the latter more focused on the development of workable, project solutions than the potentially more open-ended outcomes of the former. Their interest is in improving students' teamworking and interpersonal skills and, based on a survey of students' learning preferences, they argue that such skills may be improved when students from different disciplines work together on multi-disciplinary, project-based assignments. Indeed, such project- or case-based approaches now feature prominently in construction education, particularly when geared towards collaborative working. The Live Projects Network, for example, provides a resource base of projects intended to provide learning contexts that 'extend the institutional confines of the design studio'.[1] While not exclusively focused on interdisciplinary collaboration, the Network provides a sense of the diversity of approaches to the use of project-based learning, particularly in relation to developing participatory approaches to the design of buildings and urban spaces (see also Harris and Widder 2014, for example).

In a review of approaches to interdisciplinary construction education across four UK Higher Education Institutions (HEIs), MacLaren *et al.* (2017) argue that it is not enough to try to teach teamworking in interdisciplinary groups; rather, students need to experience the collaborative environment in multidisciplinary project settings. Such experiential learning has long been a feature of construction education, particularly in relation to site-based workers (Floe 2019) and with some application in more professional disciplines also (Forster *et al.* 2017). The more immediate question here, however, is the extent to which such learning is focused on particular concepts of collaboration. MacLaren *et al.* observe that common features of seven group-learning case studies examined include the implicit adoption of a working definition of collaboration from Wood and Gray (1991) that describes a generic form of working together (2017: 183, 195–197). This is despite the focus of many of the project-based scenarios adopted being on the development of innovative solutions rather than more conventional responses to project/study briefs that might, perhaps, have required a more purposefully creative, interdisciplinary form of collaboration described, for example, by Schöttle *et al.* (2014).

Of course, different project settings arguably provide different opportunities for group learning with a focus on collaboration. The particular context of sustainable design, also requiring novel solutions to interdisciplinary challenges, has featured explicitly in group-based learning approaches. This can, according to Brncich *et al.* (2011: 23), provide 'an effective framework for sustainable design and construction education'. In the case study featured, they focus on some of the perceived benefits to students of the group approach in raising their awareness of the roles of other disciplines and the importance of communication in team working, as well as learning about important technical and performance aspects of sustainable design. Valdes-Vasquez and Clevenger (2015) also use a sustainable building project in an approach to developing interdisciplinary classroom activity designed to promote more specific collaborative working skills such as communication, conflict resolution, decision-making and problem-solving. Their interest is in collaboration that is focused on the development of improved design solutions: 'building delivery is not produced from one person's thinking process; rather, it is the result of the technical collective knowledge from different disciplines' (Valdes-Vasquez and Clevenger 2015: 81).

The potential of computer-based developments to transform professional working is currently the subject of considerable debate and enquiry (Susskind and Susskind 2015). In construction, the particular adoption of Building Information Modelling (BIM) across its many guises to support more integrated, collaborative working has also been the focus of a number of approaches to group learning. MacDonald and Mills (2013), for example, suggest that university curricula have not developed sufficiently rapidly to respond to growing requirements for collaborative working capabilities in response to developments such as Integrated Project Delivery (IPD). They propose a new framework that seeks to integrate principles of collaboration with BIM technologies through a focus on the development of 'both technical (I.T. and discipline-specific) and interpersonal (collaborative and teamwork) skills' (MacDonald and Mills 2013: 99). To support this, they advocate revisions to professional institutions' accreditation requirements so they address more collaborative working skills.

Zhao *et al.* (2015) focus rather more on the potential of BIM to provide a learning environment for collaborative skill development and working. They characterise a good deal of BIM coverage in university curricula in the USA as being concerned with technology application, lacking a focus on collaborative working. They argue that equipping students with BIM knowledge is not all that is needed to meet industry's needs for skilled construction professionals. Using a simulation of 'real world' working conditions requiring students to work within a BIM environment, they found that students from different engineering and construction disciplines felt they improved their collaborative workings skills in communication, coordination, cooperation and goal setting (Zhao *et al.* 2015; 114–7).

Comiskey *et al.* (2017), while pointing to the relative immaturity of BIM-focused material in degree programs in the UK and Ireland, focus on the use of data sharing platforms to support collaborative learning in project-based activity among students from different disciplines. While the potential for team-based problem solving to help those involved appreciate the contributions others can make is recognised, learning outcomes seem rather more focused on technology application and use than they are on the development of collaborating skill. Similarly, Vassigh *et al.* (2020) examine the role of virtual- and augmented-reality (AR and VR) technologies to simulate 'real world' project conditions for collaborative learning approaches in which students across different construction disciplines work together. Using a sustainable building design scenario, their focus – as with Comiskey *et al.* (2017) – is as much on the potential for learning collaboratively about the technical and performance aspects of the sustainable design challenge as it is about learning how to engage in more collaborative working. Nonetheless, the use of the AR and VR technologies was seen to promote group discussion and interaction, and a generally positive student attitude towards working together.

Education for collaboration in construction – some challenges and dilemmas

The foregoing vignettes of group-based learning approaches in construction highlight some important issues for any consideration of the role of tertiary education in supporting collaborative working among construction professionals. First, they provide evidence of significant activity in the development of group-based learning approaches over a number of years to suggest that ongoing criticism that education providers either are ignoring the need for collaborative working or lack the ability to do anything about it is not well-founded. That said, however, approaches to group-based learning, despite their range and diversity, can hardly be considered mainstream across the broad landscape of construction tertiary education. And further, while many group-based approaches adopt a form of 'project-based' learning (Chan and Sher 2014), there is considerable diversity and, indeed ambiguity in terms of their purpose and the different forms of collaboration they are meant to support. It is not always clear, for example, whether such approaches are intended to enhance core disciplinary knowledge on the one hand, or skill in collaborative working on the other, or indeed both. To put it another way, are they about learning collaboratively or learning to collaborate?

Learning collaboratively or learning to collaborate?

It is possible to distinguish the potential learning outcomes of the kind of group-based learning discussed earlier in two important ways. In the first of these, students learn more *about their disciplinary or topic domain* and related areas, and at the same time develop a better *awareness of others' inputs and capabilities* (Manthe and Smallwood 2007; Brncich *et al.* 2011; Comiskey *et al.* 2017). In other words, learning outcomes are primarily single-discipline focused. While students learn from each other alongside their peers from other disciplines, they exchange information and knowledge in a primarily cooperative learning activity that is focused on enhancing their understanding of the role and context of their own discipline.

By contrast, learning outcomes may be focused more on *how to work collaboratively*. In this approach the focus is more on the development of skills in aspects of collaborative working including communication, the development of interpersonal working relationships, and leadership (MacDonald and Mills 2013; Chan and Sher 2014; MacLaren *et al.* 2017; Vassigh *et al.* 2020). In these approaches, students are typically engaged in more of a joint endeavour around the problem area as part of a mainly collaborative activity geared towards developing additional know-how in joint problem solving and in working together.

These outcomes are not mutually exclusive and may not always be so easily distinguished either. However, in terms of the collaborative working scheme outlined earlier (see Fig. 4.1), learning to work *alongside* each other may be distinguished from learning to work *with* each other. In these terms, the former may be characterised by a primarily multidisciplinary *exchange* of information between participants, whereas the latter is more concerned with the *co-creation* of knowledge as participants collaborate with each other in a joint enterprise. The problem, when these outcomes are not clearly distinguished, is that what seems to matter more is the act of bringing students from different disciplines together in the belief that this, of itself, is sufficient to open a range of collaborative learning outcomes. It is as if – as reflected in a good deal of the collaboration literature briefly reviewed earlier – the presence of appropriate conditions (here in the form of multidisciplinary, group-based learning settings focused on project 'problems') is all that is needed to enable the innate collaborative abilities of students to come to the fore. And in these terms, opportunities for both the development of disciplinary knowledge *and*, where required, new collaborative capability are expected to arise without having to pay specific attention to either of them.

But some important questions arise here. One concerns skills. By mentioning so-called collaborative skills such as communication and relationship development, many approaches to group-based learning seem to take for granted what is involved in collaboration, without explicating what these skills involve nor how they might support and help participants to engage effectively in collaborative working. And this obscures a further question, which is about the nature of the collaborative endeavour that is envisaged and may be desired. In terms of the collaborative scheme shown in Figure 4.1, is the idea to help prepare students for a future of multidisciplinary working, or of inter-/transdisciplinary collaboration?

For if these distinctions are significant, surely they also matter in terms of the skill requirements of potential collaborators. These questions go to the core of what the role of tertiary education in this context might be and raise a number of challenges, and indeed dilemmas, for providers in mainstreaming collaborative approaches that will now be discussed.

Can collaboration be taught?

Many of the approaches to group-based learning highlighted earlier draw on theories of collaborative learning to help explain and set them in context. Problem-based learning, with its early origins in the teaching of medicine and other healthcare disciplines (Savin-Baden and Major 2004) provides the basis for a number of approaches (Chan and Sher 2014; Vassigh *et al.* 2020). Group-based learning more generally has somewhat deeper origins (Dewey 1938) that are reflected in many contemporary approaches. These include a number of the construction-focused and group-based initiatives described by MacLaren *et al.* (2017) that emphasise the participatory nature of the problem/project settings, and recognise the educational value as lying in the process rather than the output. Other approaches align with recognised learning frameworks and taxonomies, such as 'Bloom's taxonomy of learning' (Bloom *et al.* 1956) that is adopted by MacDonald and Mills (2013: 99) into a framework designed to support academics in the incorporation of collaborative design principles into their curricula. But few, it would seem, are informed to a similar degree, if at all, by theories of collaborative working, or an explication of associated collaborative working practices.

A challenge, however, is that work on collaboration in construction in particular is at an early stage in its development and, as argued earlier, there is little established theory, nor much consensus on what collaborative work actually involves either. And in its absence, one problem is that, as noted briefly earlier, work on collaborative learning tends to cover a range of aspects *perceived* to support collaborative working – good communication, the development of interpersonal relationships, teamwork, conflict resolution, leadership, and a host of others – but without necessarily highlighting any of them or unravelling and explicating their respective roles in collaborative work. What is needed therefore is more scholarly work focused on what different forms of collaboration involve and what differences between them mean so that the knowledge and skill needed in their performance can be understood and developed by those providing the education intended to support them. While the need may seem obvious, it has tended to be ignored in work on group-based learning in construction and, for such work to progress, needs to be brought more fully into thinking about collaborative learning – and working – in this sector.

A further challenge, particularly in relation to the kind of interdisciplinary collaboration discussed throughout, is to understand not only how collaboration works, but what the implications of working *across* disciplinary boundaries might be, for professional institutions, businesses, and individual professionals. Questions of how normal professional boundaries (what Abbott 1988 [at 88–90] refers to as 'jurisdictions') may need to be relaxed or even reconfigured. These challenge the relevance and durability of traditional jurisdictions of professional 'practice' and the role of institutions in developing and protecting them. The role of construction businesses, many of which are organised for the delivery of professional services along disciplinary lines and are regulated by institutions (Connaughton and Meikle 2013) is also relevant. Questions of professional identity, amongst other things, arise for individual professionals also. While a more detailed exploration of these issues is beyond the scope of this chapter, they are germane to a consideration of how professional knowledge and competence is organised and managed and traded in the construction marketplace. In this context, the extent to which interdisciplinary collaboration may require new competencies, and who might be responsible for providing them raises an enduring question: is it about education or training?

Key dilemmas for education

Tertiary educators regularly confront questions of whether their role is the production of employable 'job-ready' graduates or whether they are helping to prepare young people for a longer-term lifetime of learning in a 'learning to learn' approach (for a discussion, see Oliver 2013). Many argue that they are primarily engaged in the latter, distinguishing education from training. And yet, when it comes to thinking about working collaboratively in construction, the focus is on a largely experiential approach that would seem to eschew thinking about understanding collaboration from a more abstract and theoretical perspective in favour of an emphasis on learning how to do it by experiencing the doing of it. Across many of the examples of group-learning discussed, it is possible to see a sort of 'dominant logic' in which, according to MacLaren *et al.* (2017), the preferred means of helping students understand the demands of collaborative work is self-evidently to get them to experience it in simulated multidisciplinary project settings. An irony in this, of course, is that it would seem to place educators more in the practice and training domains than in the educational one, at least in terms of the broad distinctions outlined earlier. Regardless, the more fundamental questions are about what such simulations of project life have to offer for learning outcomes and for preparing graduates for working collaboratively and these questions are ultimately about the nature and purpose of the collaborative enterprise in which graduates might be involved during their working lives.

While there seems to be wide consensus that project-based simulation is valuable and enhances the learning experience, must collaborative learning always be done by mimicking what is happening in 'real' project life? In relation to distinctions between multidisciplinary and interdisciplinary working, for example, to what extent can simulations of typical project settings prepare students for a new and different version of the project life they might encounter in a future that is more open to active collaboration involving cross-disciplinary working and responsibility sharing? Must future practice be rehearsed as something knowable that will be encountered, or can it be imagined and co-created as new sets of activities and future practices? Addressing these latter questions, and paying attention to the need for further work on the role and nature of collaborative working, educators could potentially chart a clearer path which is more about teaching to learn rather than practising to do.

In addition, important questions for educators centre around the relative efficacy of different modes and approaches. Group-learning tends to be premised on a belief that experience of this environment is valuable and improves learning outcomes. And while, in general, evidence for the benefits of collaborative learning seems mostly positive (e.g. Laal and Ghodsi 2012; Springer *et al.* 1999) findings are not entirely conclusive (e.g. Wang and Burton 2010). In the construction domain this may be because, as noted earlier, learning outcomes are not always clearly defined and are, in any event, difficult to assess. While short-term knowledge retention and understanding is typically evaluated via examinations and assignments, assessing more longer-term knowledge retention and the development of collaboration skill presents significant challenges.

A number of collaborative learning approaches in the construction domain base evaluation on student surveys undertaken during or immediately following the collaborative learning exercises (Chan and Sher 2014; Valdes-Vasquez and Clevenger 2015; Zhao *et al.* 2015; Comiskey *et al.* 2017; Vassigh *et al.* 2020) that, to varying degrees, rely on students' self-assessment and reported satisfaction. These provide at best somewhat limited indicators of learning efficacy and, in line with additional work to understand the nature of collaborative working identified earlier, there is a further need to improve an understanding of the usefulness and efficacy of different approaches to group-based learning. These could perhaps involve a more longitudinal approach (for an example in healthcare, see Pollard *et al.* 2004) to collaborative learning across different collaborative settings that would support its consolidation and further development within tertiary curricula.

Some further comments

Some important practical issues also arise in thinking about learning for collaboration and collaborative learning. The idea of 'T-shaped' professionals (Holland *et al.* 2010) that have deep knowledge of their own discipline and a broad understanding of the roles and responsibilities of others does not, on the face of it, present insurmountable challenges for tertiary providers. But given the extent to which many construction degree programs are accredited by professional institutions, and the dominance of core, mono-disciplinary content at the expense of knowledge or skill development in collaborative working (see the aforementioned text, and Table 4.1), a more practical question arises as to how to fit it all in. A further dilemma for educators is that not all students like collaborative learning – there may be particular cultural preferences across different student cohorts (Li and Campbell 2008), and there is a sense also that some more talented students may feel that less capable colleagues are benefitting at their expense in group-based work and assignments (Ford and Morice 2003). Further, students may encounter collaborative learning opportunities fairly rarely, and thus may feel less comfortable in such an environment than when working within their core discipline, which may create a greater sense of belonging (Vassigh *et al.* 2020).

While such challenges would need to be addressed in the design of group-based learning approaches, the more fundamental question of the relative efficacy of this form of learning opens up a further debate around what it could be used for. The scope is potentially wide and could cover not only a broad range of subjects that are typically taught in single-discipline modules that could be delivered in a more collaborative learning environment, but entire programs as well. Indeed, there is a growing interest in cross-disciplinary courses at both undergraduate and post-graduate level in, for example, architectural engineering in the UK[2] that provide considerable opportunities for collaborative learning. These would seem to open wider possibilities that go well beyond the provision of a small element of group learning that, at best, allows students some brief, prior experience of their likely future work settings. Instead, providing the majority of learning in a group-based, intentionally cross-disciplinary setting could start to create the sort of 'shared professional identity' among different disciplines that, as Hartenberger *et al.* (2013) argue, encourages not only information-sharing between disciplines, but also more of a sharing and strengthening of collective responsibility for project outcomes.

This then brings some of the questions for institutions identified earlier – about how generally accepted jurisdictions of professional practice to support interdisciplinary collaboration might need to be reconfigured – into sharper focus. Indeed, the key to the more widespread adoption of interdisciplinary collaboration would seem to lie in joint working between education providers and the professional institutions around an understanding of what it means to be a professional in contemporary construction, and of the benefits of collaborative working. And while there would seem to be considerable work to do here, a starting point could be recognition that, with interdisciplinary collaboration centre stage, what really matters is a shared responsibility to society generally for the outcomes of the design and construction process.

Concluding remarks

This chapter started with an overview of some of the ongoing problems endemic in design and construction processes. Collaboration between design and construction professionals – particularly in terms of inter-and trans-disciplinary collaboration in which different contributions become more integrated in developing shared problem solutions (Schöttle *et al.* 2014) – has the potential to address many of these problems. Interdisciplinary and transdisciplinary collaboration go beyond simple concepts of working together. In particular, the outcomes of such collaboration are ultimately greater than the primarily additive outcomes of multidisciplinary working, focusing more on the creation of novel, shared solutions through the sharing and integration of knowledge and skill. Further, such sharing of intellectual and creative capital can also involve sharing in the responsibility for project outcomes, and it is this that promises considerable transformative potential for construction. The industry's clients can benefit not only from more creative, innovative solutions but from an assurance of collective responsibility that avoids an endless quest for fault if things go wrong. Institutions can support the development of a shared professional identify across professions that may prove attractive to members as well as to potential entrants. Design and construction businesses also gain from working more closely together, sharing in the rewards of success, and reducing the risk of failure.

The role of education providers in helping to move working practice towards greater inter- and transdisciplinary collaboration is crucial. For some, a key challenge is for providers to break out of the predominantly monodisciplinary structure of degree-level construction education. And yet many providers have been doing that and continue to do much in the area of collaborative learning that at the very least recognises some of the realities of construction project life in which disciplines have to work together. What is less clear is the extent to which, if at all, such work is focused explicitly on developing knowledge and skill in the kind of interdisciplinary and transdisciplinary collaboration discussed here. Moreover, such ideas of collaboration are not yet widely adopted in construction practice either, nor are they well understood. But, as noted, their potential to transform how people work together to create innovative and valuable construction outcomes is considerable.

A starting point is needed for considering how tertiary construction education can support more interdisciplinary collaboration among tomorrow's professionals. This lies in developing an improved understanding of collaboration that moves away from thinking that the provision of favourable conditions for collaborative working is all that matters. Certainly, supportive conditions are important to encourage and support working together, but the nature of the collaborative enterprise and the skills needed to engage effectively in it need to be more clearly understood to provide the foundation for improved collaborative learning and, ultimately, working. This is a key role for the research and education capabilities within tertiary education providers.

An important challenge here is that interdisciplinary collaboration is not particularly easy to pinpoint or explicate. Occurring at and between disciplinary boundaries in what Klaassen (2018: 2) describes as a 'third space' in which the meeting of different perspectives stimulates co-construction of new knowledge, learning and innovation, it remains an elusive and somewhat aspirational ideal of collaborative working. And it raises challenges for institutions also, not least for what it means for established jurisdictions of disciplinary knowledge and normal professional practice boundaries. The potential, therefore – and indeed the need – for educators *and* institutions to work together is considerable. It can be achieved by focusing more on understanding and developing collaborative approaches to building design and construction in which problems might be resolved more effectively and those involved may share in the responsibility for the outcomes.

In these terms the scale of the challenge of unlocking the collaborative potential of construction professionals working creatively and responsibly in a joint endeavour is not insignificant. But neither are the potential benefits. Indeed, developing more creative and rewarding working environments for those who work in construction as well as better outcomes for those in wider society who depend on it surely provides a compelling case for concerted action by educators and institutions. Such joint, collaborative endeavour is required to start to solve the conundrum of what collaboration is needed and how education can support it.

Notes

1 https://liveprojectsnetwork.org/
2 For a selection, see one of the UK's course selector tools at www.whatuni.com/degree-courses/search?subject=architectural-engineering

References and further reading

Abbott, A. (1988) *The System of Professions: An Essay on the Division of Expert Labor* (Chicago: University of Chicago Press).

Bechky, B. (2006) Gaffers, Gofers, and Grips: Role-Based Coordination in Temporary Organizations. *Organizational Science*, **17** (1), 3–21.

Bloom, B.S., Englehart, M.D., Furst, E.J., Hill, W.H. and Krathwohl, D.R. (1956) *The Taxonomy of Educational Objectives, the Classification of Educational Goals, Handbook I: Cognitive Domain* (New York: David McKay Company).

Boudeau, C. (2013) Design Team Meetings and the Coordination of Expertise: The Roof Garden of a Hospital. *Construction Management and Economics*, **31** (1), 78–89.

Brncich, A., Shane, J.S., Strong, K.C. and Passe, U. (2011) Using Integrated Student Teams to Advance Education in Sustainable Design and Construction. *International Journal of Construction Education and Research*, **7** (1), 22–40. https://doi.org/10.1080/15578771.2010.512034.

Bruffee, K. (1993) *Collaborative Learning: Higher Education, Interdependence, and the Authority of Knowledge* (Baltimore: The Johns Hopkins University Press).

Bygballe, L., Swärd, A. and Vaagaasar, A. (2016) Coordinating in Construction Projects. *International Journal of Project Management*, **34** (8), 1479–4192.

Chan, T.W.C. and Sher, W. (2014) Exploring AEC Education Through Collaborative Learning. *Engineering, Construction and Architectural Management*, **21** (5), 532–550. https://doi.org/10.1108/ECAM-04-2013-0036.

CIOB (2018) *CIOB Undergraduate Education Framework* (Bracknell: Chartered Institute of Building).

Comiskey, D., McKane, M., Jaffrey, A., Wilson, P. and Mordue, D. (2017) An Analysis of Data Sharing Platforms in Multidisciplinary Education. *Architectural Engineering and Design Management*, **13** (4), 244–261. https://doi.org/10.1080/17452007.2017.1306483.

Comiskey, D., McKane, M., O'Shea, E., Hughes, J., McNiff, S. and Eadie, R. (2016) Collaborative & Multidiscipline Working – From Theory to Practice in 48 Hours: A Case Study from BIM Region Northern Ireland. *International Journal of 3-D Information Modeling*, **5** (2), 55–71. https://doi.org/10.4018/IJ3DIM.2016040104.

Connaughton, J. and Meikle, J. (2013) The Changing Nature of UK Construction Professional Service Firms. *Building Research & Information*, **41** (1), 95–109. https://doi.org/10.1080/09613218.2013.742366.

Cooper-Cooke, B., Sutrisna, M. and Olatunji, O.A. (2020) Innovation in the Globalised World: Educating Future Building Professionals. *Construction Economics and Building*, **20** (3), 56–61. https://doi.org/10.5130/AJCEB.v20i3.7410.

Davidson, N. (2002) Cooperative and Collaborative Learning: An Integrative Perspective. In: Thousand, J., Villa, R. and Nevin, A. (eds.) *Creativity and Collaborative Learning: A Practical Guide for Empowering Teachers and Students* (2nd edition) (Baltimore: Brookes).

Davidson, N. and Major, C.H. (2014) Boundary Crossings: Cooperative Learning, Collaborative Learning, and Problem-Based Learning. *Journal on Excellence in College Teaching*, **25** (3–4), 7–55.

Dewey, J. (1938) *Experience and Education* (New York, NY: Kappa Delta Pi – Republished by Collier, 1963).

Engineering Council (2014) *The Accreditation of Higher Education Programs: UK Standard for Professional Engineering Competence* (3rd edition) (London: Engineering Council). www.engc.org.uk/ahep.aspx.

Farmer, M. (2016) *The Farmer Review of the UK Construction Labour Model: Modernise or Die* (London: Construction Leadership Council).

Foxell, S. (2019) *Professionalism for the Built Environment* (London: Routledge).

Floe, C. (2019) *Best Practice in Experiential Learning: Evidence Review for the Construction Industry Training Board* (London: Traverse). www.citb.co.uk/global/funding/best%20practice%20in%20experiential%20learning%20-final.pdf.

Ford, M. and Morice, J. (2003) How Fair Are Group Assignments? A Survey of Students and Faculty and a Modest Proposal. *Journal of Information Technology Education*, **2** (1), 367–378.

Forster, A.M., Pilcher, N., Tennant, S., Murray, M., Craig, N. and Copping, A. (2017) The Fall and Rise of Experiential Construction and Engineering Education: Decoupling and Recoupling Practice and Theory. *Higher Education Pedagogies*, **2** (1), 79–100.

Froyd, J.E., Wankat, P.C. and Smith, K.A. (2012) Five Major Shifts in 100 Years of Engineering Education. *Proceedings of the IEE, Vol. 100, Special Centennial Issue*, 1344–1360, 13 May. https://doi.org/10.1109/JPROC.2012.2190167.

Gajendran, T. and Brewer, G. (2012) Collaboration in Public Sector Projects: Unearthing the Contextual Challenges Posed in Project Environments. *Engineering Project Organization Journal*, **2** (3), 112–126.

Gottlieb, S.C. and Haugbolle, K. (2013) Contradictions and Collaboration: Partnering In-Between Systems of Production, Values and Interests. *Construction Management and Economics*, **31** (2), 119–134.

Harriss, H. and Widder, L. (eds.) (2014) *Architecture Live Projects: Pedagogy Into Practice* (Abingdon: Routledge).

Hartenberger, U., Lorenz, D. and Lützkendorf, D. (2013) A Shared Built Environment Professional Identity Through Education and Training. *Building Research & Information*, **41** (1), 60–76. https://doi.org/10.1080/09613218.2013.736202.

Hill, S. and Lorenz, D. (2011) Rethinking Professionalism: Guardianship of Land and Resources. *Building Research & Information*, **39** (3), 314–319.

Holland, R., Messner, J., Parfitt, K., Poerschke, U., Pihlak, M. and Solnosky, R. (2010) *Integrated Design Courses Using BIM as the Technology Platform*. National Institute of Building Sciences, Annual Meeting of EcoBuild America Conference, Washington, DC, 7 December.

Hughes, D., Williams, T. and Ren, Z. (2012) Differing Perspectives on Collaboration in Construction. *Construction Innovation*, **12** (3), 355–368.

Johnson, D.W. and Johnson, R.T. (1994) Structuring Academic Controversy. In: Sharan, S. (ed.) *Handbook of Cooperative Learning Methods* (Westport, CT: Greenwood Press).

Johnson, D.W. and Johnson, R.T. (2009) An Educational Psychology Success Story: Social Interdependence Theory and Cooperative Learning. *Educational Researcher*, **38** (5), 365–379.

Kadefors, A. (2004) Trust in Project Relationships – Inside the Black Box. *International Journal of Project Management*, **22** (3), 175–182.

Klaassen, R.G. (2018) Interdisciplinary Education: A Case Study. *European Journal of Engineering Education*, **43** (6), 842–859. https://doi.org/10.1080/03043797.2018.1442417.

Laal, M. and Ghodsi, S. (2012) Benefits of Collaborative Learning. *Procedia – Social and Behavioral Sciences*, **31**, 486–490. https://doi.org/10.1016/j.sbspro.2011.12.091.

Lee, J.H., Zhou, Y. and Ashuri, B. (2020) Key Challenges to Design Professional Liability in the Design-Build Environment. *Journal of Legal Affairs and Dispute Resolution in Engineering and Construction*, **12** (3). https://doi.org/10.1061/(ASCE)LA.1943-4170.0000413.

Leite, F. (2016) Project-Based Learning in a Building Information Modeling for Construction Management Course. *ITcon*, **21**, Special issue *9th AiC BIM Academic Symposium & Job Task Analysis Review Conference*, 164–176. www.itcon.org/2016/11.

Li, M. and Campbell, J. (2008) Asian Students' Perceptions of Group Work and Group Assignments in a New Zealand Tertiary Institution. *Intercultural Education*, **19** (3), 203–216.

Lloyd-Walker, B.M., Mills, A.J. and Walker, D.H.T. (2014) Enabling Construction Innovation: The Role of a No-Blame Culture as a Collaboration Behavioural Driver in Project Alliances. *Construction Management and Economics*, **32** (3), 229–245.

London, K. and Pablo, Z. (2017) An Actor-Network Theory Approach to Developing an Expanded Conceptualization of Collaboration in Industrialized Building Housing Construction. *Construction Management and Economics*, **35** (8–9), 553–577.

MacDonald, J. and Mills, J. (2013) An IPD Approach to Construction Education. *Australasian Journal of Construction Economics and Building*, **13** (2), 93–103.

MacLaren, A.J., Wilson, M., Simmonds, R., Hamilton-Pryde, A., McCarthy, J. and Milligan, A. (2017) Educating Students for the Collaborative Workplace: Facilitating Interdisciplinary Learning in Construction Courses. *International Journal of Construction Education and Research*, **13** (3), 180–202. https://doi.org/10.1080/15578771.2016.1267667.

Manthe, M. and Smallwood, J. (2007) The Appropriateness of Built Environment Tertiary Education: Perspectives of Academics and Postgraduate Students. *Journal of Engineering, Design and Technology*, **5** (2), 102–119. https://doi.org/10.1108/17260530710833167.

Mattessich, P.W. and Monsey, B.R. (1992) *Collaboration: What Makes It Work. A Review of Research Literature on Factors Influencing Successful Collaboration. Report for Wilder Foundation* (St. Paul, MN: Amherst H. Wilder Foundation).

Menken, S. and Keestra, M. (eds.) (2016) *An Introduction to Interdisciplinary Research* (Amsterdam: Amsterdam University Press).

Mollaoglu, S., Sparkling, A. and Thomas, S. (2015) An Inquiry to Move an Underutilized Best Practice Forward: Barriers to Partnering in the Architecture, Engineering and Construction Industry. *Project Management Journal*, **46** (1), 69–83.

Monson, C., Dossick, C.S. and Neff, G. (2015) *Themes in Recent Research on AEC Collaboration*. Working Paper Proceedings (Scotland: Engineering Project Organization Conference, University of Edinburgh), 24–26 June.

Morrell, P. (2015) *Collaboration for Change: The Edge Commission Report on the Future of Professionalism* (London: The Edge Commission). www.edgedebate.com/?page_id=2829

Neuman, M. (2016) Teaching Collaborative and Interdisciplinary Service-Based Urban Design and Planning Studios. *Journal of Urban Design*, **21** (5), 596–615. https://doi.org/10.1080/13574809.2015.1100962.

Nystrom, J. (2005) The Definition of Partnering as a Wittgenstein Family-Resemblance Concept. *Construction Management and Economics*, **23** (5), 473–481.

Okhuysen, G.A. and Bechky, B.A. (2009) Coordination in Organizations. *Academy of Management Annals*, **3** (1), 463–502.

Oliver, B. (2013) Graduate Attributes as a Focus for Institution-Wide Curriculum Renewal: Innovations and Challenges. *Higher Education Research & Development*, **32** (3), 450–463. https://doi.org/10.1080/07294360.2012.682052.

Pero, M., Moretto, A., Bottani, E. and Bigliardi, B. (2017) Environmental Collaboration for Sustainability in the Construction Industry: An Exploratory Study in Italy. *Sustainability*, **9** (1), 125.

Pollard, K.C., Miers, M.E. and Gilchrist, M. (2004) Collaborative Learning for Collaborative Working? Initial Findings From a Longitudinal Study of Health and Social Care Students. *Health and Social Care in the Community*, **12** (4), 346–358.

Reichstein, T., Salter, A.J. and Gann, D.M. (2005) Last Among Equals: A Comparison of Innovation in Construction, Services and Manufacturing in the UK. *Construction Management and Economics*, **23** (6), 631–644. https://doi.org/10.1080/01446190500126940.

RIBA (2014) *RIBA Procedures for Validation and Validation Criteria for UK and International Courses and Examinations in Architecture* (2nd Revision) (London: Royal Institute of British Architects), 2 May.

RICS (2018) *RICS Requirements and Competencies Guide* (London: The Royal Institution of Chartered Surveyors).

Savery, J.R. (2006) Overview of Problem-Based Learning: Definitions and Distinctions. *Interdisciplinary Journal of Problem-based Learning*, **1** (1), 9–20.

Savin-Baden, M. and Major, C.H. (2004) *Foundations of Problem-Based Learning* (Buckingham, UK: Society for Research in Higher Education and Open University Press).

Schöttle, A., Haghsheno, S. and Gehbauer, F. (2014) Defining Cooperation and Collaboration in the Context of Lean Construction. *Conference Proceedings IGLC-22, Oslo*, 1269–1280, June.

Selman, T. and Westcott, T. (2005) *Interprofessional Issues in Construction Education* (San Diego: American Society of Civil Engineers, Construction Research Congress), 5–7 April. https://ascelibrary.org/doi/abs/10.1061/40754%28183%2944.

Sennett, R. (2013) *Together: The Rituals, Pleasures & Politics of Cooperation* (London: Penguin Books).

Smits, P.B.A., Verbeek, J.H.A.M. and de Buisonjé, C.D. (2002) Problem Based Learning in Continuing Medical Education: A Review of Controlled Evaluation Studies. *British Medical Journal*, **324**, 153–155.

Smyth, H. and Pryke, S. (2008) *Collaborative Relationships in Construction: Developing Frameworks and Networks* (West Sussex: John Wiley and Sons).

Springer, L., Stanne, M.E. and Donovan, S.S. (1999) Effects of Small-Group Learning on Undergraduates in Science, Mathematics, Engineering, and Technology: A Meta-Analysis. *Review of Educational Research*, **69** (1), 21–51.

Sunding, L. and Ekholm, A. (2015) Applying Social Sciences to Inspire Behavioural Change in the Construction Sector: An Experimental Study. *Construction Management and Economics*, **33** (9), 695–710.

Suprapto, M., Bakker, H.L.M., Mooi, H. and Hertogh, M.J.C.M. (2016) How Do Contract Types and Incentives Matter to Project Performance? *International Journal of Project Management*, **34** (6), 1071–1087.

Susskind, R. and Susskind, D. (2015) *The Future of the Professions: How Technology Will Transform the Work of Human Experts* (New York: Oxford University Press).

Tews, T., Skulmoski, G., Langston, C. and Patching, A. (2020) Innovation in Project Management Education – Let's Get Serious! *Construction Economics and Building*, **20** (3), 124–141. https://doi.org/10.5130/AJCEB.v20i3.704.

Valdes-Vasquez, R. and Clevenger, C.M. (2015) Piloting Collaborative Learning Activities in a Sustainable Construction Class. *International Journal of Construction Education and Research*, **11** (2), 79–96. https://doi.org/10.1080/15578771.2014.990122.

Vassigh, S., Davis, D., Behzadan, A.H., Mostafavi, A., Rashid, K., Alhaffar, H., Elias, A. and Gallardo, G. (2020) Teaching Building Sciences in Immersive Environments: A Prototype Design, Implementation, and Assessment. *International Journal of Construction Education and Research*, **16** (3), 180–196. https://doi.org/10.1080/15578771.2018.1525445.

Wang, F. and Burton, J.K. (2010) Collaborative Learning Problems and Identity Salience: A Mixed Methods Study. *Journal of Educational Technology Development and Exchange*, **3** (1), Article 1. https://doi.org/10.18785/jetde.0301.01

Wood, D.J. and Gray, B. (1991) Towards a Comprehensive Theory of Collaboration. *The Journal of Applied Behavioral Science*, **27** (2), 139–162.

Xue, X., Shen, Q. and Ren, Z. (2010) Critical Review of Collaborative Working in Construction Projects: Business Environment and Human Behaviors. *Journal of Management in Engineering*, **26** (4), 196–208.

Zhao, D., McCoy, A.P., Bulbul, T., Fiori, C. and Nikkhoo, P. (2015) Building Collaborative Construction Skills Through BIM-Integrated Learning Environment. *International Journal of Construction Education and Research*, **11** (2), 97–120. https://doi.org/10.1080/15578771.2014.986251.

Editorial Comment

Most construction work in developed/industrialised countries is performed under some form of contract; in spite of that, disputes are common, and the industry has a reputation for being adversarial in nature. Resolution of disputes often depends on careful interpretation of the exact words in the contract. While a verbal contract (i.e. one that is not committed to paper or stored in a computer) can be legally binding but, according to American movie producer, Sam Goldwyn, 'a verbal contract isn't worth the paper it's written on'. Shakespeare's play, *The Merchant of Venice,* revolves around contact, prepared by a notary, which sets out the terms of the deal including the very nasty penalty that would apply in the case of a breach. In the play the merchant, Antonio, is unable to pay back money borrowed and thus there is a breach, and the final outcome of the dispute between the parties depends on the precise meaning and intent of certain terms in the contract, something that would have been much more difficult to do had it been a verbal agreement.

It is most unlikely these days that anyone would go ahead with any sort of construction project larger than a small domestic job without a written contract that, at the very least, describes the job to be done and the price to be paid. Not surprisingly, the larger and/or more complex the project, the more detailed the contract.

Construction is inherently risky, and construction contracts, apart from delimiting the scope of the work to be done, and fixing the price (and, usually, the time in which the work is to be completed), are largely concerned with apportioning risk to the parties involved. Not surprisingly, clients would like to simply agree on a fixed price and pass all risk to the contractor; similarly, contractors would prefer that all risk be borne by the client so that, in the event of a risk materialising, the contractor will be able to claim additional money for added costs that accrue as a result of a risk event occurring. Typical risks to be managed include unforeseen (latent) ground conditions (such as unexpected rock excavation) and delays due to bad weather.

Contractors compete for work by tendering or bidding for projects, and price is a key factor in determining who wins each job. If a contractor accepts a particular risk then they must make a decision about how they will price that risk into their tender/bid, and that decision will be driven by various factors, including the state of the market; when work is plentiful bids go up, when it is scarce bids go down and profit margins are reduced, sometimes to zero and even lower as firms struggle to maintain cashflow and keep their businesses afloat in anticipation of better times. When bids are low contractors may seek to improve their position by exploring every opportunity to claim for additional payments due to design changes or any other circumstance that may enable them to ask for more money. Ideally, the contract will clearly state what options are available, and who carries the various risks, but, in practice, there will be claims made that require resolution through some legal mechanism, up to and including, as a last resort, litigation.

The following chapter provides an overview of some fundamental aspects of construction contracting from a legal perspective.

Editor's note: More detailed discussion of many of the points raised in the following chapter can be found in these previous publications by the same author:

Trickey, G. and Hackett, M. (2001) *The Presentation and Settlement of Contractors' Claims* (2nd edition) (Abingdon: Taylor & Francis).

Hackett, M. and Statham, G. (2016) *The Aqua Group Guide to Procurement, Tendering and Contract Administration* (2nd edition) (Chichester: John Wiley & Sons).

5 A guide to construction contracting

Mark Hackett

Introduction

For most clients, to have a building designed and erected is only a means to an end, not an end in itself. It is a means of expanding a manufacturing process, of accommodating staff, of solving housing problems, of operating a hotel or the like (Trickey and Hackett 2001).

For a construction project to be successful, careful planning is required in which the client's brief is fully and properly formulated and, importantly, communicated back to the client (possibly a layperson) to ensure that the brief, as interpreted by the design team and as reflected in its proposals, is indeed responsive to the client's expectations and requirements. If there is a disconnect between what the client wants and what the client is being given and it is discovered too late then the cure for this will usually be in the form of variations (and even re-work) which will have to be undertaken at what should have been an avoidable cost.

When things go wrong, financial claims ensue in which the stakes are so high and the adverse consequences of losing so great that the presentation and settlement of contractors' claims has become an industry in itself. This has, in turn, led to debates as to whether the way in which construction contracts are drafted is a factor in adversarial relations between contractors and clients or, more particularly, the client's professional advisers. To blame the words in a contract for creating adversarial relationships serves as a distraction to the real issue which is the need for design information to be fed to the contractor in a timely and complete fashion.

The client's brief

On its simplest analysis, lay clients will want to have their project constructed: (a) in the shortest time because they are naturally keen to put into operation the building that they have invested in; (b) at the lowest cost because in an increasingly competitive and cost conscious world there are not unlimited funds available to be devoted to projects no matter how dear they are to the client's heart; and (c) to the highest quality because people do not want inferior goods which, of their nature, are prone to failure and require costly maintenance.

DOI: 10.1201/9781003102403-5

Experienced clients of the industry (e.g. developers, utility operators, retailers and so on) have a similar outlook but it is one tempered by experience. Such experience makes it evident that time, cost and quality exist in tension with one another and that, for example, lowest cost is not compatible with highest quality. In the light of this, an experienced client will appreciate that the objective is to strike the optimum balance as between the time, cost and quality parameters of what will evolve into the project's physical representation on site. The brief represents the early articulation of how time, cost and quality are to be synthesised and a considerable degree of thought needs to be devoted to this if it is to serve the client's interests.

The client's brief might best be considered in two parts (i.e. strategic definition and the ensuing briefing), as it is in effect in the case of the RIBA Plan of Work Overview(RIBA 2020).[1] That Plan of Work is arranged in stages and the successful resolution of any stage informs the step to be taken in the next stage. The RIBA Work Stages are as follows:

- Stage 0 – Strategic Definition
- Stage 1 – Preparation and Briefing
- Stage 2 – Concept Design
- Stage 3 – Spatial Coordination
- Stage 4 – Technical Design
- Stage 5 – Manufacturing and Construction
- Stage 6 – Handover
- Stage 7 – Use

For the purposes of this chapter, the emphasis is on:

(a) the brief (Stages 0 and 1 of the RIBA Plan of Work)
(b) conflicts which may arise during the construction phase (Stage 5 of the RIBA plan of work)
(c) whether there is a correlation between the procurement strategy/contract selection and the incidence of disputes

The initial focus should be on developing the brief so that the design team knows exactly what is required and is not left to guess a requirement whose correction will come late and, most likely, at great cost. It therefore follows that if the brief is to set the project's objectives, then any shortcomings therein are likely to manifest themselves in shortcomings in the subsequent stages. Whether dealing with a lay client or an experienced client, the initial decision to build rests with the client rather than the design team. Accordingly, the client will need to be acquainted with the means of drawing up a strategic definition of what is to be expected from the completed project. For the experienced client, this will be a more straightforward process and one which may be assisted by some input from design team professionals encountered on prior projects. For the lay client, this might feel like entering into uncharted territory, but the strategic definition is, in effect, a business case and the sort of client who has outgrown existing business premises ought also to be the sort of client well-versed in producing and delivering successfully upon business cases.

The strategic definition requires, as a core objective, the client to establish the business case, determine the strategic brief and consider other project objectives. The principal aim is to consider substantive issues prior to developing an initial project brief. At this stage of the process, issues regarding the assembly of the client's project team and the establishment of a preliminary programme regarding the duration of the project might be considered (Hackett and Statham 2016: 14).

When considering the strategic brief, one is expected to assess the options for the client with regard to issues such as its functional requirements, required quality, whole life costs and environmental requirements and/or constraints. When combined with the client's business case, the strategic brief will enable the client to consider if the project should comprise a refurbishment, an extension to an existing building or the construction of a new building (Hackett and Statham 2016: 14).

From a client's perspective, the cost, quality and time paradigm might be considered as being the highest quality, at the lowest cost and in the shortest time. Unfortunately, this is not always possible, and a compromise has to be sought, based on the client's priorities. If particular circumstances dictate that one of the factors must take precedence, at least one of the other two factors will 'suffer' or carry less weight or emphasis. The decision as to the weight or emphasis to be given to each factor must lie with the client (Hackett and Statham 2016: 30) albeit taking on the advice of the design team since its collective experience can be brought to bear in mitigating the effects of one emphasis (e.g. cost) being detrimental to another (e.g. quality).

If quality is of paramount importance, then adequate time must be allowed for the design and the specification to be perfected. In this regard, cost could rise on both counts. If speed of completion is paramount, then quality and cost may both have to suffer. If lowest cost is the priority, time may not be prejudiced but quality could suffer (Hackett and Statham 2016: 30). The permutations are endless, but they must be engaged with and tested if the client is to secure the optimum balance as between cost, quality and time – and that is a balance which is unique for each project.

The effect of these different priorities is relative and there is no reason as to why, with proper planning and management, those elements with a lower priority cannot be adequately controlled (Hackett and Statham 2016: 30).

As will be appreciated, multiple factors need to be considered in formulating a brief and these factors can and do operate in tension with one another. There are then external factors to consider such as enhancing the built environment,[2] public procurement directives[3] and sustainability.[4] Table 5.1 provides a high-level view as to how securing one objective can adversely affect another if not addressed with care:

As the design progresses (it starts in outline form and becomes progressively more detailed thereafter), it should be regularly tested against the brief to validate whether the client's objectives continue to be met. One way in which to do this

Table 5.1 Competing project objectives

Objective	*Positive*	*Negative*
Drive the cost as low as possible.	Minimises capital outlay.	May produce an eyesore (inconsistent with making a contribution to the built environment) with high-running costs (inconsistent with sustainability responsibilities).
Completion to be achieved in the shortest possible time.	The building will be ready sooner whether that be a householder keen to occupy their new home or a commercial client keen to put the building to use and see a return on investment.	Accelerated working can lead both to an increase in labour costs (e.g. due to extra gang shifts) and to a decline in quality.
Trophy/statement architecture (based on a design led brief).	The client's requirements as to quality will be met.	The time taken to construct the project may take longer to complete with a commensurate increase in cost which is then compounded by the cost premium associated with higher quality materials being used.
Utility architecture (based on a budget led brief).	The outturn cost is more likely to align with the client's initial budget.	The finished quality of the building may fall short of serving the intended performance requirements.
Leading-edge architecture incorporating the latest technology and sustainability features.	The building will have an element of future-proofing, lower running costs and strong sustainability credentials.	The rate of technological advances may make for a greater incidence of variations as the project progresses on site and this will be against the backdrop of an already high capital outlay.

is to have regular design reviews and value management workshops in which the delivery programme is reviewed against the client's objectives. It is not uncommon to find in such reviews that elements of design enhancements have crept in which add to the cost of the project but without necessarily adding to the ability of the completed project to perform its intended function. A value management workshop is a good place in which to probe whether a design feature is a 'nice to have' or desirable element (hence something which could be omitted without compromising the completed project's intended function) or something which is indispensably necessary (hence something which, if omitted, would adversely affect the functioning of the completed project).

Needless to say, it is at the discretion of clients as to whether they will allocate funds to desirable elements, but it is a good discipline for construction advisers to assist clients in making that an informed decision rather than something which emerges as a consequence of 'design creep' which many accept rather than guard against.

The distinction between that which is desirable and that which is indispensably necessary is often not fully appreciated until tested against cost considerations. One is reminded here of an architect who would say to his clients: '*Tell me what you want or what you can afford but please don't tell me both.*'[5] Hopefully rigorous monitoring and evaluation will enable clients to find alternative solutions such that desirable elements of the design can be incorporated without compromising the client's budget, thus following the invocation of a great figure from another sphere who said: '*We haven't the money so we've got to think.*'[6]

Getting the brief right from the outset requires the client to have developed a coherent business case which is capable of being absorbed by the design team. The client must also appreciate that there will be engagement with members of the design team who possess a wide range of skills which must be coordinated such that the right emphasis is placed on any one or more of the relevant skill sets at any time. To be able to do this requires an understanding of construction procurement and techniques, and even lay clients should invest some time into reading around the subject if they are to be able to integrate with and get the best from the design team (Hackett 2015).

Procurement strategy

Procurement is the process of acquiring goods and/or services from a supplier or service provider. It is, in essence, a simple transaction but today's construction projects are often considerable in scope, involve several designers, require specialist contractors, are executed at arm's length and take considerable time to complete (Hackett and Statham 2016: 29).

In addition, a degree of complexity arises from the requirement to comply with numerous regulations, engage professional consultants, achieve value for money, demonstrate accountability, regulate complicated contractual relationships and achieve a timescale largely dictated by the client's specific business objectives. Nevertheless, irrespective of however complex modern construction projects may be, they are all based on the comparatively simple established principles of addressing cost, quality and time (Hackett and Statham 2016: 29).

It is advisable to apprise the client of the procurement options, recommend the most appropriate route, describe how the priorities are protected and outline the various implications. Only when the method of procurement has been determined can the pre-contract programme be finalised. This also represents the point in time when the specific services required of each member of the design team can be defined, fee proposals confirmed and agreements finalised (Hackett and Statham 2016: 30).

It must be stressed that the particular circumstances of each project may result in different conclusions as to the procurement route that best satisfies the priorities of time, cost and quality. For example, a high level of provisional sums within a contract could undermine the cost certainty otherwise afforded by a particular procurement route.

The procurement strategy provides the vehicle in which the project is taken from concept through to delivery and governs such things as:

- the appointment, composition and duties of the design team
- risk allocation and responsibility
- responsibility for the design
- input or otherwise of specialist sub-contractors
- the project programme.

Historically, the design from which a contractor would work would be supplied by the design team in the form of a prescriptive specification which the contractor would be obliged to comply with and not depart from and this procurement route is, not surprisingly, called 'Traditional'. Recognising that contractors have great experience of more economic design solutions than might be obtained from architects (not least because architects may be more focused on visual and aesthetic considerations than with the guts of the building), there have been significant developments with 'Design and Build' where the work to be performed is set out in a performance specification thus providing contractors with latitude to implement their own design solutions provided, of course, that they meet the criteria set out in the performance specification. There are also 'Management Contracting' routes in which the overall project is divided into packages such that design and construction activities may be overlapped thus producing savings on the program duration.

There are also somewhat different means of dealing with financial entitlement which lie on the spectrum between lump sum contracts (where the sum due to the contractor is agreed before construction commences) and cost reimbursable contracts (where the contractor is reimbursed for the work as it is undertaken and in respect of which the final costs will not be determined until the construction has been completed). On its simplest analysis, a lump sum is the most attractive for clients because they will know of their financial commitment from the outset and can budget accordingly. However, to obtain a lump sum commitment from a contractor also requires that the client must first of all be able to set out in comprehensive and detailed terms all of the requirements on the contractor with regard to cost, quality and time. Needless to say, to be able to set out that sort of information is both a somewhat time-consuming process and a client may be impatient to start on site before all aspects of the design have been fully formulated. It is in the light of this that, on the other end of the spectrum, a cost reimbursable contract might be chosen but, of its nature, the client may not know until it is too late that the project embarked upon is no longer financially viable due to unforeseen costs being incurred.

The procurement strategy adopted must be compatible with the client's objectives because if it is not then, to stay with the vehicle analogy, there will be a bumpy ride ahead with various breakdowns. It is therefore essential that the vehicle selected is suitable for the conditions to be navigated and this may involve a hybridised version of the principal procurement strategies outlined earlier. These issues are addressed in the following examples:

- If the client needs a high level of control over the finished design, then the Traditional approach is appropriate since the contractor may not depart from what the client requires.
- If dealing with an industrial facility or the unseen elements of a building (e.g. the substructure and structural frame), a design and build approach could be appropriate because the contractor can address the most economical solution (but consistent with the stated performance criteria) since aesthetics (and the costs associated with achieving them) are a lower priority.
- If a particularly early start on site is required then a Management Contract approach could be adopted because the excavations and substructure could, for example, be carried out on site whilst elements like the cladding or services are still in the design or tender stages.

Project execution

Having finally made the decision to build, probably after months if not years of debate, most clients are understandably impatient for a result and, having believed their financial commitment to have been established at the outset, most are understandably keen for their project to be contracted for – and delivered to – budget.

It is unrealistic to assume that, on every project, all of the design and specification work will have been completed before going to tender; thus it would be fanciful to produce a standard form of contract that is based on this assumption. However, where this can be achieved on any project the contractual clarity, and the absence of any excuses for bad management that will ensue, can be very rewarding. For such documentation to be complete not only must the design team have fully considered and completed all aspects of the building design but they must have also:

- based the design on a comprehensive site investigation (so as to consider, amongst other things, work on or near existing buildings, access restrictions, ground conditions, etc.);
- eliminated discrepancies (e.g. at interfaces between trades particularly so with specialist elements where multiple designers may be involved);
- resolved the details of any sub-contractors providing a specialist design input (e.g. cladding, lifts, building management systems, etc.); and
- co-ordinated the work of any other contractors who are directly appointed by the client.

Not only will this have removed the majority of the scope for claims but it will also have created a clear backdrop against which the effect of any variations, if they occur, can be precisely measured. All too often the line between a variation and the continuing development of the initial design is impossible to draw.

Desirable though it may appear to be, it would be unrealistic to attempt to ban variations or design changes. A building project is usually a complicated, unique venture erected largely in the open, on ground the condition of which is never fully predictable, and in weather conditions which are even less so. There is a considerable time span between making the initial decision to build and the completion of the project on site. During this period technology, fashion and the client's requirements invariably change. Standard construction contracts recognise these possibilities and attempt to ensure that such changes in circumstances are valued fairly in accordance with rules which are known at tender stage, rather than leaving the establishment of those rules until the problems occur at which time one party would most probably be in a more advantageous position than the other.

If contractors are to be able to quote a proper price for the project and be kept to their contractual terms then they must always know and have clearly explained the contractual obligations. Initially, the basis for these will be set out in the contract documents. Those obligations may include target dates by which design information is to be released around which contractors will plan their site operations. Where delays in information, or the extent or timing of variations, eliminate these target dates then they should be re-established quickly and fairly using the contractual mechanisms provided. The contractors' duties will then always be clear; they must achieve those objectives and they should ensure that adequate management is provided to do so. On a contract which is beset with delays and disruption and where the targets of time and monetary entitlement are temporarily undetermined, it is understandable, if not contractually supportable, for the contractors to turn their backs on the contractual obligations and divert energies into building a claim rather than building the project.

This is not to say that variations or further instructions should not be given but, where they are necessary, the effect on both time and monetary entitlement should be quickly evaluated so that the contractual objectives may be repaired. This is where many building projects fail and a tension then exists in which the contractor is likely to seek, in addition to the cost of the varied work itself, costs in respect of prolongation and disruption. The latter might be regarded as 'wasted costs' in the client's eyes because even though they may be an unavoidable consequence of implementing a variation the outlay in respect of them does nothing to enhance the finished quality of the project. As such, wasted costs often become viewed with such suspicion that attempts to negotiate and agree upon what is properly due can become fractious and descend into a dispute between the parties.

Ongoing design reviews and cross-checking the state of the design against the client's requirements mean that if a variation is necessary then the need for it can be implemented sooner rather than later in the construction process. As a broad proposition, the opportunity to vary work diminishes as the project progresses (at least that is the case if re-work is to be avoided) and the later in time one calls for a variation the costlier it is likely to be.

Clients will observe that, from an early stage of the design process, detailed estimates of time and money are given; changes in cost are monitored as the design is developed; tenders are usually called for on the basis of detailed measurement and the rates for valuing any variations are established as part of the tendering process in advance of letting the contract. Against this background of care, inevitably purchased at the client's expense, it must be bewildering to discover, as sometimes is the case, that progress on site has lapsed and there appears to be little that can be done about it; that the final cost and completion date are impossible to predict; and that the final settlement is arrived at out of exasperation rather than by evaluation. Any client, and in particular a lay client, will have formed the view that after extensive interaction with the design team at the brief stage, there should be clarity as to the ranking of cost, quality and time considerations hence the delicate striking of a balance between these things ought not to unravel during the execution phase. That such planning could unravel at all may point to the fact that the briefing phase was not conducted with the diligence and care it was thought to have been. Should a brief unravel, and then be triaged through a variation process, additional costs will arise which: (a) the contractor must seek to recover as a matter of commercial reality; (b) the design team will be embarrassed about because it may point to shortcomings in its own procurement practices; and (c) the client will be dissatisfied with meeting because they are costs which do not add to the value of the completed project.

The construction contract

There are frequent references to the adversarial or confrontational behaviour in the construction industry.[7] For example, if you search the internet for examples of adversarial contracts you are at once faced with several examples of such from commerce in general but the construction industry in particular. Examples are as follows:

- Bossom (1934) described the construction industry as wasteful, expensive and unsatisfactory for clients.
- The Banwell Report (Banwell 1964) recommended the need for the construction industry to conduct its relationships in a less adversarial manner.
- The Latham Report (Latham 1994) referenced the construction industry as being fragmented, ineffective and adversarial.
- The Egan Report (Egan 1998) described a situation in which the construction industry: (a) has too many clients dissatisfied with its performance; (b) could learn from radical changes in other industries; and (c) suffers a lack of integrated processes. It was also, and rather importantly, observed that low profitability and insufficient investment (whether by way of capital, research and development or training) were other shortcomings in the construction industry but one can rather understand how the former might feed the latter and vice-versa.
- The National Audit Office (NAO 2001) observed the general incidence of adversarial relationships within the construction industry.

- Regan *et al.* (2015) asserted that 'adversarial contracting methods are used for most public infrastructure procurement . . .'
- Banaszak *et al.* (2020) identified 'traditional, adversarial contracting practices' in which parties 'rush through fixed-price contract negotiations, opening the door for purposely understated timelines, long delays, and massive budget overruns.'

Even one of the UK's top spies at the time of having her headquarters refurbished, observed it to be '. . . clear that dealing with the building industry was just as tricky as dealing with the KGB' (Rimington 2002).

It was earlier observed that construction projects involve unique requirements which are to be constructed outdoors in weather and ground conditions which are never fully predictable. This might usefully be contrasted to a situation where dealing with those types of industries which operate in factory conditions[8] in which: (a) considerable investment is put into prototypes before replicating items on a mass scale;[9] (b) being indoors the fabricators are spared the elements and (c) the consumer base is more accepting of a 'standard' solution.

Considering what has been said, it might reasonably be concluded that the only way of completely avoiding the prospect of confrontation in a building contract is to let the contractor to build what it wants to and to be paid for whatever is asked for. As soon as the client insists on a specific commitment, i.e. '*this is the building I want*', or particularly if they go further and say '. . . *and I want it completed by a specified date and for a specified price*', the seeds of confrontation are inevitably sown. It is not forms of contract or methods of procurement that are of themselves confrontational; it is the parties to a contract who, having been required to honour commitments, create the confrontation by either failing to achieve them or actively seeking to relieve themselves of them. Certainly, ill-prepared contract documents, or wanton changes in them thereafter, will produce fertile ground in which claims will propagate; but to seek to blame particular forms of contract or methods of procurement for the widespread recourse to claims is to ignore the real issues.

The real issues almost invariably arise in connection with design information which is variously:

- late (in the sense that it is not supplied at a rate compatible with the contractor's rate of progress on site); and/or
- inadequate (in the sense that it does not provide the particularity necessary for the contractor to be able to procure materials and assemble the parts); and/or
- additional (in the sense that the work needs to be varied whether because the original design did not meet the client's brief or because the client's brief needs to be changed).

Where design information is late and/or inadequate and/or additional, this will probably lead to additional costs and delays. Contractors will know that if they are late without excuse then they will bear additional costs in funding the resources detained on site and may also be liable for delay damages. In contrast, if contractors are excusably late (in the sense that a typical construction will recognise that if contractors are faced with design information which is late and/or inadequate and/or additional then they will be paid the cost consequences of this and be absolved of liability for delay damages) then clients will know that they have to compensate the contractor whilst also forsaking access to the revenue stream from the project until the delay has run its course.

In all of the earlier circumstances, the allocation of responsibility for design information which is late and/or inadequate and/or additional has considerable cost implications for the contractor, the client or both of them. This can lead to a situation of people allocating responsibility in accordance with what best serves their financial interests rather than in accordance with the objective standards of the contract to which the parties subscribed. To this extent, it is human nature rather than the words in the contract, which is the driver for adversarial conduct.

The objective of contract draftsmanship is often expressed as being to deal fairly with the interests of the contracting parties and, whilst this is laudable, it must be recognised that the concept of fairness is subjective and all the harder to crystallise in one's draftsmanship as a result. Contracts which are drafted to achieve fairness may become excessive in length and over-sophisticated in procedure due to the many competing factors which have to be addressed; ironically, such contracts may be self-defeating since their complexity will increase the incidence of discrepancies or clauses which are genuinely capable of alternative meaning. Alternatively, the objectives of contract draftsmanship may be to achieve clarity and certainty though the ensuing document may not be 'fair' in the sense of balancing the parties' expectations under the contract, but it will at least provide clear and relatively incontestable procedures to be followed whether they are palatable or not. In this regard, it is worthy of note that if the parties cannot themselves agree upon what their contract means then they may find themselves having the point decided for them in consequence of formal legal proceedings. Such formal legal proceedings will be conducted before a tribunal[10] and that tribunal will then be required to find, as a matter of law, what the contested contractual terms mean. At this juncture, the tribunal does not (and rather sensibly should not) concern itself with what might look 'fair' (which would be pointless in circumstances where each party will by then have convinced itself that its position is the one which bears all the hallmarks of what is fair) because it will then consider what, in objective terms and by reference to the language of the contract, obligations and benefits the parties negotiated for themselves. At that point, the Tribunal will take a cold hard look at what the parties agreed and do so without regard to the heated state in which the parties may represent their viewpoints in vehement and diametrically opposed terms.

In response to a situation where the terms of a contract might be perceived as giving rise to confrontational or adversarial exchanges, one suggestion is that this situation should be remedied by eliminating clauses from the contract which promote confrontation and replace them with clauses which instead promote co-operation. Such a view fails to recognise the commercial reality behind construction projects namely that clients will endeavour to maintain the balance of cost, time and quality for which they initially bargained whilst contractors will endeavour to maximise their recovery with the least outlay. Whereas the interests of each party are quite different, they might co-exist peacefully until such time as the balance of cost, time or quality is redistributed; the danger then exists that the parties will push for their diametrically opposed interests.

In a sense, it is the very contracts which have endeavoured to avoid adversarial relationships which have created the most problems. For every concession which it is thought fair and reasonable to embody into the contract, there is a further reason for contractors to depart from their original obligations; there will certainly be those occasions where any such departure is wholly justified but there are also innumerable instances where contractors will exploit a clause for exploitation's sake to see if they can benefit by it. A contract will not make any significant contribution towards a non-adversarial approach by providing endless remedies for defaults since these remedies then risk being pursued to the detriment of progressing the project itself. Instead, the contract's contribution will be made by incorporating clauses which are clear and simple to implement thus removing suspicion that one party may unfairly benefit from any adjustment thereunder. To this extent, that which is 'fair' (and nobody really knows how this might be objectively decided) might better be achieved if the contract is clearly expressed (such that the price for performance thereunder plainly reflects what was asked for) rather than be left to a time when a challenging circumstance arises in which financial consequences rather than the terms of the contract cloud people's thinking.

With any construction contract, the potential will exist for adversarial relations if only because contractors are likely to have secured the work in competition (which could be at a price which owes more to securing the work than at a price which reflects their obligations) and may feel bound to exploit each amendment to cost, time and quality as they are presented. The unique nature of each construction project will lead to such amendments since every possible eventuality is unlikely to have been foreseen in the contract documentation. To this extent, financial consequences will follow, and human nature is likely to dictate that which is fair being gauged by reference to how little the client has to pay or how much the contractor can make from it. In this context, it is difficult to imagine circumstances where an objectively fair outcome can be agreed upon between parties whose focus is directed to their balance sheets rather than the contract agreed between them. The sheer pressure of an adverse financial outcome (in respect of which only a few people will ever find themselves unaccountable) involves occupying a situation in which the company's officers tasked with resolving the situation (be that someone for the contractor taking money in settlement of a claim or someone for the client agreeing to pay out money) may face commercial pressures which blind them as to how the matter should be dealt with according to the terms of the contract.

Contractors' claims

If, through default, a client or the design team (for whom the client is responsible) prevents the contractor from completing the project by the contracted completion date and there are no express contractual provisions enabling the client to set a new date for completion, the contractor's obligation is transformed from one of having to complete by the contracted date to one of completing in a reasonable time. In other words, time is 'at large' and clients lose any right that they may otherwise have to levy liquidated damages for delay. It may be seen, therefore, that far from favouring only contractors, the existence and proper execution of an extension of time clause within a contract is of the utmost importance to clients in that it preserves their other contractual rights should they or their representatives cause the project to be delayed by default.

Events which delay and/or disrupt any project will have been caused by one of the following:

- the contractor
- the client or its representatives (the latter being those for whom the client is contractually responsible)
- events outside the control of both the contractor and the client (often referred to in the industry as *shared risk events* or *neutral events*).

If a contractor makes a significant claim, then it should not be rejected out of hand. Instead, it should be investigated and, if it is justified, be treated in the manner dictated by the terms of the contract – which will invariably involve an attempt to put the contractor in the position that it would have been had it not been delayed or disrupted. In this regard, if a contractor has its resources delayed and/or disrupted in circumstances for which the client is responsible then the contractor will be entitled to be paid damages[11] or loss and expense.[12] Under English law (and something which the writer has also seen applied in several other jurisdictions), if someone has been wronged and is, by reason of that, entitled to be compensated, the ensuing compensation will be calculated on the basis that '*where a party sustains a loss by reason of a breach of contract, he is, so far as money can do it, to be placed in the same situation with regard to damages as if the contract had been properly performed.*'[13]

Construction contracts will typically include provisions such that the contractor is required to give timely notice in the event that it is being or is likely to be delayed or disrupted in the progression of the project and/or that the contractor is incurring or is likely to incur extra costs in progressing the project. Despite such provisions, parties all too often convince themselves that it is to their advantage, for a number of reasons, to delay the resolution of claims for as long as possible. Occasionally this is due to a mistaken belief that matters might become clearer as the contract progresses or due to the preoccupation with other seemingly more pressing priorities. For example, the architect who is notified of a delay occurring in the construction of the foundations may understandably be tempted to claim that the inadequacy of the contractor's particulars and estimates provided under any such notification provisions prevents the architect from determining a new completion date. The architect might, therefore, leave judgment on extensions of time for, say, six months because of an impression that its effect upon the remainder of the work will have become evident by that time. In practice, nothing is likely to be further from the truth; the occurrence of yet more variations or instructions, delay due to bad weather or mismanagement will all serve to have obscured the issues rather than to have clarified them.

Another reason offered in support of delaying the granting of extensions of time or ascertaining entitlement to further monies is that the contractor may catch up. The contractor may well catch up but that in itself will not remove any entitlement to an extension of time or additional monies to which the contractor had become entitled upon the delaying event having first become apparent; the test is to compare what actually happened with what would have happened had the delay or disruption not occurred.

So rather than procrastinate, one must engage. In order for a strategy of timely resolution to work, the contractor must first give the requisite notice but some contractors, with misplaced gentility, still feel that to give notice of delay or of the incidence of loss is to cause discord and that as long as such delays or disruption do not occur too often then they are best ignored. Ironically, an overly diplomatic reaction to harm caused may make it more difficult to seek a remedy to the situation in the event that the harm continues – the person causing the harm might even have the temerity to suggest that any earlier failure to signal an objection amounted to acquiescence. It is a fundamental misunderstanding of contractual relations if either party feels that the submission of a notice of this kind is somehow letting the side down. On the contrary, it is, if nothing else, a timely notice which enables the design team so to organise itself as to mitigate the continuing effects of such delays and the like and to require proper records of the facts to be kept. Put another way, notifying a problem in a calm and factual manner affords the opportunity to restore the contractual obligations without creating antagonism in contractual relations.

Correspondence between the contractor and the design team is also relevant when looking back over a project in order to establish the facts. There is always the temptation, particularly for the architect, not to respond to correspondence. The architect is, after all, responsible for the management of the project from the client's point of view and this can be both time consuming and pressing. The contractor, on the other hand, may be tempted by a direct monetary interest in pursuing relentlessly any potential for a claim. However tedious, therefore, all correspondence must be dealt with at the time and in the proper degree of detail. Often such correspondence, from either side, is written with the potential adjudicator, arbitrator or judge in mind rather than to achieve current objectives, and such a tactic may rebound heavily if the response is not fully thought through and backed up by evidence. If, instead, the correspondence is written in dispassionate, logical and factual terms (duly backed by relevant evidence), this is more likely than mere bombast to persuade the addressee as to the seriousness with which a response should be composed.

In the same way that an estate agent may say that the three most important factors in the selection of a property are 'location, location, location', those involved in construction disputes will tell you that the three most important tools are 'records, records and records'. The absence of contemporary records is probably the single most important reason for the settlement of claims degenerating into an expensive game of chance. The only way to ascertain the proper amount due is to bring matters out in to the open as they occur, when allegations can be fairly tested against the prevailing facts and when the real impact of any delay or disruption can be seen. In contrast, uncertainty is quite the wrong environment in which to expect even efficient organisations to perform well; and it is fertile ground for the inefficient and imaginative alike. The monies at stake and the complexity of the issues on which they turn can often mean that the resolution of the claim will take time, and that one of the parties (often the contractor) will threaten litigation before the issues have been fully debated.

It is impossible to set out in advance all of the rules for running a meeting aimed at reaching a negotiated settlement but there are some general points which ought always to be kept in mind:

- The overriding objective is to achieve a figure which best represents the proper amount due to the contractor under the contract. The negotiation should be conducted in a professional manner and with regard to the parties' respective rights and obligations under the contract. The negotiation should not be allowed to descend into a bartering exercise in which bluff and counterbluff are the order of the day.
- The fear that if a negotiation fails formal proceedings will follow is something which hangs over both sides and should, therefore, play only a neutral role in any negotiation.
- Constantly giving way to the threat of legal costs will set the pattern for settlements that will always favour the party making those threats and may even embolden that party to make further threats.
- Never be induced into believing that because the other side has 'made a concession' it is now your turn to do so. However, if an obvious error is revealed in your case then concede it quickly and graciously. Beware those who, having grossly overstated their case, dress up a modification as a genuine attempt to settle.
- Do not be fooled into thinking that silence is golden. There is a widely held belief that the first one to break the deadlock in a negotiation will lose the battle. In fact, being able to calmly but relentlessly pursue the facts is the sign of a well-documented case.
- Do not get stuck on any one particular issue. Failure to agree a particular item should not prevent progression to the next. Once all items have been considered, total up the unagreed items and ascribe to them the amounts that you are prepared to accept. It may well be that the resulting total is also acceptable to the other side, or at least very much nearer to their aspirations than might otherwise have been thought.

Claim tactics

No commentary upon the subject of claims would be complete without some reference to the tactics sometimes resorted to by one or both parties in either the presentation, response to or negotiation of a claim.

Not all of the following devices are employed all of the time; nor does everyone resort to them some of the time. In the same way that a medical textbook, crammed with the real or potential physical maladies of the body, is not typical of the majority of the human race, the following paragraphs are not included because they are typical of the construction industry. Nevertheless, for those who do have to address claims, it is as well to be acquainted with some of the tactics which have been encountered over the years (Trickey and Hackett 2001).

The fatality of the first figure or the relentless wedge

This rarely fails. Either party gives the other an exaggerated view of their expectations; the client or architect pitches the figure in payment of claims very low, or the damages for delay very high, or both and/or the contractor pitches the alleged loss and/or expense very high. Of course, the contractor would not normally copy the client with correspondence addressed to the architect, but a copy of a claim submission sent to the client 'for information' may be a very worthwhile investment for the contractor. The party in receipt of such figures seeks consolation in the thought that at least they are aware of the worst position and are curiously grateful for the removal of uncertainty. The receiving party in effect steels itself for a disappointing outcome and the more that party becomes resigned to this the less objectivity and resolve there is for supporting what might be a long struggle through to a just settlement; this is the thin end of a relentless wedge.

Architects and quantity surveyors faced with the need to deal with a claim submission from a contractor should in no way be influenced by the global figures that have been put forward. The contractor's duty in the ascertainment of loss and/or expense is simply to convey the facts and to make available such cost records as are called for but almost without exception one finds that contractors are only too keen to evaluate their own entitlement as they judge that it suits their cause. Albeit that it is the role of the appointed design team member (invariably the quantity surveyor where UK contracts are involved) to ascertain that which is properly due, the client will often be unable to resist the temptation of peeping at the contractor's submission and being seduced into accepting that it has some relevance rather than waiting for (what should be) the design team's wholly objective ascertainment.

It is for the design team to consider the contractor's submission objectively and, therefore, to have careful regard to the extent that any of the facts supplied are relevant. However, if it is judged that the contractor's monetary evaluation is exaggerated, it is unrealistic to pretend that this will not have had some impact upon the client (who will almost certainly wish to be kept informed of developments) and who may be faced with a dilemma if, by contrast, faced also with what might appear to be a very low assessment in response from the design team. To prevent such a credibility gap appearing then, in the first instance, an immediate and approximate ascertainment is necessary from the design team in order to counter any prejudices forming in the client's mind as to the validity of the contractor's figure.

The Solomon syndrome or splitting the difference

At the root of too many negotiations is the bland acceptance that, if two contrary views are expressed on a claim, there must be some merit in both and the real truth lies somewhere in between; moreover, the nearer the middle that the answer is judged to be, the more sensitive and sophisticated has been the adjudicator. This, of course, is nonsense. But the more widely it is practised the more it is exploited, with each side stretching ever wider the boundaries of their imagination in order to establish a base in as extreme a position as can be tolerated without totally losing credibility.

This predisposition to 'split the difference' can be particularly distressing when attempting to reconcile the design team's view on entitlement to reimbursement of additional monies arising from delay and/or disruption. The roles of the contractor and the design team in the run up to any such confrontation are quite different and this difference is often not recognised by the client nor, regrettably, by the client's legal advisors. In arriving at its ascertainment as to the contractor's entitlement, the design team will have been under a duty to operate the terms of the contract as best it can as between the client and the contractor. The contractor, on the other hand, has no such constraint and is perfectly entitled to express every item in dispute in a way which is to its best advantage. Given such a situation, it is quite clearly wrong to assume that the view of each side has equal merit, because the design team has been representing the contract whereas the contractor has been representing itself. Any temptation to split the difference should therefore be resisted.

The contractor and the design team are not at opposite ends of a tug-of-war rope; the latter are (or at least should be) standing where they judge the centre marker to be.

The clandestine cost plus

In this type of claim, the contractor's material will present undeniable facts relating to the running of the contract, trace the remedy to the correct clause in the contract and then finish with a calculation of claim entitlement which turns out to be no more than a comparison between: (a) what the contractor alleges the costs to be; and (b) what is claimed to have been recovered through the contract to date. Proceeding in this manner is, in effect, to seek reimbursement of all costs incurred in progressing the project.

However just the contractor's cause might be, this is no way in which to proceed. For the sake of decency, such an approach to a claim is often camouflaged by a distracting detour through fascinating formulae, e.g. converting the alleged labour content of the final account into a percentage of the total value and describing this as the *tender output*, relating the alleged labour costs incurred to the same value and describing this as the *achieved output*, comparing the two and describing the difference between them as the *disruptive loss*.

This approach is unacceptable because it (a) makes no attempt to relate cause to effect; and (b) embraces all causes whether they be the responsibility of the contractor or the client. A nice touch is often added by making a modest deduction for fluctuations charged elsewhere by the introduction of sophisticated but irrelevant calculations . . . at the end of the day it is no more nor less than a claim for reimbursement of all costs howsoever caused!

The veneer of precision

No one is impressed by a claim of such generality as US$100,000.00; one's suspicions are at once aroused as to its authenticity. On the other hand, a total of USD 99,787.32 does concentrate the mind wonderfully. Peeling back the veneer, however, might show some underlying detail like: (a) Joe Bloggs & Co's documented and indexed claim for USD 9,787.32; and (b) a 'say' allowance for claims which *might* be forthcoming from other sub-contractors in the amount of USD 90,000.00.

The heart-rending narrative

Building is at best a difficult process and an accurate record of the unfolding of even a well-run contract with a few variations would strike fear into the hearts of many from other industries. The inevitably complex and almost unique nature of any construction contract provides limitless opportunities for distorting the picture of events. Consequently, many a claim is presented in wonderfully tabbed files crammed with myriad diagrams, graphs and statistics and eloquent narrative but little which directs itself to causation which is the very thing needed to link together (a) the costs which the contractor most likely incurred in trying circumstances, and (b) the client's responsibility for those of the circumstances which have proven to be more trying than should already have been envisaged by the contractor when formulating the contract price.

Never forget that in the final analysis an adjudicator, arbitrator or the courts, may have to review the facts; if you genuinely believe in the strength of your case, there is no reason why others should feel differently. Bear in mind that the views being so vociferously put forward by the other side might, in any event, be flying in the face of received advice. For example, consider those who raise their voices and talk over others so as to prevent an orderly enquiry into the 'facts' upon which they rely.

Conclusion

Building is a complex process and requires great forethought in its planning and a regimented approach to its delivery. That said, no amount of planning is ever likely to wholly eradicate instances of design information which is late and/or inadequate and/or additional, and these are the factors which beget claims. Without doubt, a number of claims are often successfully pursued which need never have arisen, and many others are settled at levels which are unrealistic given the real facts surrounding them. Such occasions leave one or sometimes both parties dissatisfied and in time they seek extreme contractual procedures for future projects. This cannot be in the long-term interests of the construction industry.

It is often claimed that alternative procedures to the competitive lump sum contract prove that claims are minimised. Almost invariably this is because such procedures enable the contractor to build a margin of safety into the pricing. This of course does not minimise claims, it masks them, and the contractor keeps the margin whether disrupted or not.

It does not follow that a profligate price begets a higher profit and a healthier industry; on the contrary, such a 'relaxed' price will simply mean the removal of any incentive to strive for efficiency and the resultant picture will be no more edifying than other occasions when a corset is removed. If the view is to be created that the construction process is one which represents good value for money, then lump sum contracts expressed relative to a complete and integrated design must be the bedrock. And for this to be acceptable to clients, for whom reliability of time and cost is usually of paramount importance, the incidence and settlement of claims must not be allowed to reach unacceptable levels.

Buildings provide the infrastructure which support our daily activities yet are so much taken for granted that people can be forgiven for overlooking, or more simply, not understanding their complexity. The challenges in the delivery process will always create tensions but the industry's efforts would be better focused on developing comprehensive and integrated designs which are delivered in a seamless, collaborative and timely fashion rather than worrying too much over the words in a contract.

At the end of the day, it is not the words in a contract but, rather, the behaviour of the parties which dictates outcome. A collaborative approach to a project may mean that the words go untested. If, however, the parties end up in a situation where financial damage is likely to be experienced by one or both then human nature may well become an impetus towards adversarial relations rather than the words of the contract itself. The words in the oft blamed contracts ought not to allow our attentions to be diverted from a collaborative, precise and mutually beneficial approach which could (and should) be pursued in which the parties are equally aware of their contractual rights but do not feel compelled to express this in response to every event on site.

Notes

1 This is a publication produced by the Royal Institute of British Architects in 2020 and which, in its Foreword, describes the document as *the definitive design and process management tool for the UK construction industry*. It is also gaining international traction.

2 Clients may have certain aspirations in this regard, but local authorities tasked with granting planning permission will have expectations of their own.

3 Where public funds are being committed, procurement guidelines will apply to ensure transparency and fair treatment in the contract award process. A well-established example of this is public procurement law directed by the European Union and passed into the law of its member states. For further reading with regard to EU Procurement, refer to Hackett and Statham (2016), in particular Chapter 16 authored by Michael Bowsher QC.

4 Sustainability touches on: (i) environmental protection; (ii) economic growth; and (iii) social equity. For further reading with regard to sustainability in construction, see Hackett and Statham (2016), in particular Chapter 33 authored by Professor John Connaughton of the University of Reading.

5 This quotation is popularly attributed to Sir Aston Webb (1849–1930), a highly distinguished architect of his era, but the writer can find no evidence that he actually said it.
6 Sir Ernest Rutherford (1871–1937), Nobel Prize winning physicist, quoted by RV Jones in the Bulletin of the Institute of Physics (1962), 13, at 102.
7 This author is based in the UK, from where many of the following examples are taken, but has worked across the globe on international arbitrations and has noted a universal pattern as it relates to the construction industry.
8 Such as in the case of a car manufacturer. It is notable that Sir John Egan, the lead author of 'Rethinking Construction' the report, was the Chairman of Jaguar Cars which he turned from a struggling business (financially that is because the quality of the cars was never in doubt) to a profitable one whose stock value had increased fivefold during his ten year term.
9 In the case of construction, the unique requirements of each project make that project a prototype in its own right and not necessarily one which will ever be replicated and therefore learnt from.
10 Which could variously include a Judge (in the case of litigation), an Arbitrator (in the case of arbitration) or an Adjudicator (such as in the case of adjudication as it applies in the UK).
11 Being a general form of compensation for losses incurred as a result of events for which the contractor bears no responsibility.
12 Being a term used in UK construction contracts but whose effect is essentially the same as recovering those costs which would not have been incurred but for adverse circumstances encountered and the responsibility for which rests with the client.
13 Robinson vs. Harman (1848) 1 Ex Rep 850.

References and further reading

Banaszak, J., Billows, J., Blankestijn, R., Dussud, M. and Pritchard, R. (2020) *Collaborative Contracting: Moving From Pilot to Scale-Up* (London: McKinsey & Company).

Banwell, H. (1964) *The Placing and Management of Contracts for Building and Civil Engineering Work* (London: HMSO).

Bossom, A. (1934) *Building to the Skies: The Romance of the Skyscraper* (New York: Studio Limited).

Egan, J. (1998) *Rethinking Construction* (London: Dept of the Environment, Transport and the Regions).

Hackett, M. (2015) Commentary on Chapter 4: 'Project Feasibility, Business Case and Funding' (S. Gray). In: Ullathorne, P. (ed.) *Being an Effective Construction Client: Working on Commercial and Public Projects* (Newcastle: RIBA Publishing).

Hackett, M. and Statham, G. (2016) *The Aqua Group Guide to Procurement, Tendering and Contract Administration* (2nd edition) (Chichester: John Wiley & Sons).

Latham, M. (1994) *Constructing the Team* (London: HMSO).

NAO (2001) *Modernising Construction: Report by the Comptroller and Auditor General.* National Audit Office (London: The Stationery Office).

Regan, M., Love, P. and Smith, J. (2015) Public Infrastructure Procurement: A Review of Adversarial and Non-Adversarial Contracting Methods. *Journal of Public Procurement,* **15** (4), 405–438. https://doi.org/10.1108/JOPP-15-04-2015-B001.

RIBA (2020) *RIBA Plan of Work* (Royal Institute of British Architects). www.architecture.com/knowledge-and-resources/resources-landing-page/riba-plan-of-work.

Rimington, S. (2002) *Open Secret: The Autobiography of the Former Director-General of MI5* (New York: Penguin).

Trickey, G. and Hackett, M. (2001) *The Presentation and Settlement of Contractors' Claims* (2nd edition) (Abingdon: Taylor & Francis).

Editorial Comment

Procurement is a broad term that may be defined at its most general as 'the process of obtaining supplies'. More specifically, in a business context, it can be defined as 'the process by which an organization buys the products or services it needs from other organizations'.[1]

Contractors have procurement departments that manage the ordering and acquisition of the materials and components required for the construction of buildings and other types of built projects (e.g. roads, bridges) while government offices have people who procure all manner of supplies from furniture to paper clips. In the context of the built environment, it may be described as the delivery method or system used to take a construction project from inception to completion. Procurement then specifically refers to the contractual framework that enables a client to engage a range of participants in order to have a built facility designed, constructed and handed over ready for occupation or use.

At its simplest this may be nothing more than a client making a verbal agreement with a builder to add a room to an existing house or build a fence. At the other end of the spectrum there are joint venture projects that involve consortia of several major contractors working with large consulting firms (cost consultants, engineers, law firms, financiers, insurance brokers and so on) and hundreds, even thousands, of trade and specialist subcontractors. There are complex contractual arrangements that are intended to formalise virtually all aspects of the process of turning the client's wishes into a built project, incorporating protocols and systems for payment, risk allocation, lines of reporting, financial controls and much more.

A typical feature of building procurement (and the same is true when public money is involved, e.g. in the procurement of supplies for government departments and offices) is that the selection of suppliers is made through competitive tendering, and the management of tendering processes is a key concern for those managing a project.

Different delivery systems have different management and contractual frameworks that determine exactly how a project will be run, and different circumstances will mean that often some forms of procurement are perceived as being more appropriate than others as clients hope to achieve the best outcomes for themselves when they decide to commission a new built facility. Often clients will lean towards systems that they hope will minimise claims and disputes during construction and to do this they may attempt to lock other players into rigid contractual arrangements such as fixed price contracts which limit the scope for contractors to make claims during the contract period. Other aspects that influence their choices include responsibility for design management and innovation, and more recently, digital information management and ownership of intellectual property such as digital building models.

Different circumstances require different approaches and there is certainly no 'one size fits all' procurement method that is suitable in all cases. The various methods have their particular characteristics, and their own advantages and disadvantages; selection of a method for a particular project is likely to require clients to balance risk and benefit as they look for the system that best suits their project and their desired outcomes. In this chapter the author examines a range of procurement methods in use today and provides insights into the pros and cons of the different approaches.

6 Construction project procurement

A critical review

Steve Rowlinson

Introduction

What is construction project procurement? In the construction industry, the real estate industry, manufacturing industries and all forms of government it is about producing a facility that fits the needs of the commissioning organisation, the client. Clients have a goal that is something other than a building, facility, hospital or other public good. What the industry produces is a 'frame' within which business, services and utilities fulfil demand. So, the question 'how does a client go about efficient and effective project procurement?' is very difficult to answer to answer. It is laden with perceptions and objectives and strategies that, for all of the participants, are different each time and from one another. Participants, *inter alia*, are the client, financier, facilities manager, building manager, asset manager, contractor, subcontractors and a whole range of professional organisations including architects, surveyors, engineers, environmentalists, landscapers.

Construction procurement is a process that has many facets such as strategic, economic, structural, environmental and institutional. Project procurement nowadays is considered as a cradle to grave activity. A facility that meets the needs of the client and the various stakeholders has to be produced but must be capable of decommissioning. recycling and be acceptable to society. It must add value to society and so must take stakeholders' interests into account.

This chapter takes a critical look at how construction project procurement has changed over the past 40 years. The generic forms of construction procurement are introduced using a theoretical framework based on organisational theory. The role of clients, and their demands and expectations, are then introduced as drivers of change. The public client, that provides 40 percent or more of construction industry demand, is introduced and the role and nature of a specific procurement system, public private partnership (PPP), is explored. Following on from this is a discussion of the rise of collaborative forms of procurement such as partnering and alliancing. The current move towards integrated project delivery (IPD) that has emerged due to the power of digitisation of the design, construction and facility management processes since the turn of the millennium is highlighted. The chapter concludes by exploring the drivers of change in the way the industry operates

DOI: 10.1201/9781003102403-6

and how major clients are imposing change through their expectations and knowledge of logistics and digitisation.

W092 Procurement Systems

Studies of project procurement have come a long way since the setting up of the International Building Council (CIB) Working Commission W092 Procurement Systems in the 1980s. Our needs as a society, as an industry, and our social responsibilities have developed multi-fold since then. W092 was set up by CIB in 1989; its aims and objectives were as follows:

- To research into the social, economic and legal aspects of contractual arrangements that are deployed in the procurement of construction projects
- To establish the practical aims and objectives of contractual arrangements within the context of procurement
- To formulate recommendations for the selection and effective implementation of project procurement systems

The commission undertook groundbreaking research into the use of different procurement systems in countries around the world. However, it was obvious that the impact of culture, economics and institutions made a big difference in the way that project procurement was undertaken in each country. Project procurement is a multi-faceted and multi-dimensional issue. Simply put, the key players are the clients of the industry, the industry itself, and the social, economic and cultural systems within which procurement takes place. That, of course, makes the choice of an appropriate procurement system a vexed issue and one that is culturally bound and institutionally difficult. (Walker and Rowlinson 2007).

Assessing the performance of different procurement systems is fraught with problems. Clients have different and specific aims in procuring construction works. The approaches to procurement, the procurement systems, are very different in each jurisdiction. The capabilities of the construction industries and the roles that the various professionals play are also different in these jurisdictions. Hence, the choice of procurement system is very much situationally dependent (Love *et al.* 2019).

This chapter addresses the common approaches to procurement and the nature and differences of how these are implemented in various jurisdictions. A broad overview of the way apparently similar procurement systems are used in different jurisdictions and how they have commonality and very specific differences is presented. A framework within which procurement systems can be analysed and used for different types of facility that have different economic, social, environmental and strategic purposes is presented by way of examples. There is no easy answer to what procurement should look like in this third millennium. It has become a complex and political issue that addresses sustainability and social issues.

Modern procurement methods

Traditional views of procurement and how these have changed due to political, social, economic, technological and stakeholder issues that impact procurement are presented. The aim is not just to discuss and describe best current practice but critique and postulate how current systems can be improved and the whole process can be integrated in terms of delivery. These issues are addressed from both a theoretical standpoint and through practical examples from around the world.

Procurement has a historical context. The systems prevalent in the United Kingdom are still practised in some former colonies. This reliance on custom and practice in the way that approaches are adopted to procurement is based on a shortsighted view of how design, construction and facilities management interact with one another and fit together. The European system, of which Germany and France are good examples, took a different route, in part because the European countries have different legal systems rather than the common law system of the United Kingdom. It is an important point to bear in mind, that the institutional framework within which construction works are undertaken has a profound influence on our procurement systems (Ju and Rowlinson 2020). The United Kingdom and Australia have made significant and different changes to public sector procurement over the past 25 years in an effort to deliver value and sustainability.

Changes in procurement systems have often been driven by changes in technology. For instance, the development of prestressed concrete in France by Freyssinet and others led to a change in the way procurement took place. The engineer had a leading role in the design process and the architect made use of the engineers' knowledge of new technology to develop new and leaner built forms. Nowadays there is a focus once more on prefabrication, volumetric construction and modular integrated construction (MiC). In the early 1960s the British construction industry focused on prefabrication for building schools and housing as demand for rapid construction as a growing economy and population forced change on the industry and society (Orlowski 1969). Unfortunately, many of these prefabrication systems failed to meet expectations due to quality and technology issues. These issues are being addressed with modern systems that make use of digital construction and logistics to improve quality and delivery.

In Europe the architect was central to most procurement systems, to the extent that the architect became the leader of the project team and also dealt with much of the administration of the contract (Walker and Rowlinson 2007). This is often a barrier to integration of the design and construction processes and, so, a lack of buildability in design. Singapore has addressed this issue in recent times by introducing a Buildability Score for Design (BCA 2021).

Nowadays, digital technology is driving change, particularly the sharing of information through a common data environment (CDE). For over 20 years the construction industry has used building information modelling (BIM). Only recently has BIM started to come of age. BIM is information management. Modelling gives the impression of a three-dimensional workspace. However, buildings and structures are much more complex than that and have five, six, seven or more dimensions such as time, cost, sustainability, facilities management and building management. These must be embedded in the modern procurement system.

Construction project procurement is now complex. Not only are the professions of architect and structural engineer involved but also the environmental engineer, LEED accreditor, facilities manager, asset manager and a whole host of other professions that are part of the design, construction and facilities management processes. Consequently, the construction industry has to develop new ways of thinking and dealing with procurement. However, putting this into context, the development of project procurement systems since the 1960s is briefly reviewed.

Generic forms of procurement

Presented in the following is a taxonomy of different approaches to construction project procurement. It is a broad taxonomy as procurement is undertaken differently around the world but there are many similarities in the basic approaches. These are discussed in detail in Rowlinson and McDermott (1999) and the categorisation is based on concepts in management and business theory based on organisational form, selection criteria, bidding process, payment methods, competition and collaborative practices.

Traditional approach

The traditional approach to procurement, sometimes referred to as 'the animals came in two by two', focuses on dividing up all of the tasks in the production of a built asset to the various professions within the industry. It is the least integrated approach to procurement and generally takes longest to deliver the facility. It is typical of procurement in Great Britain and its past colonies and is a conservative approach.

The architect is the leader of the team, and the quantity surveyor prepares estimates and bills of quantities that are then used in the bidding process by the contractor and subcontractors. There are generally between three and six contractors bidding, providing the client with a competitive tender.

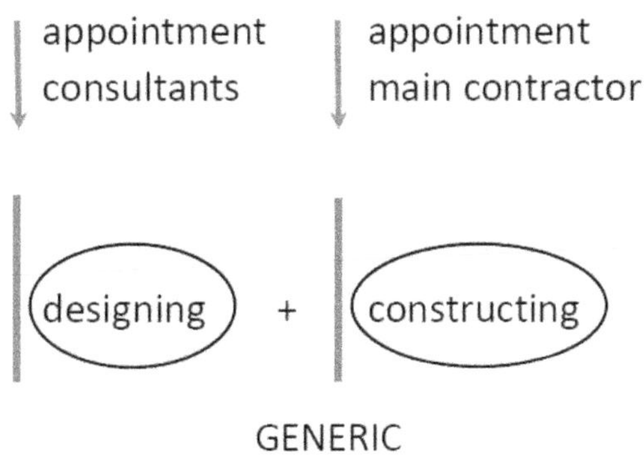

Figure 6.1 The traditional approach

However, as each professional works independently of one other there is scope for errors and omissions in the documents and specifications produced for the project. These are likely to lead to claims from the contractor for compensation for variations, delays and other issues. It is not surprising that other procurement methods have become more popular than this traditional approach.

Design build approach

The design build approach is in many ways the most integrated approach to project delivery. Generally, the design, costing, construction, and delivery come under one organisation. This approach has been popular in USA and goes under many names such as turnkey, system building, EPC (engineering procurement and construction – popular in China and USA) and design and construct (D&C). Within this approach there is opportunity for all involved to collaborate on designs that are both buildable and economic. However, these opportunities are not always grasped, and individual commercial interests often take precedence, potentially leading to disputes.

Adopting design build can lead to a lack of specific expertise in the design and construction team. Unlike the traditional approach, where the client can choose the best architect, structural engineer and contractor, the design build process is a package deal where all of the knowledge and experience and skill lies within one organisation. In order to address this issue one of the many variants of the design build approach allows for the novation of an architect's design to the builder.

In general, the fully integrated approach to design build is a highly unlikely scenario and often the design build organisation will buy in expertise in structural engineering and architecture and so the concept of the one-stop shop with the design build organisation becomes flawed. Design build organisations appear as three separate types; pure design build, where all of the expertise and skill lies within one organisation, a partially integrated approach where much of the expertise lies within one organisation but certain key skills and professions are bought in on a project by project or regular basis, and the fragmented approach whereby

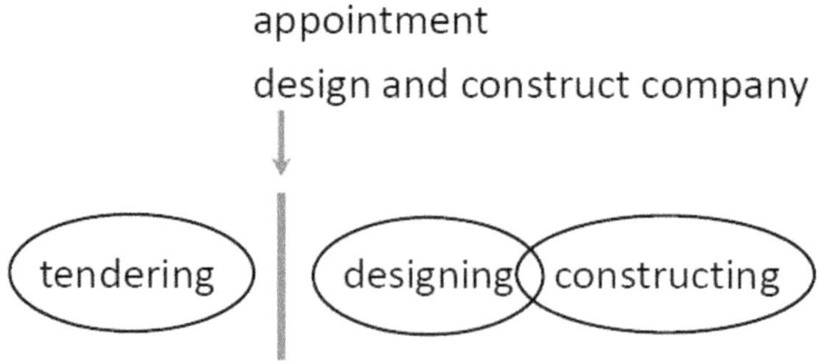

Figure 6.2 The design build approach

virtually all of the professional services are bought in by a contractor and the integration of the process is limited. The types of design build organisation are clearly identified in Rowlinson (1987). Generally, the design build approach can bring savings in both time and cost compared with the traditional approach. However, depending on the type of design build organisation there can be criticisms of this approach in terms of the quality of the facility produced. For this reason, many design build firms focus on one or two sectors of the market only and develop expertise in these particular markets.

The divided contract approach

The final approach to procurement can be classed as a divided contract approach. This is also referred to as the 'management approach' and is based around a construction manager or a managing contractor that pulls together the resources and expertise required by a client in order to develop a project. The contractor's role is one more akin to project manager as the majority of the construction is undertaken by specialist subcontractors that often also contribute to design development or detailing.

The divided contract approach brings an extra level of project management that oversees packaging of the works and tendering for each of the packages. The major advantage of this approach is that design can take place concurrently with construction thus saving time which can be of great benefit to the client. An example of this was the construction of Central Plaza in Hong Kong by Sun Hung Kai properties.

This project broke ground 40 days after the land was won at tender, and the project went ahead on the basis of constructing the 78-storey, 374-metre tower in three separate sections. The lower section was constructed and fitted out whilst the second section then rose above the first and, again, was fitted out while the third section was being constructed. Not only did the design and construction

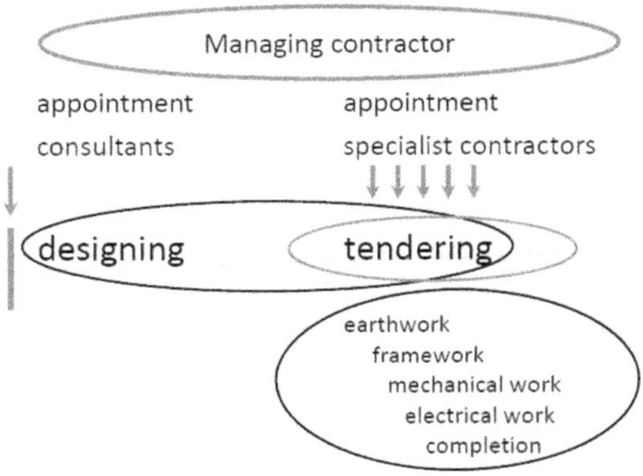

Figure 6.3 The divided contract approach

processes overlap but the client was able to let parts of the building whilst higher floors were being constructed. In terms of cash flow, and return on investment, this was an excellent approach to property development. In order to facilitate this a transfer plate podium was constructed on the fourth floor of the building as the design above had not been completed whilst the piles and diaphragm walls were being constructed.

Although the cost of design and construction may have been more expensive than undertaking a traditional approach, the ability to let parts of the building whilst the rest was being constructed meant that the return on investment and the cash flow far outweighed the extra costs in terms of the overall profitability of the project in a rising property market.

Another example of the divided contract approach along with the use of pre-fabrication was the construction of the Hong Kong and Shanghai Bank in Hong Kong. Cooke (1987) explains the novel method of constructing the building and the use of a local construction manager to facilitate the subcontracting process. Vosloo (2020) explains the architecture in terms of hanging skyscrapers and discusses the sustainability of the building. The nature of the banking business is addressed by Quantrill (2018) in which the value of the building and its enclosure is explained in terms of its relationship to the business of banking. The Hong Kong and Shanghai Bank building is an excellent example of project procurement leading to highly innovative designs which are sustainable in terms of environmental, economic and business terms.

Modern construction contracting and the client

In recent years construction contractors have focused more on the management of the building process, as in the divided contract approaches, and do less physical construction works with their own labour force, relying on specialist subcontractors to do this work. For this approach to be successful, it is necessary to carefully set up the company business model. Indeed, if this is not done with vision and care, what is left is the hollowed-out shell approach which is now common in the UK and is discussed by Smyth (2018: 28–30). In his critique of the UK construction industry's business model, he states:

> *The overall consequences among the major main contractors for the period were:*
>
> - *Main contractors were largely managers rather than constructors, and a degree of professionalisation had developed at operational level*
> - *Geographical and sector diversification continued in the early part of the period, yet was in reverse towards the end of the period 1995–2020*
> - *Management focused on sectors and projects, overlooking the need to become more strategic, reform and undertake investment for the long term.*
>
> *British-owned main contractors still did not copy the construction professional service firms. These consultants were growing, and London had become a world*

cluster for the professions. They had restructured around high-value specialised service provision.

On the contrary, the reliance on return on capital employed (ROCE) as a measure of profitability is now outdated given the worldwide professionalisation of the international contracting business. As Smyth points out:

Clients had developed programme management to support all their projects and were developing strategic approaches at the front-end of each project. British-owned main contractors did not mirror this and follow suit. As such, contractors continued to react to conditions, rather than strategically manage their prime markets. The period is characterised by two main features:

- *Growing misalignment between the service offered and the combination of client demands and project complexity*
- *The continued loss of market share among British-owned main contractors over the period, despite initial market growth and exacerbated by a severe downturn as a result of the financial crisis in 2008.*

The lessons learned from this are that modern contractors must 'move with the times', and the old methods of winning work and outdated business models fail to appeal to modern, sophisticated clients and stakeholders. Construction procurement systems are being shaped by client demands for sustainability, facilities management and digital construction. 'Modernise or die' is the current mantra of the UK construction industry; the controversial *Farmer Report* is discussed at the end of this chapter.

Public private partnerships – PPPs

The World Bank (2020: 6) defines PPPs as:

a mechanism for government to procure and implement public infrastructure and/or services using the resources and expertise of the private sector. Where governments are facing ageing or lack of infrastructure and require more efficient services, a partnership with the private sector can help foster new solutions and bring finance.

PPPs are the most political procurement form. Many countries rushed into launching PPPs without putting in place an appropriate institutional framework. Often, the introduction of PPPs was based on political expediency rather than a well-thought-out procurement strategy.

PPPs have been controversial since their inception. Looking back historically, perhaps the first successful modern-day PPP was the Hong Kong Cross Harbour Tunnel in the late 1960s. However, the Brunels, father and son, constructed a foot tunnel under the Thames a century earlier and that could also be considered a PPP. Also, the rail lines from London to the West Country in England were project managed and financed by Isambard Kingdom Brunel and his backers and these could also be considered PPPs. Incidentally, Brunel was a multi-talented engineer and an entrepreneur, and was capable of designing bridges, tunnels and ships (Brindle 2013). Very rarely nowadays do we get such rounded talents in construction.

In the 1970s and 80s Japanese contractors leveraged funding from banks at virtually zero interest rate. This was part of a national policy to extend the reach of the Japanese economy into Asia and Australasia. Thus, the aid projects and the publicly tendered projects could be seen as part of an informal PPP that promoted Japanese companies and products. Japan entered the PPP market in Australia and in Hong Kong through infrastructure projects such as the Melbourne Link (Obayashi) and the Eastern Harbour Tunnel in Hong Kong (Matsumoto *et al.* 1990). At this point in time the leverage was not just the low interest rates available in Japan but the setting up of a special purpose vehicle to look after the asset in its planning, design construction, use and facility management phases.

However, the Japanese contractors began to get cold feet during the late 1980s and 90s as the economy turned down and the competition became more intense. The innate conservatism of the senior Japanese company directors led to a withdrawal from what were quite lucrative markets. Around this time, Japan and economies worldwide suffered a major problem in the construction industry, what the Japanese called *Dangō* or bid rigging/prearranged business agreements (Jefferies and Rowlinson 2016). This an oligopoly and competition problem and is one of the areas of governance that really needs to see a change in the way construction projects are tendered and procured in order to ensure that it does not happen.

In the UK, the Thatcher government attempted to make use of its so-called Private Finance Initiative (PFI) to deliver value for money from public infrastructure spending. However, some suspected that there were political dimensions to this initiative. The aim was to effectively keep public expenditure out of public borrowing although the public explanation was to secure the so-called benefits of private sector efficiency. Another explanation was touted as being leverage against the common agricultural policy of the European Union. It was almost a tit-for-tat approach: 'we will pay for your sheep, but you pay for our roads and railways'.

At the same time, the banks in Australia saw PPPs as cash cows for the special purpose vehicles (SPVs) in which they could invest and later divest themselves of after maybe five years and make a very good return on their investment whilst also accruing interest from their partners in the SPV. Defining the structure and governance of PPPs is difficult as each country has its own approach to them. For instance, Australia categorises PPPs into two separate types (see Fig. 6.4). The distinction is between predictable, quantifiable revenue streams of hard infrastructure (economic) and much more complicated financial and demand forecasts for public services (social).

Economic PPPs tend to be easier to assess than social PPPs. For instance, the number of users of roads or railways is much easier to predict from current usage data compared to an aged care facility or hospital that depends on future demographic

Economic PPPs	Social PPPs
• Roads	• Hospitals
• Railways	• Schools
• Bridges	• Prisons
• Motorways	• Public Housing
• Tunnels	• Aged Care
• Ports	• Tertiary Education

Figure 6.4 PPP infrastructure types

data, developments in medical technology and drugs and other less predictable changes. Hence, economic PPPs in general can be less risky investments.

In the past two decades China embarked on a massive rail program throughout the country, and this was a learning curve for the rail construction industry. The system is now in place, though still being developed further, and is exemplary. The internal development of expertise in express rail construction allowed the Chinese construction companies and rolling stock suppliers to develop their own niche in worldwide markets, particularly in developing nations. Many projects in China, not just rail, were undertaken through PuPuPs – public public partnerships. We might say that the Great Wall was an example of a PPP in that the Emperor enlisted workers and craftsmen from nearby provinces to work on this mega-project for China. In some ways, following the Japanese example, infrastructure was being built as part of a national policy to provide funding and technology to less developed nations. Through this approach, driven by its Command Economy and as laid out in its Five-Year Plans by the Central Committee of the Chinese Government, Chinese construction contractors now play a leading role in international construction and *Engineering News Record* lists six Chinese contractors in its top 20 international contractors.

From this brief review of the development of PPPs it can be seen that the SPVs, the governance structures of PPPs, appear in many different forms with many different parties involved, so no standard form of agreement has been developed. The key issues blocking standardisation are the sources of finance and the sourcing of engineering expertise. One country that did not dive headfirst into the PPP approach was Singapore, apart from desalination plants, waste disposal plants, and the controversial Singapore Sports Hub (Trager and Ping 2014). The Singapore view was that there was no need for PPPs as it manages its economy very effectively and so has ample cash to produce infrastructure to stimulate the development of its economy. Wealth generated from business and taxes enabled Singapore to eschew PPPs until 2003 and it went about controlled development of infrastructure and other strategic assets in order to safeguard the integrity of its resources and stimulate economic growth. The Singapore government thought carefully about whether it needed to push itself into a difficult position with PPPs when it could have a much more collaborative contract with integrated project delivery and perhaps a contract like NEC (ICE 2022) to actually organise projects effectively. An example of this far-sighted, open-minded view was the Sands on the Bay project, where the government decided not to accept the lowest bid to develop the site. Government made it clear that it was going to accept the best proposal, in terms of what was to be built on the site and not the lowest bid. This is an example of an enlightened view of how to stimulate industry and the economy and the development is undoubtedly a huge success.

An issue that plagues the efficiency of PPPs is the special purpose vehicle (SPV) that must be set up for each project. The SPV is the bidder so multiple bidders need to spend time, money and negotiating skills to set up this entity without any guarantees of winning the project. Looking at a typical relationship network for a PPP SPV in Figure 6.5, it is obvious that setting up a PPP requires negotiation and collaboration among a diverse range of interested parties. Indeed, developing a legal agreement between the stakeholders is a difficult and expensive process. Figure 6.6 makes this very clear. There are so many bonds and trusts and contracts and supply agreements that the whole process is complicated, time consuming and expensive. Hence, the business case for a PPP must be carefully assessed and all transaction costs clearly identified. This makes the case for a national PPP authority and for clear guidelines and procedures essential.

To summarise, a key criticism of PPPs is the lack of accountability of the partners within the SPV and there is little or no evidence that PPPs provide value for money or that benefits exceed costs. If an entity within the SPV is able to divest interest in a PPP before the end of the franchise it may cash in on the initial advantage and profitability and leave the liabilities with subsequent partners and the public. This is one of the many criticisms of PPPs. Though effective in some areas, governments have to, as former British prime minister Harold MacMillan described it, 'be wary of selling off the family silver.'

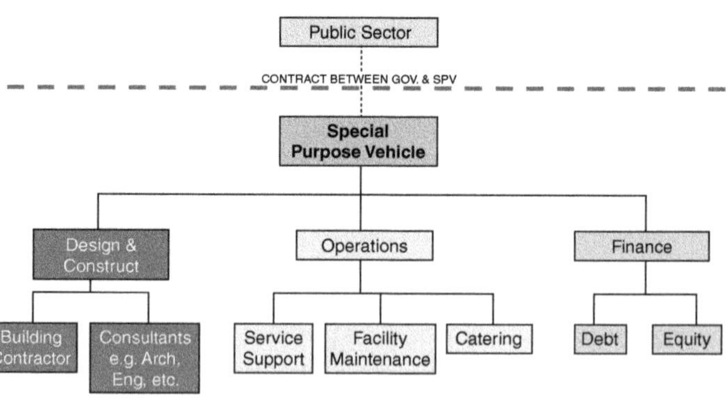

Figure 6.5 Typical PPP network of relationships

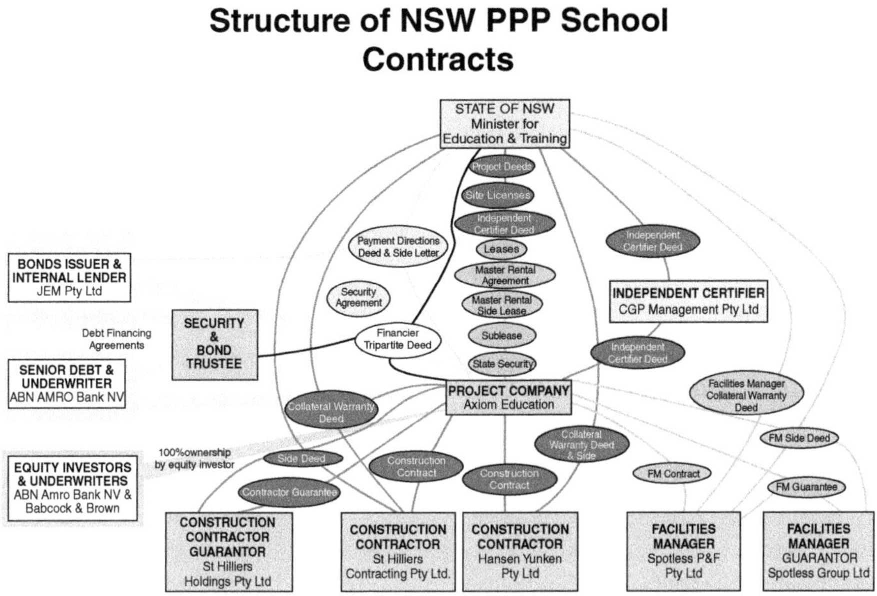

Figure 6.6 Typical legal structure of a special purpose vehicle (SPV)

On top of this, there is the issue of what type of PPP is easily or readily accessible in terms of its return over the 20 or 30 years or whatever is the period of the franchise. The issues of facility and asset management are very important considerations in determining the viability of the PPP, as is the distinction between social and economic PPPs as discussed earlier.

Collaborative approaches to construction procurement

Briefly, looking at other 'best for society' approaches to procurement, 20 years ago in the UK framework agreements were introduced with the intention of stimulating both social development and industry demand, thus stimulating capacity building. The point of the framework agreement was that several contractors, subcontractors, designers, architects got together in a consortium to bid in competition with other consortia with the promise of several projects over a number of years. This allowed them to work together and to share information to collaborate to improve the processes of project delivery to help 'support public sector bodies to deliver their strategic targets'. This was a driver for change, but also social development, which is a very important issue for the industry. Although it was politically driven, and despite changes of government, it still operates as a stakeholder engagement form of procurement in certain regions.

However, it was not all plain sailing; Tennant and Fernie (2012: 209) provide an interesting critique of frameworks using a neoclassical lens to expose business tensions around individualism, instrumentalism and equilibrium. They state that 'a number of private sector clients [are] rejecting framework agreements in favour of traditional competitive procurement practices.' Public sector investment continues to sponsor collaborative working practices via framework agreements but there have been an increasing number of clients demanding 'more for less'. Such attitudes will continue to influence which construction procurement system is preferred. As noted earlier, there is no 'one size fits all' approach to construction procurement.

Transparency

In Japan, and other countries worldwide, collusion and corruption between contractors and authorities are major barriers to an efficient and effective construction industry. Indeed, the Japanese refer to such activities as *Dangō* (Woodall 1993), or bid rigging or prearranged business agreements. This is an anti-oligopoly or cartel issue and has caused several government officials and elected members in Japan to be dismissed from their posts in recent years. This is an area that has plagued construction worldwide and the sector needs to see a change in the way that the organisation of tendering and procurement is conducted. Transparency is essential for effective, efficient and sustainable construction procurement systems; see, for example, the work by the Infrastructure Transparency Initiative (Infrastructure Transparency Initiative 2021).

Alliancing

In Australia, Walker and Rowlinson (2019) reported the use of alliancing on the Wivenhoe Dam extension and drew the conclusion that alliancing provides the opportunity and facility to align the goals and aspirations of all contributors and stakeholders in a project. This same conclusion is also reported by Lloyd-Walker and Walker (2015) in relation to a level crossing program implemented in Melbourne.

As with the other relational contracting approaches, trust between partners plays an important role in alliancing. To reinforce the trust element, a no-dispute clause was introduced to the alliance agreement in this case study project. However, a no-dispute approach is not possible without a positive approach to relationship management. So, is 'no litigation alliance' fundamentally tautological? Without a clear relational vision that leads to both soft and hard infrastructure that assist decision making and relationship building, a no-litigation contract cannot exist. This is the basic premise behind alliancing.

Relational contracting approaches were introduced into the construction industry in the 20th century. Among them, alliancing is regarded as a long-term business strategy that links together clients, contractors, and the supply chain (Rowlinson *et al.* 2006). Commitment and action by the project alliance board, comprising parent organisations, has a strong impact on the team in developing an alliance culture. Thus, alliancing has a high chance of failure when there is inadequate buy-in and support from top management. Consequently, the management of the alliance is a key element in determining success, as discussed in the following.

Scholars have categorised alliancing into two main types: 'strategic' and 'project' alliancing. Strategic alliancing is often defined as establishing interorganisational relations and engaging in collaborative behaviour for a specific purpose (Love and Gunasekaran 1999) whereas project alliancing is described as project delivery strategies whereby the project participants work together to share risks and outcomes (Hutchinson and Gallagher 2003; Manivong and Chaaya 2000). Aiming to maximise performance, proactively manage risk, reduce cost, and achieve outstanding results in attainment of client'goals, the objectives of the client and the commercial participants are aligned. A clear, if almost unattainable, definition of project alliancing was given by Hutchinson and Gallagher (2003) as 'an integrated high performance team selected on a best person for the job basis; sharing all project risks with incentives to achieve game breaking performance in prealigned project objectives; within a framework of no fault, no blame and no dispute; characterised by uncompromising commitments to trust, collaboration, innovation and mutual support; all in order to achieve outstanding results.'

The diversified approach to construction projects, which has been enabled by the formation of alliances, has received mixed responses from the industry. According to Jefferies *et al.* (2001), while many are wary of the concept and the new ideas associated with it, though with limited knowledge about the concept, some are willing to enter an alliance, desiring to perform as a participant and develop business expertise.

An alliance should provide a framework and a collaborative environment to adapt behaviours to project objectives, to share resources and experiences, and expose the 'hidden risks'in a project. In order to identify and report critical success factors of alliancing, an alliance project between public and private organisations in Queensland, Australia is presented (in the following) as a case study (Cheung *et al.* 2005).

A typical alliance

The case study featured an alliance project between a multifaceted public sector organisation (the client) and several private sector organisations to upgrade three wastewater treatment plants located in three different suburbs. The public client chose the alliance approach, with the aim of developing mutually beneficial relationships among all parties to deliver outstanding project outcomes over a number of projects. Depending on how the project outcomes compared with pre-agreed targets, all parties took collective ownership of the associated risks of project delivery with reasonable risk sharing at fixed pre-agreed ratios of the financial outcomes (pain or gain).

This is a commercial alignment that is consistent with the 'no-blame, best for the project' philosophy of alliancing that directs all participants towards setting and achieving common objectives, difficult though this may be when using current adversarial standard forms of contract, NEC excepted as it is designed to promote collaboration. In line with this approach, a no-claim clause was included in the contract. This unique government project involved large capital outlays and many private sector organisations, and its success rested on both flexibility and innovation.

The selection of the alliance project team was crucial in ensuring that, in meeting project challenges, the strengths of the individual alliance partners were properly utilised.

The selection criteria included:

- Capability and capacity to complete the full scope of works
- Proposed approach to each project
- Affinity for project alliance culture
- Relationship management capabilities.

In Australia, alliancing is increasingly popular and successful and another example is the recently completed Level Crossing Alliance in Melbourne. Taking traffic from at grade to underneath or over level crossings was an approach to improving road safety, ensuring that trains and motor vehicles are separated so that incidents or accidents cannot occur at crossings. An alliance was set up as it was recognised that such a large-scale venture involving hundreds of locations required community and stakeholder management as well as continuous refinement of techniques and technologies used. Learnings from one project were carried over on to the next project and all of the participants were encouraged to learn from this. Achieving such objectives does come with a cost, and so a need was identified to employ alliance managers to keep the various parties actively involved in this particular type of program of projects.

To encourage dissemination and subsequent learning, project learnings and innovations were written down and recorded on multi-media and stored in a common data environment. This was designed to facilitate contractors', consultants' and subcontractors' learning and skill development. There also needs to be a promise of continuing workload and future value in lessons learned, and such an approach has the potential to lead to a much more mature and profitable and efficient industry (see Walker and Rowlinson 2019).

Rowlinson *et al.* (2006) discuss the structure of a project alliance board, and their analysis indicates that there are considerable overhead costs in a continuing alliance that include the alliance manager, alliance communication manager, risk and opportunities manager, alliance coach and alliance psychologist, as shown in Table 6.1. In some ways this structure mirrors what we find in BIM-enabled projects nowadays.

Industry, particularly in Australia, took a big picture view of what was needed to improve construction project procurement and alliancing appeared to suit the Australian culture of open and frank discussion, prompting a move forward into the field of alliancing as described earlier. The alliance approach allowed for

Table 6.1 Alliance organisation structure – Project Alliance Board (PAB) (Rowlinson *et al.* 2006)

Project Alliance Board (PAB)	Senior executive from all alliance partners	Provide governance Set policy and determine delegation Monitor performance of the AMT High-level leadership/support
Alliance Management Team (AMT)	Alliance Manager Deputy Alliance Manager Project Managers from each site Design co-ordinator Alliance communication coordinator Environment manager Risk/Opportunities and Innovation Manager Alliance Coach Alliance Psychologist Services Manager	Provides overall management for all three projects Ensures effective integration into public sector organisation operations Performance management
Integrated Project Team	Project staff at operational level	Individual project work

multiple contractors, subcontractors and consultants and subconsultants to work together over a period with a number of similar projects and so enhance, piece by piece, the process of design, construction and facility management. There was a common ground in which information was shared and learnings were passed on between the members of the team.

Unfortunately, this development took place before BIM and common data environments (CDE) had properly developed and so the ideas and concepts behind alliances were held back by the lack of digitisation in the industry and the lack of a trusting and collaborative environment that could be maintained throughout the project and a series of projects. Thus, we might say that partnering was the initial concept that was identified as long ago as the Bovis A5 contract that was used with the British supermarket chain Marks & Spencer in the 1950s. It took a couple of decades before the rest of the industry realised that for ongoing programs of projects a partnering approach was both worthwhile and necessary. It then took perhaps another decade to understand that the concept of the alliance could address many of the issues of collaboration, cooperation and trust. The era of digitisation and digital twins is here and the ability to virtually construct projects in a collaborative environment exists. This environment is now being driven further by the incorporation of prefabrication, MiC and volumetric construction into the initial phases of project development. Hence, we can see that more than 50 years have passed in developing the current approaches to collaboration, cooperation and trust. Indeed, some of the structures as described earlier may well suit

IPD, CDE enabled procurement systems and enhance the 'big room' approach to project management by addressing more directly the collaboration aspects of the process. A virtual FM based view as to the future of project design, construction and use will also facilitate the assessment of value for money and sustainability.

Partnering

The alliance approach grew from some of the weaknesses that are inherent in partnering. For a detailed analysis of the weaknesses see Bresnen (2007), who outlines his *Seven Deadly Sins of Partnering*. Partnering was often seen as a one-off approach to getting all contracted parties to work together as a team (Bennett and Jayes 1998) That team included clients, designers, contractors, subcontractors and suppliers. Partnering workshops normally took place after a tender had been accepted with the aim being to get the members of each organisation to work and think as a team together in a 'best for project' approach, however, such an approach is not ideally suited to the nature of the construction industry. Design, costing, building, construction and facilities management were traditionally addressed as separate items in the list of tasks to be completed and there was no element of integrated project delivery really within partnering systems. Partnering was treated as an add-on to the traditional approach to procurement. There had been little or no process change. Nevertheless, partnering does bring about an awareness of the need for shared values and 'best for project' thinking and, in some measure, paved the way for alliancing to develop these concepts further.

Contractor driven innovation in procurement

Looking at innovation from a different perspective, the traditional way of constructing in Hong Kong, and in most other parts of the world, involved masses of false work supporting formwork which eventually supported the beam, column or slab that was to be constructed. Gammon Construction took on a traditional, consultant designed structure and turned it on its head. The 'Methods' team decided to design a formwork and false work system which came all in one that could be craned into place and would be much safer. This system had a whole series of innovations in terms of how the forms were put together and how they were

Table 6.2 Seven pillars, seven paradoxes and seven deadly sins of partnering (Bresnen 2007)

Pillar	*Paradoxical effect*	*Deadly sin*
Strategy	Wishful thinking about strategy and behaviour	Sloth
Membership	Fostering of relationships built on exclusivity	Lust
Equity	Encouraging exploitation and opportunism	Avarice
Integration	Reinforcing a desire for control	Gluttony
Benchmarks	Setting of inappropriate targets	Envy
Processes	Over-engineering of processes	Wrath
Feedback	Failing to capture knowledge and learning	Pride

struck and working platforms were built into the formwork. For the cargo terminal under construction, prototypes of these shutters were used for training the workers, so that they were conversant with the operations when they started production in earnest. This idea was moved forward into other projects with the goal of eliminating risk. A BIM model was used to show workers how to strike the formwork without having to touch the formwork. A modularised access system was in place from the columns to support the beam formwork and so this was a well thought through innovation, that used digital technologies such as BIM to put together models of how this work could be done before the structure was physically built. In fact, in the mechanised system, there were only seven workers and 50 man-hours labour needed compared to the traditional system. Half of the workforce completed the work in a quarter of the time, so the system was efficient, effective and safe (claimed to be eight times safer on the basis of reduction in man-hours).

This approach to procurement where contractors suggest alternative methods and designs that are cheaper, quicker and safer is changing attitudes to project delivery towards early contractor involvement and IPD. This approach also affects a key issue in procurement, that is the safety of the individual worker; the higher the level of mechanisation, the smaller number of people on site, the quicker the task is completed and so less risk to each worker. That is the claim, but that claim, as yet is not substantiated. The issue lies in ensuring that the mechanised and prefabricated systems are safe in terms of design and use (Walker and Rowlinson 2019).

Considering this process on a larger scale, the Shanghai Green Building company produces modular panels and combined services units for the construction of housing, offices, and even hospitals. These residential and business units consist of standard elements that are designed to be put together on site and they are manufactured in a 200-hectare factory in Shaoxing in China. Across China there are companies such as this, setting up in different provinces and achieving a five-year plan of building 40 million housing units in five years. What is supplied is a high-quality product that is delivered and put together on site in such a way that the workforce is vastly reduced. Working conditions in the factory are much safer compared with a construction site and the factory is highly mechanised and impressive, running with the precision of a car manufacturing production line. This approach to the provision of vast quantities of social housing is promoted and supported by the central and provincial governments. It is enabled by sophisticated digital technologies in both design and production in the factory and robust tracking of the individual components for assembly and quality assurance.

Prefabricated pre-finished volumetric construction (PPVC) is one of the innovations that Singapore has introduced. Dragages Singapore, a member of the French Bouyges conglomerate, built a Crown Plaza hotel at Singapore Airport using PPVC and that project was completed in record time. All room modules were placed on top of one another and side by side, fully finished and equipped from the factory. Such approaches challenge the conventional procurement systems and the current roles of the professions in the construction industry.

There are both specific requirements and an institutional background of governance to this move towards prefabricated prefinished volumetric construction (PPVC) in Singapore (BCA 2020) and these are dealt with through the Building and Construction Authority (BCA). The BCA sets the ground rules for construction projects in Singapore and these rules provide a sound basis for a level playing field, a distinguishing characteristic of construction in Singapore. There is a minimum level of construction on site, mostly 'connection' operations, and so vulnerable workers are taken away from traditional danger areas. The limitation, of course, is the size of these units that are lifted into place. In one of the early attempts to deal with this, the contractor had to use A-frames to lift some of these massive concrete units and that, in a sense, took away from both the cost advantage and the time advantage of having to put this process together. This has to be addressed by considering the size of each module and using different materials in order to ensure this innovation is a success.

The quest for the Holy Grail

How does the architectural engineering and construction (AEC) industry make construction project procurement more efficient, effective, safe and sustainable, and develop an industry that people want to work in? An industry that adds value to projects and society? BIM is not the answer, information management is part of the answer and IPD is what should be the industry's goal. Careful thought, planning and policy about the future state and shape of the industry is needed. From start to finish of a project, decisions must be made about what should be constructed through to actually operating, maintaining and finally dismantling the facility.

Infrastructure, whether it be a road or a factory or an office is an asset to be used over a lifetime. Putting all the relevant information together at the front end of the process, as done in IPD, is more efficient, effective and economic in the long term. This is the sea change in procurement that the construction industry must accept. Unless the industry embraces change this challenge will allow logistics companies and management consultants to take over the role of constructors. The industry already relies on outsourcing most of its production, so it faces a threat from those who plan and manage in detail and operate with discipline. As pointed out by Smyth (2018) the UK construction industry has already been slow to adapt and change its business model and has lost competitive advantage to European counterparts. Change is necessary in the industry worldwide and a few major companies will lead the way.

Why is a change needed? There is an institutional monster, the set of institutions that govern construction industries worldwide. There are many different organisations all of which have either power or influence that can increase or decrease construction efficiency, quality and safety. For the construction industry to perform procurement effectively it needs to shed the silo mentality of the professions and embrace a change to IPD. The constructor, designer, building services engineer, and subcontractors must come together at the beginning of the project, even before a price has been agreed, in order to ensure that the project is delivered in a form that is usable, economic and that meets the client's and stakeholders' needs. This change will not come easily, and it depends on a whole range of collaborative and trusting behaviours being recognised and further developed. Constant information sharing and testing understanding is essential for this process change to succeed.

The construction industry culture is adversarial rather than collaborative and that is not just true of Hong Kong or the UK or Australia; it is true all over the world. Teamwork does not come without trust; sharing information leads to trust; inspiration comes from speaking with people. Supporting the constant exchange of ideas leads to success, so everybody coming to the table must be involved at the beginning. This includes the facilities managers, the asset managers, and the workers and users whose safety and health must be paramount. This must be designed in.

What is needed is to change attitudes and have an awareness of the impact of design on many aspects of construction and use of construction works, a DfX approach. Thus, we have Design for Safety, Design for Assembly, Design for Disassembly, Design for Use, Design for Manufacturing. Indeed, there is a need to change design and construction thinking and move into a new paradigm that is inclusive and reflects the future. Unfortunately, it is highly unlikely that the older generation of construction professionals will lead this change, let alone buy into this. It has to be dealt with through education systems, bringing young people through the industry who actually understand and care about this change taking place. Our professionals must be comfortable with digital technologies and the extra dimensions it gives to their work and their contributions to the work of others.

Another obstacle to culture change in construction project procurement is the forms of contract, for public or private clients. Current contracts are generally adversarial and do not promote collaboration, trust, information sharing or mutual understanding. The *New Engineering Contract*, version 4, promotes discussion and collaboration and resolution of disputes before they happen. It has a whole series of compensation events that link to the construction program. This promotes collaboration and teamwork, which is what the industry needs if it is to survive the external threats to its future.

The Farmer Report

Farmer's report (2016) in the UK dealt with the state of the industry and proposed reforms to the way the industry operates. It is important as it promotes a move to industrialisation through off-site manufacturing. Farmer approached the problems in the construction industry from three perspectives. He looked at the industry and focused not only on construction but also on design organisations and then moved on to address clients and how clients should reform their approach. The focus of his report was what he called pre-manufacture which others would probably call prefabrication or offsite construction. He addressed all three of these with the view that the industry was unproductive, inefficient and badly organised. However, his proposals could be critically appraised and Green (2016) did just that. The evolution of the role of the client in driving change in construction project delivery is discussed in Walker and Rowlinson (2019).

Essentially, Farmer presents what he considers the critical symptoms of failure and poor performance as: low productivity, dysfunctional training and funding, and inappropriate construction project procurement models. According to Farmer, in 2016 the workforce was too small, had inappropriate demographics and exhibited structural fragmentation leading to a lack of collaboration and trust in the industry. He identified a lack of leadership that manifested itself in terms of lack of research and development and investments in innovation. He also stated that profit margins, adversarial pricing models and financial fragility led to the industry having a particularly poor image. These symptoms are not unique to the UK; they exist in many countries.

In his critique of this analysis and raft of initiatives aimed at solving the industry problems, Green (2016) pointed out a series of issues in the report. For instance, the problem with training stemmed from the fact that contractors in the industry had to ensure they could expand or contract according to fluctuations in demand by offsetting risks onto the supply chain. By this he pointed out that 40 percent of the workforce is self-employed. Hence, there is no mechanism for integrated training and building skills. He usefully points out that Farmer appeared to be demanding that the industry change in order to improve its relationships with its clients.

Agreeing with Smyth (2018), Green pointed out the industry has become used to functioning as a hollowed-out shell that needs multiple layers of subcontractors in order to survive in the current markets. Thus, the problem is exacerbated at the lower end of the chain where so many of the workforce are self-employed.

Green attacks the suppositions behind Farmer's propositions. Innovation in the housing sector through pre-manufactured solutions is possible but 50 percent of UK construction output is repairs, maintenance, additions and alterations (RMAA). RMAA cannot be undertaken using the pre-manufactured paradigm. Green goes on to point out that the suggested approach is likely to reduce the level of traditional skilling even further.

Donaghy, in the previous decade, was even more damning of the business models adopted:

> '*the construction industry is modelled to provide maximum flexibility. Consequently, most functions are contracted out and at least 40% of workers are self-employed . . . The advantages are obvious in that it reduces overheads. Some but not all argue that it improves profitability and productivity. The disadvantages are that it becomes more difficult for a safety culture to flourish, worker engagement is weak, employment security and continuity is minimal, and skills training is at best patchy.*
> (Donaghy 2009: 21)

Green also points out that 'contractors make their money on the basis of *contract trading*'. Thus, the contractor is not interested in productivity but merely the price he pays to his subcontractors, having been paid already for work by the client. If manufacturing of components is taken offsite then the productivity of the industry will apparently drop even further. Such works cannot easily be modularised. As

Smyth pointed out, Green also notes that the design and engineering services companies have changed in order to ensure that their incomes are protected by moving with the market and following clients' directions. To say that the construction industry operates on low profit margins is to ignore the fact that the industry, in its current state, returns a good return on capital employed (RoCE). This makes it an attractive sector for many firms to work in.

The way forward

If the business model is to move from a project-based to a product-based model, the value chain must be refocused and those people who supply components, which go together to make a building, need to be involved at the outset of the project. Also, components need to be more sophisticated so that they can be used and reused and fixed in place efficiently and effectively. They must be able to be assembled on sites and use product based digital processes and digital twins to guide assemblers, owners and users. A data environment that captures the whole constructed project information, the digital twin that should be built, not at the end of the project, but at the beginning is needed.

For IPD to be successful the facility and asset managers and clients must make inputs before the construction process starts. Hence, more time is needed at the front end of the procurement process, but this must not be wasted time. Many projects go through three phases of design elaboration before a final design is completed. This long, drawn out process leads to the client attempting to compress the construction program to an unacceptably short time. More sensible tendering procedures are also needed so information management is the key, a CDE that focuses on integrating unconnected contributors or being, preferably, part of an integrated project delivery system.

As for business, nowadays construction firms need to ask whether they need such large offices. Employees using flexible working is common so the industry should be questioning the need for current office space. Overheads can be reduced by divesting property. That opportunity to reduce location costs might be a boost in the arm for the industry post-pandemic. Indeed, a recent report commissioned by Sir Robert McAlpine (*Construction Manager* 2021) indicates that a 50 percent increase in flexible working could lead to over 50,000 new jobs, and an economic gain of GBP 55 billion to the UK economy.

Note

1 https://dictionary.cambridge.org/dictionary/english/procurement

References and further reading

BCA (2020) *Prefabricated Prefinished Volumetric Construction (PPVC)* (Singapore: Building and Construction Authority). www.bca.gov.sg/buildsg/productivity/design-for-manufacturing-and-assembly-dfma/prefabricated-prefinished-volumetric-construction-ppvc.

BCA (2021) *Buildability (Buildable Design and Constructability)* (Singapore: Building and Construction Authority). www.bca.gov.sg/buildsg/productivity/buildability-buildable-design-and-constructability.

Bennett, J. and Jayes, S. (1998) *The Seven Pillars of Partnering: A Guide to Second Generation Partnering* (London: Thomas Telford).

Bresnen, M. (2007) Deconstructing Partnering in Project-Based Organisation: Seven Pillars, Seven Paradoxes and Seven Deadly Sins. *International Journal of Project Management*, **25** (4), 365–374.

Brindle, S. (2013) *Brunel: The Man Who Built the World* (Hachette, UK: Phoenix Press).

Cheung, F.Y.K., Rowlinson, S., Jefferies, M. and Lau, E. (2005) Relationship Contracting in Australia. *Journal of Construction Procurement*, **11** (2), 123.

Construction Manager (2021) Increased Flexible Working Could Unlock £457m for Construction. *CIOB Construction Manager*, 26 November. https://constructionmanagermagazine.com/increased-flexible-working-could-unlock-457m-for-construction/.

Cooke, A. (1987) *Hong Kong and Shanghai Bank, 1979–1986*. www.Fosterandpartners.Com/Projects/Hong-%20kong-And-Shanghai-Bank-Headquarters/.

Donaghy, R. (2009) *One Death Is Too Many: Report to the Secretary of State for Work and Pensions* (London: The Stationery Office).

Farmer, M. (2016) *The Farmer Review of the UK Construction Labour Model: Modernise or Die – A Critique of the Farmer Review of UK Construction* (London: Construction Leadership Council).

Green, S. (2016) Modernise . . . or Not. *Construction Research and Innovation*, **7** (4), 24–27.

HM Treasury (2019) *Private Finance Initiative and Private Finance 2 Projects: 2018 Summary Data* (London). https://assets.publishing.service.gov.uk/government/uploads/system/uploads/attachment_data/file/805117/PFI_and_PF2_FINAL_PDF1.pdf.

Hong Kong Housing Authority (2011) *Going an Extra Mile to Innovate the Procurement System*. www.housingauthority.gov.hk/en/about-us/publications-and-statistics/housing-dimensions/article/20110318/going-an-extra-mile-to-innovate-the-procurement-system.html.

Hutchinson, A. and Gallagher, J. (2003) *Project Alliances: An Overview* (Australia: Alchimie Pty. Ltd and Philips Fox).

ICE (2022) *NEC Contracts* (Institution of Civil Engineers). www.ice.org.uk/knowledge-and-resources/professional-practice/nec-contracts-and-ice-conditions-of-contract.

Infrastructure Transparency Initiative (2021) https://infrastructuretransparency.org. Accessed 5 November 2021.

Jefferies, M., Gameson, R., Chen, S. and Elliott, T. (2001) The Justification and Implementation of Project Alliances – Reflections on the Wandoo B Development. *Journal of Construction Procurement*, **7**, 31–41.

Jefferies, M. and Rowlinson, S. (eds.) (2016) *New Forms of Procurement: PPP and Relational Contracting in the 21st Century* (Abingdon: Routledge).

Ju, C. and Rowlinson, S. (2020) The Evolution of Safety Legislation in Hong Kong: Actors, Structures and Institutions. *Safety Science*, **124**, 104606. https://doi.org/10.1016/j.ssci.2020.104606.

Love, P.E. and Gunasekaran, A. (1999) Learning Alliances: A Customer-Supplier Focus for Continuous Improvement in Manufacturing. *Industrial and Commercial Training*, **31**, 88–96.

Love, P.E., Ika, L.A. and Ahiaga-Dagbui, D.D. (2019) On De-Bunking 'Fake News' in a Post Truth Era: Why Does the Planning Fallacy Explanation for Cost Overruns Fall Short? *Transportation Research Part A: Policy and Practice*, **126**, 397–408.

Manivong, K. and Chaaya, M. (2000) *Life Cycle Project Management*. Paper presented at the Proceedings of the 4th International Conference of the International Research Network on Organising by Projects, UTS, Sydney.

Matsumoto, Y., Oakervee, D., Thomson, A. and Morton, D. (1990) Eastern Harbour Crossing, Hong Kong. In: *Immersed Tunnel Techniques* (pp. 387–395) (London: Thomas Telford Publishing).

Orlowski, S. (1969) *Research Study of Modular Design of School Buildings in Europe. April 30, 1968 – May 20, 1968* (Toronto: Ontario Dept. of Education). https://eric.ed.gov/?id=ED035219.

Procure Partnerships Framework (2021) *Procure North West Framework User Guide*. www.procurepartnerships.co.uk/wp-content/uploads/2020/08/Procure-Partnerships-Framework-North-West.pdf.

Quantrill, A. (2018) The Value of Enclosure and the Business of Banking. In: *Grey Room* (pp. 116–137) (MIT Press). www.greyroom.org/issues/71/88/the-value-of-enclosure-and-the-business-of-banking/.

Rowlinson, S. (1987) *Design/Build: Its Development and Present Status* (Ascot: Chartered Institute of Building).

Rowlinson, S., Cheung, F.Y., Simons, R. and Rafferty, A. (2006) Alliancing in Australia – No-Litigation Contracts: A Tautology? *Journal of Professional Issues in Engineering Education and Practice*, **132** (1), 77–81.

Rowlinson, S. and McDermott, P. (eds.) (1999) *Procurement Systems: A Guide to Best Practice in Construction* (London: E&FN Spon).

Simon, E. (1944) Ministry of Works. In: *The Placing and Management of Building Contracts. Report of the Central Council for Works and Buildings to the Minister of Works (Chairman Sir Ernest Simon)* (London: HMSO).

Smyth, H.J. (2018) *Castles in the Air? The Evolution of British Main Contractors* (London: The Bartlett School). www.ucl.ac.uk/bartlett/construction/castles-in-the-air.

Tennant, S. and Fernie, S. (2012) The Commercial Currency of Construction Framework Agreements. *Building Research & Information*, **40** (2), 209–220.

Trager, A.M. and Ping, T.J. (2014) Making Public Private Partnerships Work: Implications for Singapore and the Region. *Ethos*, **13**, May. www.csc.gov.sg/articles/making-public-private-partnerships-work-implications-for-singapore-and-the-region.

Vosloo, C. (2020) Early Sustainable Architecture in Hanging Skyscrapers – a Comparison of Two Financial Office Buildings. *Acta Structilia*, **27** (1), 144–177.

Walker, D.H.T. and Lloyd-Walker, B. (2015) *Collaborative Project Procurement Arrangements* (London: Project Management Institute).

Walker, D.H.T. and Rowlinson, S. (eds.) (2007) *Procurement Systems – a Cross Industry Project Management Perspective* (London: Taylor & Francis).

Walker, D.H.T. and Rowlinson, S. (eds.) (2019) *Routledge Handbook of Integrated Project Delivery* (Abingdon: Routledge).

Woodall, B. (1993) The Logic of Collusive Action: The Political Roots of Japan's *Dangō* System. *Comparative Politics*, **25** (3), 297–312.

World Bank (2020) *About Public-Private Partnerships*. Public-Private Partnership Legal Resource Center (Washington, DC: World Bank). https://ppp.worldbank.org/public-private-partnership/about-public-private-partnerships.

Editorial Comment

In the construction world, estimating is a fundamental factor in the planning and execution of building projects. *Early estimating* refers to the prediction of whole-of-project out-turn costs, essentially it is an attempt to establish how much a client will have to pay for their project, so an estimated cost (or price) includes the builder's profit and general overheads, project overheads (preliminaries) and the cost to the builder of the various components of building cost: labour, materials, plant and equipment and so on. It may or may not include other costs such as legal fees, land purchase, clearing, demolition and design fees. For construction firms, estimating usually means predicting the net cost, to the builder, of building, and as such it is restricted to the cost of direct inputs (e.g. labour, materials) plus job specific overheads or preliminaries, i.e. costs directly associated with a particular project such as fencing, site security and scaffolding which do not form part of finished structure or facility. With a predicted net cost established, margins are added to cover profit and a portion of company overheads to arrive at a bid or tender sum.

With either sort of estimating the procedure has generally involved measuring projects to some level of detail and then applying unit rates to the measured quantities, whether they be relatively coarse measurements of floor areas or detailed measurements at the level of individual items of work. Whatever the type of estimate, and however they are prepared, many estimates share a common trait: they are often wrong, and sometimes they are spectacularly wrong, most usually because the final cost exceeds the predicted cost.

There are many factors that may contribute to the unreliability typical of estimates of construction cost; they include:

- Incomplete and/or inaccurate documentation
- Lack of time to measure and price
- Design variations
- Changing market conditions
- Human error
- Optimism bias
- Price competition
- Latent conditions

With regard to these factors the common thread is uncertainty: for example, estimators are usually attempting to predict the cost to build across the life of a contract, which could range from a few months to several years, commencing at an often unknown future date. Much can change during the life of a project; inflation may push labour and material costs up, interest rates may fluctuate and affect the cost of finance, and so on. Allowances for these uncertainties are made based on little more than experience and hope, as predicting trends in markets has made some people rich while sending many others into bankruptcy.

In this chapter the authors look at traditional estimating methods and suggest some alternatives based on statistical methods which are more akin to cost modelling as opposed to detailed measurement and pricing. In the Digital Era, the technology and tools that could allow a shift from the unreliable (and time-consuming) traditional estimating processes are available, and may provide more effective ways of predicting building costs in the future.

7 Estimating challenges in the Digital era

Pat O'Donnell and Inna Kolyshkina

Introduction

It is important, at the outset, to clarify the meaning of the two key terms in the title of this chapter. Dictionary.com (2021) offers the following definitions for the verb *estimate*:

- roughly calculate or judge the value, number, quantity, or extent of . . .
- to form an approximate judgment or opinion regarding the worth, amount, size, weight, etc., of
- calculate approximately
- to estimate the cost of . . .
- to form an opinion of; judge.

The Digital Era, otherwise referred to as the Information Age, is a historic period in the 21st century characterised by the rapid shift from traditional industry that the Industrial Revolution brought through industrialisation, to an economy based on information technology. Shepherd (2004: 1) states that it is characterised 'by technology which increases the speed and breadth of knowledge turnover within the economy and society'. Rogers (2017) suggests that, in the digital era, growth is about more than technology, and that 'to bring a legacy business into the digital era, managers need to unlearn all the old rules.'

Contractors' estimating offices are legacy businesses operating in an environment with disruptive digital technologies available that potentially make the traditional ways of doing things obsolete yet, in the authors' experience over the last 30 years, the processes of estimating the cost of buildings have changed little. Change has been evolutionary with adoption of new digital tools such as estimating software that interfaces with Computer Aided Design (CAD) software to speed up measuring the quantities of work yet estimates look much the same as they did 30 years ago. The construction industry has a reputation for being conservative, with a lot of inertia when it comes to accepting change and adopting new technologies.

Despite the promise of change, there have been no real disruptions that would mean that the quantity surveying and estimating professions are no longer required. Theoretically, multidimensional Building Information Modelling[1] (BIM) systems

DOI: 10.1201/9781003102403-7

and data science could make estimators redundant. In reality, there is great demand for competent estimators and quantity surveyors, and they are paid reasonably well.

An additional complication with estimating, occurs with procurement models where the lifecycle costs need to be estimated for an extended term of 20 to 40 years in many cases. The lifecycle costs include maintenance, capital replacement, energy, handover upgrades at the end of the term and residual value. This is often the case with procurement models such as:

- Build, Own, Operate and Transfer (BOOT).
- Private Finance Initiative (PFI).
- Public private partnership (PPP).
- Infrastructure owners wanting a long-term asset such as a power station.

This chapter looks at issues relating to estimating in the digital era, offers suggestions as to why change has not been more radical, and makes some predictions about what is likely to occur in the future.

Estimating unknowns

In construction, estimates are predictions of future costs for future construction operations, for a defined project, including fixed and variable costs.

- The fixed costs can include capital costs for land, plant and equipment, statutory fees, mobilisation/demobilisation and the like.
- Variable costs are dependent on time, quantities of work and values for things like trade work, margins, taxes and insurances.

The estimator has information available in digital format for the design of the building or facility with varying degrees of detail. There are, however, many unknowns that impact cost that are not part of the design, and this is a fundamental barrier to being able to automatically generate the estimate from a 5D BIM system. 3D BIM is the three dimensional design, 4D includes scheduling/programming and 5D includes estimating (RIB n.d.).

Unknown cost factors include:

- The weather during construction
- Latent site conditions such as the presence of rock, unstable foundations, contamination and groundwater
- Labour and plant productivity rates for the specific project being estimated – these vary depending on the industrial climate, the level of skill and training, the capacity and type of machinery used, and the complexity of the design, among other things
- Usually designs are incomplete at the time of preparing the estimate – the estimator has to allow for the costs associated with missing detail which requires a great deal of experience with similar types of projects

- The prices for purchasing materials and equipment vary depending on market forces, and over a multiyear project, the cost of plant purchased from overseas, for example, will vary depending on foreign exchange rates.

The fundamental nature of estimating is that there are many variables that impact the final cost for a project and the skill of the estimator is being able to predict the likely outcomes within a reasonable range of accuracy. Estimating accuracy is dealt with in more detail later in the chapter but as a guide, basic residential buildings can be estimated very accurately, within, say, a 5 percent range, by those experienced in this field. The cost of complex infrastructure projects such as road tunnels is much more difficult to predict, with cost overruns in excess of 20 percent not uncommon.

Risk allocation

Owners tend to pass on as much risk to the contractor as possible, however construction companies try to mitigate the risk of actual costs exceeding the estimate, by pushing back some of the risk to the client or owner. This is usually done using allowances and qualifications in the tender submission and mark-ups to the contract documents. Some examples of contract provisions that pass risk back to the client from the contractor include:

- Provisional allowance for rock excavation
- Excluding the cost of remediation of unknown contamination
- Compensable delay definition including delays beyond the builder's control
- Stipulating a fixed foreign exchange rate for offshore plant.

Construction contracts allocate risk between the parties and estimators need to frame their estimates taking into account the risk allocation.

Estimate constraints

Project stage and user

An important consideration is who the estimate is being prepared for and the stage of the project. Estimates are used across a variety of industry sectors; it is not the intention to provide an exhaustive coverage for all sectors, however, a sample selection is provided to outline the key participants and consultants involved. A summary of the different project stages and types of estimates[2] for the different users is shown in Table 7.1.

Time constraints

One of the challenges for estimating is the time that practitioners have available to prepare an estimate. Often there is only a concept design and a performance specification provided on which the tender is to be prepared. For a contractor

Table 7.1 Estimating summary

Project stage	*User*	*Type of project*	*Producer*	*Type of estimate*	*Available information*
Conception	Client	Building	QS	Unit rate e.g. $\$/m^2$	Funding limits
		Civil	Engineer/QS	Unit rate e.g. $/km/lane	Capacity requirements
		Process plant	Engineer	Unit rate e.g. $/MW	Benchmark cost information
Brief	Client	Building	QS	Functional area cost plan	Functional area cost rates
	Designers	Civil	Estimator/QS	Budget estimate $/component	Benchmark cost information
		Process plant	Engineer	Cost models	Benchmark cost information
Sketch design	Client	Building	QS	Elemental cost plan	Elemental cost rates
	Designers	Civil	Estimator	Approximate resource-based estimate	Approx. quantities and resources
	Contractors	Process plant	Engineer	Approximate system-based estimate	Approx. system requirements
	Financiers				
Detailed design/ Tender	Client	Building	QS	Detailed trade-based estimate	Subcontractor & supplier pricing
	Designers	Civil	Estimator	Detailed resource-based estimate	Material and resource pricing
	Contractors	Process plant	Supplier	Detailed system-based estimate	System supplier pricing
	Financiers				
Construction	Client	Building	QS	Cost to complete estimates	Tender price schedule
	Project managers	Civil	Engineer/QS	Variation estimates	Actual cost information
	Contractors	Process plant	Engineer		Schedule of rates information
	Financiers				
Operations	Owners	Building	QS/FM	Lifecycle cost plans	Benchmark cost data
	Facility/asset managers	Civil	AM	Maintenance and capital replacement	Actual cost data
		Process plant	AM		Asset Management Database

QS – Quantity Surveyor, FM – Facility Manager, AM – Asset Manager to manage their risk and have some certainty, they need to work up reasonably detailed designs and then prepare the estimate. For BOOT, PFI and PPP projects, it is often necessary for the designs to extend to choosing plant and equipment brands and types, to estimate the lifecycle costs. The process includes:

- Inviting tenders from sub-contractors
- Preparing a construction program
- Reviewing and marking up the contract
- Viewing and analysing the site to work out logistics, materials handling, storage, cranage, security, work methods and similar
- Preparing a construction management plan with traffic management plan, hoardings layout, crane operations and similar
- Obtaining geotechnical reports including bore logs to determine engineering requirements, ground water issues, possible contamination and potential issues for neighbouring infrastructure
- Preparing schematic designs for the services and key plant items
- Preparing the tender submission which must cover a wide range of issues including safety, quality, stakeholder relations and communications as well as the tender price.

The estimate requires information regarding each of these components to be finalised. Tendering is also an iterative process as new ideas emerge, and designs change, or new information is discovered at the last moment. Usually there is a process to try and cut costs and this includes refining the design and repricing. The estimating office is often in a perpetual state of panic trying to meet tender deadlines and the practitioners have no time to learn new technology and processes. The lead author here has personally experienced strong resistance to any change in processes as there is always a learning curve that will delay the immediate delivery program of current jobs.

The next section deals with the concept of economy of effort and examines ways estimators traditionally manage the conflicting requirements for accuracy and limited time without adopting new technology.

Economy of effort

One of the important considerations is how long and what resources are needed to produce an accurate estimate. Estimating is generally considered an overhead and successful businesses try to keep overheads low. Consideration needs to be given to efficiently predicting the project costs and the diminishing accuracy improvements that will be achieved by increasing the estimating resources necessary for more detailed estimates.

With so many variables impacting the final cost, as discussed earlier, the estimate is never perfectly accurate. Instead, as more detail is investigated, the accuracy will tend towards a limit as shown in the hypothetical graph in Figure 7.1. As the curve flattens out, there is a diminishing marginal improvement in accuracy

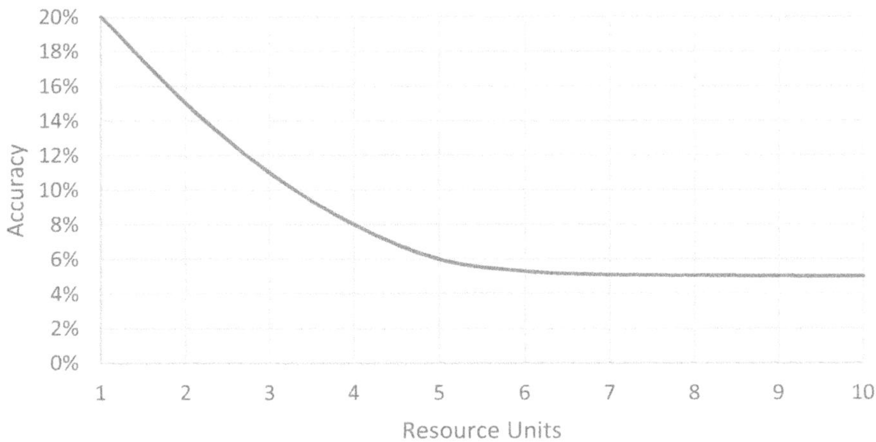

Figure 7.1 Accuracy vs. resources

for each additional resource unit added to the estimating process. The estimating process needs to be optimised for the desired accuracy outcome.

For an early budget estimate, clients need a guide to the costs for making their business case and obtaining funding but there is limited value in investing large amounts of resources to obtain detailed estimates when the scope of the project is evolving. On the other hand, a contractor preparing a tender estimate for a fixed price contract will value a more detailed and thorough estimating process so they can have a high degree of confidence in the predicted cost outcomes. The requirements for accuracy need to be balanced against the cost of achieving that accuracy.

Cost modelling including regression models

Cost modelling is a way of predicting costs using some form of simulation process rather than working out in detail all the quantities of material, hours of labour, plant requirements and other costs. Simplified parameters are chosen that can be used to model or predict the cost of a project. Computer applications have been developed that produce cost predictions with parameter inputs. Care needs to be taken regarding the outputs in the absence of professional oversight.

The following graph is a simplified Unit Rate Estimate for a regional shopping centre where two rates are used showing high and low costs.

Effectively, the unit rate estimate (such as the $/m^2 of floor area cost estimate shown in Figure 7.2), is a simple linear regression model using the formula:

$y = a + bx$, where:

- y is the cost outcome
- a is the y axis intercept (in Figure 7.2, 0)

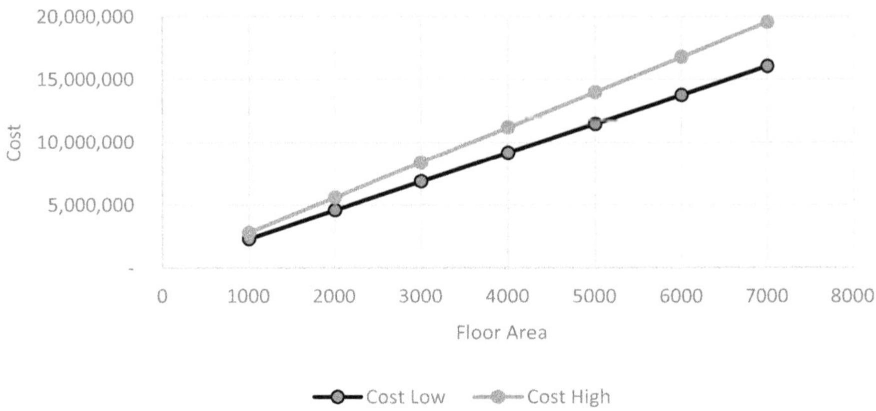

Figure 7.2 Unit rate estimate

- *b* is the slope of the line (in Figure 7.2, the cost per m^2)
- *x* is the input parameter (in Figure 7.2, the floor area)

Linear regression models have an advantage in that they can take into account fixed costs in the *a* parameter. Costs are never all variable as assumed in the unit rate model. In addition to the cost per m^2 there will be costs for site establishment, project management, planning approvals, demolition, decontamination, legal fees and the like that will shift the whole curve up and make the Y access intercept greater than 0.

Many different variations of regression models have been used to predict tender prices and building costs. The models can have many parameters (variables) and may produce curvilinear prediction lines (curves) or stepped curves where, for example, each additional floor level will provide a step in the rate per square metre.

More complex linear models can use data analysis of big data sets to produce the model algorithm using the following general form with *p* independent variables:

$$y_i = \beta_1 x_{i1} + \beta_2 x_{i2} + \ldots + \beta_p x_{ip} + e_i$$

where x_{ij} is the *i*-th observation on the *j*-th independent variable (Wikipedia 2021)

Any of the variables that influence cost can be modelled using regression if data is available. Some important variables for determining cost or price include:

- floor area
- site area
- site conditions including topography, geotechnical conditions, groundwater
- building height
- number of basements

- number of floors
- roof area
- construction type/materials
- climate
- economic factors such as unemployment, inflation, GDP growth

Regression models, using historical data for more accurate parameter inputs, potentially provide disruptive digital technology for estimating that is very different to the traditional methods of measuring and pricing quantities of work. Potentially the estimator is not needed, just the data analyst and a modeller with good mathematical skills.

Regression models have been around for decades, and the average desktop computer can handle complex regression models. While they are used by academics doing research work (see, for example, O'Donnell 2001; Lowe *et al.* 2006; Thomas and Thomas 2016), there seems to be little evidence of regression models being used for real world estimating.

The construction industry is conservative and business owners trust what they know, which is the experience of estimators in the estimating office. Risking the business on a complicated mathematical model that the CEO does not understand, is not likely to happen any time soon. Developers also need finance to do developments and the financial institutions require tried and tested methods for estimating the costs for developments they are financing. Banks tend to be very reluctant to lend for a project unless the estimate is prepared in strict compliance with their guidelines.

Early development of cost modelling using data science techniques

O'Donnell (2001) investigated a revolutionary new method of cost modelling using data from previous projects. There was a need to be able to provide more detailed and accurate cost planning in the early stages of projects, particularly for large complex projects like hospitals. The Functional Area Cost Evaluation Technique (FACET) system was developed by O'Donnell whilst working for a large construction contractor. FACET provided detailed early budgets, utilising data captured from previous similar projects, to model both the quantities of work and the costs for future projects. The modelling of quantities, using regression analysis of historical data, grouped into functional areas and elements, is unique to this system. By modelling detailed quantities there is no issue with inflation affecting the cost data. It is relatively simple to reprice using current local rates. (O'Donnell 2002a). FACET has been used to cost plan many major projects.

FACET overview

The FACET system provides detailed cost plans in a variety of summary formats, based on input of functional areas which form part of the brief. Input is also required for external work and site-specific items.

The summary format outputs comprise:

- functional area summary costs
- elemental summary costs
- trade grouping summary costs
- detailed trade item costs.

Each summary output has quantities and rates to facilitate benchmarking against other projects using industry standard groupings.

The FACET model uses two-stage regression analysis of elemental and trade quantity data from previous similar buildings containing the same or similar functional areas.

- First regression analysis is used to produce regression models to generate quantities for each element for the new project
- Second regression analysis is used to produce regression models to generate quantities for each trade item within each element for the new project.

The costs for both elements and trade items are also generated with the costs adjusted for inflation using a building price index.[3] The costs generated by the analysis can be overridden by the operator with local current rates inserted instead.

Data collection, management and selection

FACET is based on a three-dimensional database that stores quantity and cost information from previous projects. Data from previous completed projects is classified according to a standardised system of trade, element, functional area and building type codes. When a new project is to be costed using the system, data is selected by use of a building type code and the user inputs appropriate location and time price indices to enable the system to adjust all the costs before the selected data is transferred to the analysis module.

The accuracy and reliability of building price indices is questionable when applied to specific locations and trade items in a building estimate. This is an important reason for facilitating repricing by an experienced estimator in the FACET system.

The database structure is a series of related tables as shown in Figure 7.3.

- BPI is the applicable Building Price Index for the project
- LI is the applicable Location Index for the project
- FA is the Functional Area. The system was used initially for hospitals and used the Functional Areas defined in the Australian Health Facility Guidelines (AusHFG 2021).

The biggest challenge with using the FACET system is that the available data from previous projects must be translated to the standard classification system developed for FACET. The data available can be 'messy' and unstructured, requiring

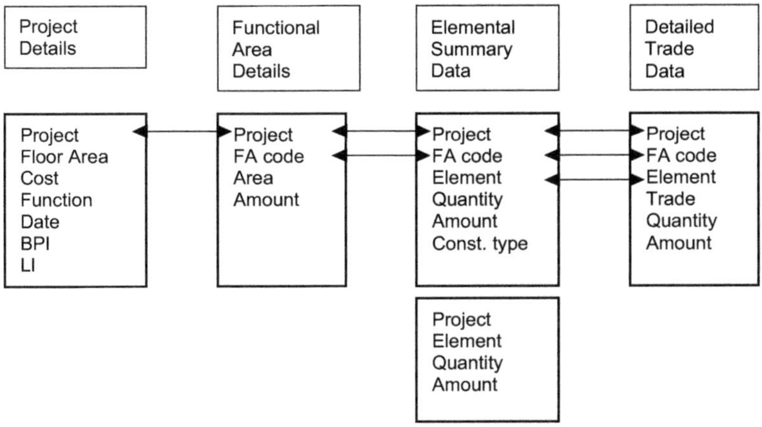

Figure 7.3 Database structure

significant manual intervention to interpret and classify. Different people describe the same work items in different ways and even use different units to quantify the work. For example, concrete slabs may be measured in cubic metres or square metres.

Twenty years ago, when the FACET system was developed, text mining[4] was in its infancy and could not compete with an experienced cost planner in interpreting the messy unstructured historical data. Now as we progress into the new digital era, modern machine learning techniques, including text mining as well as BIM systems, provide exciting opportunities for revisiting cost modelling systems such as FACET.

Data science and estimating

Machine learning

Data science is an emerging area that uses advanced modelling techniques to predict future outcomes using large data sets from multiple sources ('big data'). Advanced machine learning (ML) techniques derive maximum value from organisational and external data while effectively addressing various data limitations. ML has been proven to be effective in cases of noisy, sparse and otherwise suboptimal data, and is widely used in many industries with great success. For example, these techniques are widely used in the public sector as well as in banking, insurance, marketing and utilities.

Estimating for construction costs can be informed by very large volumes of internal data from previous projects, combined with external data sources such as those related to climate, economic market conditions, labour statistics and the like.

Recently ML techniques have been gaining popularity in the construction industry (Bilal *et al.* 2016; Witten *et al.* 2017); however, construction management still relies heavily on empirical information even when aided by modern informatics tools like BIM (Bilal *et al.* 2016; Miettinen and Paavola 2018). Data science, including ML, has the potential to become an important part of the digital revolution impacting building and construction in the areas of estimating and risk management.

Many researchers argue that more efficient utilisation of available data could enhance decision making and hence project performance in the industry (e.g. Kifokerisa *et al.* 2019; Kumar and Reinartz 2018). Chen and Jupp (2019) report that data within BIM can aid in such efforts while Bilal *et al.* (2016), as well as Miettinen and Paavola (2018) state that ML can help in utilising BIM data.

Data science/machine learning provides the tools to derive valuable information from unstructured data such as free text, or images such as site photos. Free text could be site diaries or maintenance notes and text mining methods facilitate resolution of issues arising from any inconsistencies such as misspellings, abbreviations or use of synonyms, and when the same object is referred to in different ways.

Another important consideration is ML's ability to allow consideration of myriad factors that can potentially influence the outcome of interest and identify the most important factors that drive the outcome, as well as quantify the extent of influence of each such factor. An estimating ML model could identify those variables that have the greatest influence on cost or time and provide management with insights on how to optimise resource usage on the areas with the greatest potential to minimise cost and risk of cost overruns.

ML methodology is a key success factor

As some organisations have found, the application of ML techniques to organisational data does not work without issues. Recent reports estimate that between 70 and 85 percent of data science projects fail (NewVantage Partners 2019). A key reason for that is the lack of proper process and methodology (NewVantage Partners 2019; Gartner 2018).

A key aspect of a successful methodology is ensuring sufficient interpretability and auditability of the ML solution. Some ML techniques, such as deep neural nets and ensembles, are known to be 'black boxes' and this makes it difficult to interpret the output and understand how the system arrived at a specific outcome.

One way to overcome this issue is to use the CRISP-ML methodology (Kolyshkina and Simoff 2019). It is an upgrade of the popular CRISP-DM methodology created in late 1990s to early 2000s (Smart Vision 2021). It is application neutral and incorporates the latest developments in the ML and artificial intelligence (AI) areas. It is built to ensure the level of interpretability meets the requirements of the organisation and that the necessary level of ML system interpretability is met by going through each project stage (Kolyshkina and Simoff 2019).

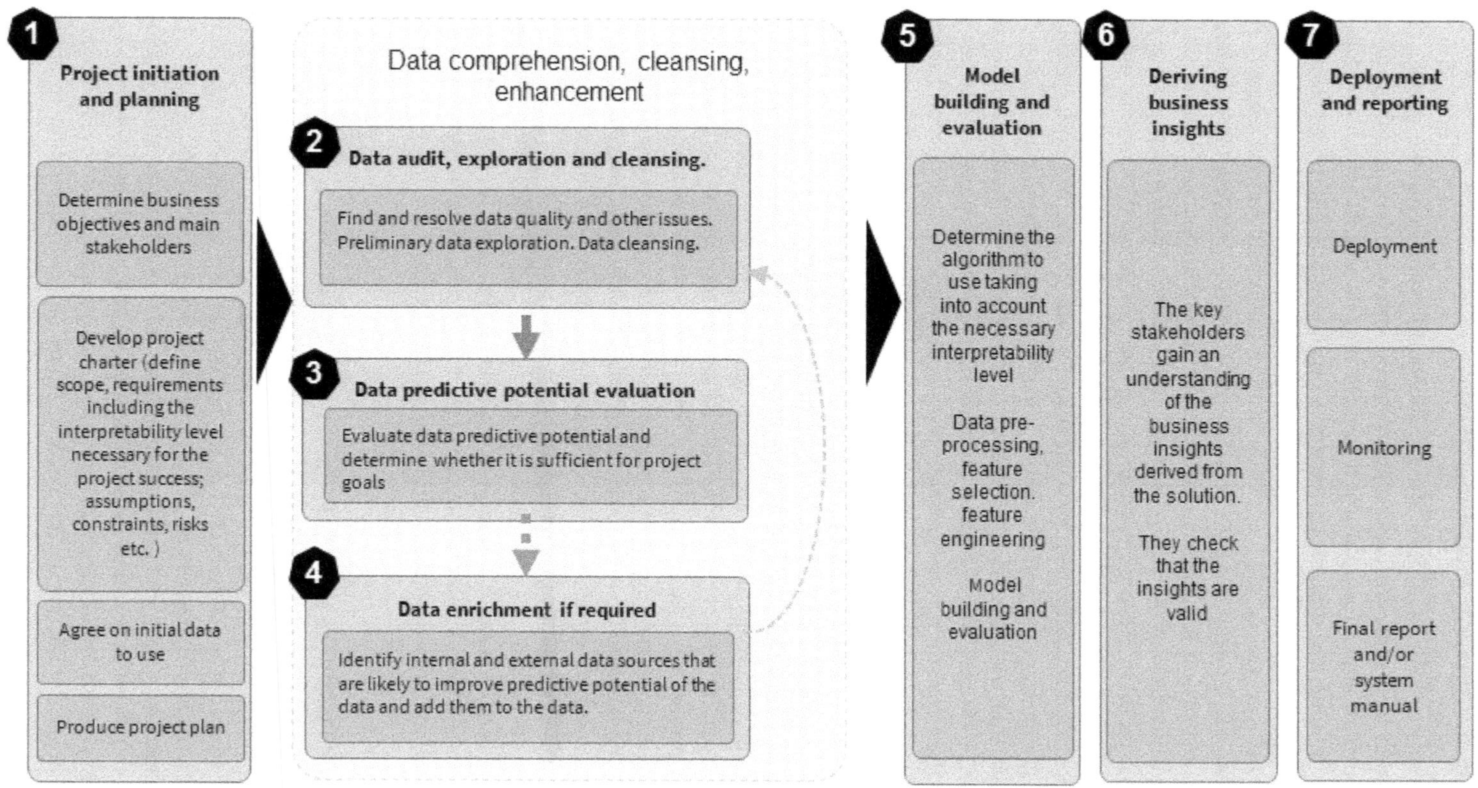

Figure 7.4 Crisp-ML flowchart

Building Information Modelling (BIM)

Measurement of the quantities and producing an estimate for the project have traditionally been done with information shared across the programming and estimating departments. The construction program provides the time schedule for work to be done and integrating data between estimating and programming has long been an important requirement for the construction industry. BIM systems provide full integration. BIM has evolved from Computer Aided Design (CAD) to become 3D modelling design and now BIM systems can produce simulations of the construction sequencing and full construction programs (4D), as well as providing quantities of all the surfaces, volumes, masses, components, service runs and fitments (5D). All the information required for detailed programs and estimates is available in the BIM system.

Construction is catching up with the digital world and the way both estimating and programming is done in the future is likely to change significantly. This does not mean that the expertise of the estimator is no longer required. BIM offers advanced tools that will eliminate the tedious tasks of measuring quantities and free up the estimator to concentrate on engaging with the suppliers and subcontractors to get the best pricing. BIM also provides advanced tools for optimising the designs, identifying clashes and analysing the construction sequence to achieve efficiencies in material handling, safety in design and more efficient programming.

One example is Line of Balance programming. Any construction project has multiple activities that need to be completed in parallel. To build efficiently, the resources need to be used efficiently with steady production. Having gaps in the work of groups of resources is inefficient. Resources allocated to each activity that occur in parallel must be balanced so that the work for each group of activities progresses at the same rate. If one activity is finished early and the workers must wait for another activity to finish before they can work again, there is down time and extra cost built into the project. Balancing the progress of the different activities using line of balance programming can be done using BIM.

Estimators tend to assume parallel progress, and this is one reason why delays and cost overruns happen and there is always a need to allow for contingencies for unexpected events. A delay to any one of the trades on any level, has the potential to delay all the following trades, and overall completion.

Asset management: a case study

In the introduction it was identified that lifecycle costs also have to be estimated for some long-term projects including for any infrastructure owner that has long-term assets. Estimating is not restricted to predicting the cost to build but can include predicting facility management (FM) or asset management (AM) costs.[5] Asset Management is a field where the authors have been involved in doing a detailed study using machine learning (ML) techniques to help estimate future repair and replacement costs for a large electricity infrastructure organisation in Australia that needed help in predicting assets at risk of failure and identifying how to optimise their limited maintenance budget. The organisation owns assets such as substations and a power distribution network spread over a vast geographical area and has experienced equipment failures that lead to blackouts impacting the community. It is also subject to economic penalties from the regulator for failure to meet the required key performance indicators. This completed ML project is used as a case study to demonstrate how digital era techniques can be used to enhance estimating.

An important consideration is the whole of life cost for an asset which includes the final capital procurement cost as well as ongoing costs throughout the life of the asset (Kolyshkina and O'Donnell 2018). Procurement methods such as BOOT, PFI and PPP involve a consortium taking on the full lifecycle costs and risks. This requires estimates of the recurring maintenance costs as well as cycles of replacement and capital upgrades. Planning and managing the limited maintenance resources are important considerations.

The following case study illustrates how an asset-owning organisation used machine learning to optimise lifecycle planning, budgeting and strategic management. It describes the application of ML techniques to historical data in a successful implementation where organisational data was used and matched against external data including geophysical (weather, location, and distance from the sea, among others) to build predictive models giving insights that aided effective, evidence-based business decisions. Included is an outline of the ML techniques used as well as data-related issues and how they were resolved.

An important aspect was identifying the assets likely to incur the highest cost and failure rates and to focus effort on these. Another important driver was to demonstrate compliance with regulatory requirements. The process needed to be easy to understand and implement (Kolyshkina and O'Donnell 2018).

Approach overview

A broad set of factors that had potential to influence the outcomes was identified. This was done in close consultation with the organisation's subject matter experts. These are the maintenance engineers who had the expertise to identify which components of their distribution network caused the biggest problems in the past and what they thought were the likely causes. It was valuable to capture the organisation's prior learning to help focus the study.

Rich asset attribute data was combined with historical maintenance records and external geographical, environmental and socioeconomic data. The asset attribute data included details of the asset being analysed such as the type of equipment, model, size, capacity (such as voltage and amperage), installation date, maintenance history, which facility the equipment was located in and the team that did the maintenance, for example. The resulting large and rich data set comprised hundreds of variables and millions of records and formed the basis of the advanced analysis undertaken.

Powerful ML techniques were applied to the data in order to identify the key factors that affected the main outcomes (i.e. cost and failure recurrence rates) and to quantify the extent of influence of each factor. The resulting solution was easy to understand for the end users as well as to implement within existing organisational systems (Kolyshkina and O'Donnell 2018).

Data challenges and resolution

Data categories included in the study were:

- in-house asset records such as asset type, model, manufacturer, location, installation date, replacement and overhaul dates
- operational statistics and maintenance records over 15 years
- management variables such as different policies/strategies in different regions
- environmental data including temperature, rainfall, humidity, wind and seasonal variability
- technical characteristics such as voltage, size and capacity
- information about the type, size and technical characteristics of the facility where the asset component was located.

External data, including environmental, geographical and socioeconomic factors describing the location of the facility (for example, distance from the city and proximity to the coast), were matched against asset performance history. Some data such as maintenance notes contained free text. Thorough data audits and detailed exploratory data analysis were conducted to ensure raw data quality and consistency. The data was then reorganised into a layout suitable for modelling. All maintenance cost figures were adjusted for inflation using the relevant Consumer Price Index (CPI) data from the Australian Bureau of Statistics. Cost data spanning a 15-year period was used and had to be converted to a common base date for valid comparisons.

It was discovered that the analysis was potentially affected by a high level of random variation and sparseness in the data. Some asset types had small numbers of available failure records. Such data issues are typical for asset data.

To accurately build a model, it is necessary to have enough data to estimate the parameters included in the model. The data contained many variables with a large number of categories such as equipment model code, the code of the geographic location of the asset and many more, so the number of parameters could potentially exceed the number of available data records.

For certain asset types, failures occur very rarely and therefore only a very small number of non-zero failure data points were available for use in modelling, which potentially complicated analysis and required special consideration for modelling such asset types.

Due to these data issues, the use of a traditional linear regression model was not feasible. An ML-based modelling approach was devised to deliver superior predictive capabilities by enhancing classical statistical methods with contemporary ML techniques that derive maximum value out of data, while effectively addressing various data limitations. These methods have been proven to be effective in cases of noisy, sparse and otherwise suboptimal data and are widely used in other industries with great success.

There are a number of methods available, including:

- Random forests (Liaw and Wiener 2002)
- Stochastic gradient boosting (Friedman 2002; Lemmens and Croux 2006; Zhang 2004)
- Least absolute shrinkage and selection operator; LASSO (Tibshirani 1996)
- Multivariate adaptive regression splines (MARS) (Lee and Chen 2005; Lee *et al.* 2006)
- Regression trees (Strobl *et al.* 2009; Wu *et al.* 2007; Bishop 2006)
- Synthetic minority over-sampling technique (SMOTE) (Chawla *et al.* 2002)
- Random over sampling (Menardi and Torelli 2012).

Random variability or missing variables

Conventional opinion suggests that age is the main driver of maintenance cost and failure rates. The bathtub curve (Xie and Lai 1996) is often used to describe the generalised cost and failure trend over time (Li and English 1997). It would therefore be reasonable to expect that a significant amount of variation in the data would be explained by a curve of this type. However, exploratory data analysis identified a high level of random variation in both cost and failure data when related to asset age. At any age point, the variability of each outcome is quite high, varying from zero to the maximum of the range. This suggests that while asset age does play a role in explaining maintenance cost generation and failure occurrence, there are other important factors apart from age that have a strong influence on the outcomes of interest.

Further analysis identified that a considerable amount of the outcome variability was explained by other variables including:

- asset subtype, model and manufacturer
- climate, geographic and socioeconomic characteristics of the facility such as proximity to the coast and the degree of remoteness
- operation and performance history of the individual asset
- location and technical environment.

Results and findings

The models allowed the key drivers of the maintenance cost and asset failures to be discovered and the extent of each effect quantified. The solution therefore identified what levers were available to strategically plan the asset management operations to minimise overall cost and maximise reliability. In addition, the models enabled the individual assets at highest risk of cost generation or failure to be identified.

Overall, for the majority of asset types, the main factors driving cost and failure were:

- asset age
- number and pattern of historical failures

- performance of the equipment directly connected with the asset
- history of maintenance in previous years
- asset model, manufacturer and type
- asset operational characteristics (such as frequency of switching and power output levels)
- characteristics of the facility where the asset was located, including facility type, size, distance from the city, region, average age of assets at a given facility
- weather conditions at the asset location.

For some asset types, asset age proved to be less important in predicting failure rates than asset model and manufacturer.

Identifying at-risk assets

One of the objectives of the project was to identify the assets that were likely to incur the highest cost and failure occurrence in the future and thus enable the organisation to develop strategies for remediation or replacement of these assets. The models allowed identification of such at-risk assets. This is illustrated by the gains chart (Rolling 2017; Hamilton 2012) in Figure 7.5.

The horizontal axis of the gains chart shows cases ranked by the model-predicted outcome such as cost amount or failure count from the highest to the lowest. The axis is divided into deciles. The vertical axis shows the cumulative percentage of

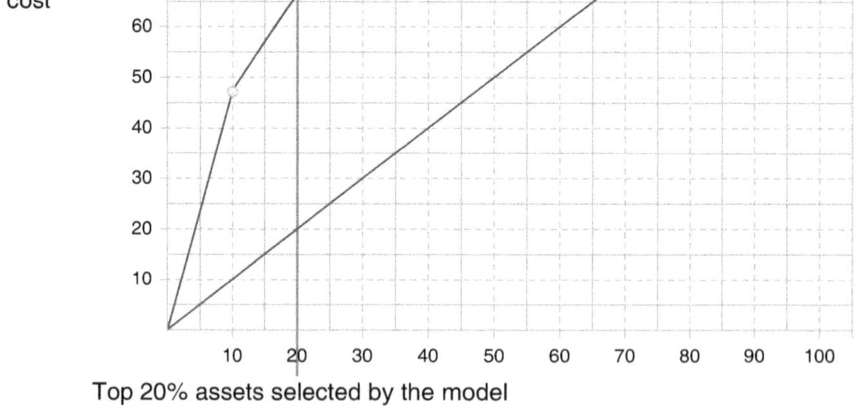

Figure 7.5 Gains chart

total actual outcome (cost, or failure count) captured by each decile. In this way, the assets responsible for the highest proportion of cost or failures are ranked at the left side of the plot. The overall shape of the gains chart reflects the proportion of assets responsible for known proportions of total cost or failures. The steeper the curve, the better the model identifies at-risk assets.

General applications for the methods used

The approach described is generally applicable for any organisation that owns assets to facilitate estimating future costs and failure intervals. In this example, making evidence-based, informed decisions regarding effective interventions and maintenance planning to reduce asset failure rates can help to optimise maintenance expenditure and resource allocation. High risk assets that require immediate attention can also be identified.

Some examples of organisations where such applications would be of benefit include:

- hospitals with a large number of plant assets including generating plant, substations/transformers, boilers, pumps, fans, air conditioning plant, hydraulic systems, electrical distribution systems, imaging equipment, etc.
- commercial buildings with mechanical, electrical, hydraulic, transport and fire protection systems
- electrical generating and distribution infrastructure including generating plant, lines and poles, substations, circuit breakers, switching equipment, etc.
- water treatment and distribution infrastructure with pipework, valves, pumps, water treatment plant, etc.
- road and rail tunnels with power supply systems, substations, lighting, ventilation and fire safety systems.

Risk and opportunity

As discussed earlier, there are many variables in estimating and the actual cost is not known until after the work has been completed. Estimating is usually done on a net basis, aiming to arrive at the likely cost for the completed work and anything that can go wrong will increase the cost and lead to a cost overrun. The most efficient methods of construction are assumed in a competitive tender situation, to enhance chances of winning the tender, so large cost savings are difficult to achieve. The estimated cost represented by '1' on the chart (Figure 7.6) tends to have an optimism bias (Australian Transport and Infrastructure Council 2018) and the probability distribution of actual costs will have an extended right hand tail.

The distribution shown in Fig. 7.6 is a theoretical representation of the lead author's experience of major construction projects in Australia. This view is reinforced by some examples from the literature in the following and research presented at the ICEC World Congress in 2002 (O'Donnell 2002b).

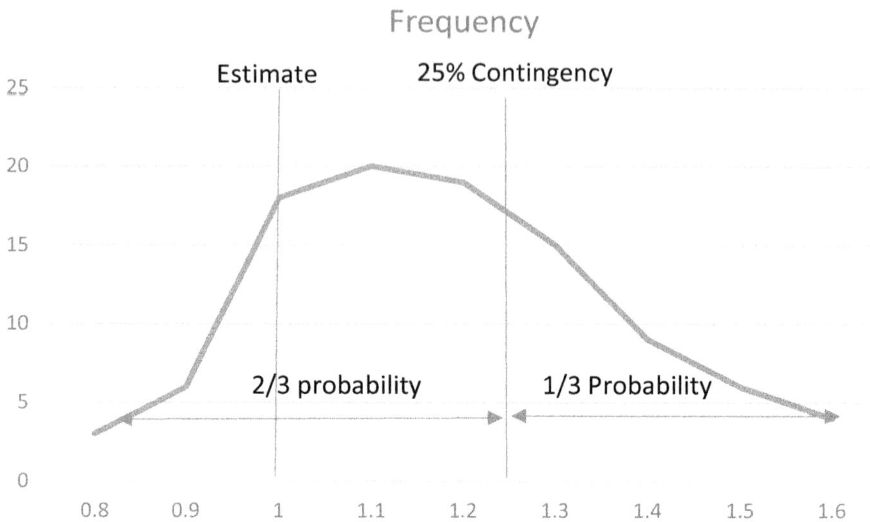

Figure 7.6 Typical distribution of cost outcomes on large construction projects

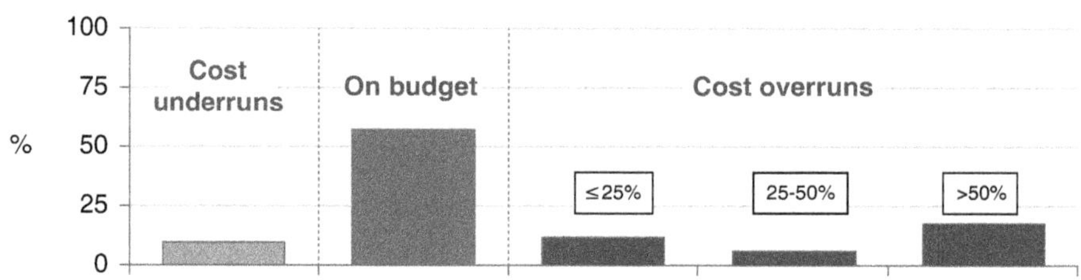

Figure 7.7 Transport projects: frequency of cost differences to budget as a proportion of all projects (Terrill 2016)

Some cost savings will be found during construction but most likely a contingency allowance is needed to add to the estimate to cover the risk of unforeseen costs. The theoretical example aforementioned may seem exaggerated, but it does represent the likely outcome on a large and complex transport engineering project. A 2016 study by the Grattan Institute, for example, analysed 836 projects valued at $20 million or more built in Australia since 2001 (Terrill 2016). It found that cost blowouts accounted for nearly a quarter of the total budgets of these projects. Alarmingly 17 percent of projects overran their original budget by more than 50 percent.

The extended righthand tail is further illustrated in the graph (Figure 7.7) published in the 2016 Grattan Institute Report on Cost overruns in transport infrastructure (Terrill 2016). A lot of projects are on budget, a small number are under budget and a few projects have large cost overruns.

A 25 percent contingency could well be required to cover the risk of cost overruns on a complex civil engineering project as discussed further in the following.

According to Flyvbjerg (2017), over the last 20–30 years, 70–90 percent of megaprojects experienced cost overruns.

Risk mitigation

Risk can be managed or mitigated on construction projects by a variety of means including transfer through the construction contracts, insurance and management processes. The principle of optimising risk allocation is to allocate the risk to the party that is in the best position to control and manage it. The business model for construction contractors is based on managing risk. Having commercial control over a wider range of variables is important in managing risk (O'Donnell 2002b).

BOOT, PFI and PPP projects are often cited as providing increased value through optimising the risk allocation.[6] Currie (2000) gives examples of savings achievable through the private sector taking over work formerly done by the public sector, with a range of 15 to 40 percent for service type contracts such as refuse collection. The savings on some contracts for prisons averaged 13 to 18 percent, and road contract savings averaged about 10 percent (Currie 2000).

Estimating accuracy

Historic cost overruns

For many years there has been concern about cost overruns on construction projects, particularly in the public sector (Morrison 1984; Bowen and Edwards 1985; Arthur Andersen and Enterprise LSE 2000; Currie 2000; O'Donnell 2002b, for example). This calls into question the traditional method of estimating based on design documents being measured and cost rates applied to each item to come up with a series of cost amounts that are summed to give a total.

Recent evidence (Terrill 2016; Flyvbjerg 2017 for example) suggest that little has changed. The industry is still using the same methods and getting the same results. Minor changes in the way quantities are measured using CAD systems does not improve accuracy of the estimates.

This sort of approach can be summed up by the well-known definition of insanity, i.e. '. . . doing the same thing over and over and expecting different results.'[7]

Twenty years ago in the UK, capital cost overruns allowances recommended by the Audit Commission, based on calculated average cost overruns in local authority projects, was 15 percent. (Arthur Andersen and Enterprise LSE 2000).

Some historical references for accuracy of budgets and estimates include:

- Newton (1992) found that the coefficient of variation between budgets and final costs in the UK was 15–20 percent, improving to 13–18 percent at detailed design stage
- Bowen and Edwards (1985) showed that the standard deviation of cost differences between the estimate and actual costs was about 24.8 percent compared with expectations ranging between 2 and 7 percent

- Morrison (1984) reported that quantity surveying offices on 557 projects in the UK had a coefficient of 15.45 percent
- Parker (1994) reported that in the US '. . . statistics show that their bid costs vary from a mean of 29 percent to a maximum of 66 percent from the original budget'
- Victorian Auditor-General's Office (1997) found that aggregate project costs could be 20 to 40 percent higher or lower than the precontract project cost estimates for road projects in Victoria.

O'Donnell (2002b) carried out research that supported the literature, showing that capital cost overruns on 120 public sector projects in Australia were similar in magnitude. Very few projects had a final contract sum under the tender value. The cost overruns were greater on larger, more complex projects and less on smaller projects; civil engineering cost overruns were a lot higher than building project cost overruns. The distribution of the overruns is shown in Figure 7.8 with the typical shape of a very short left hand tail, a significant skew to the right and a long right hand tail.

Aggregating risk

Arriving at an overall cost overrun risk (%) often involves aggregating the components of the project to arrive at a total risk percentage. This would only be valid if each risk were separable, no cross-correlation existed between the risks, and if the distributions were normal. With collinearity issues (cross-correlations between the individual risks) and distributions that are not normal (i.e. skewed) then a more sophisticated method is needed to determine the overall distribution and the aggregate risk. (O'Donnell 2002b)

The literature and Figure 7.8 show that distributions are heavily skewed to the right and are therefore not normal. There is also a positive correlation between capital cost overrun and time overrun. If a project runs significantly over time, it will almost certainly be over budget as well. This tends to increase the overall risk value and the cross-correlations need to be included in the modelling/simulation process.

The overall distribution incorporating the individual risk distributions and the cross-correlations can be modelled using simulation techniques such as Monte Carlo. Historical data can be used to run simulations. Using data and data science can provide much better insights into the true variability inherent with complex projects.

Contingencies for risks

The authors' experience in Australia is that a typical contingency allowance during construction for a simple development like an apartment building is 5 percent. Financial institutions generally want to see a contingency around 5 percent added to the budget for a fully designed metropolitan building they are providing finance for.

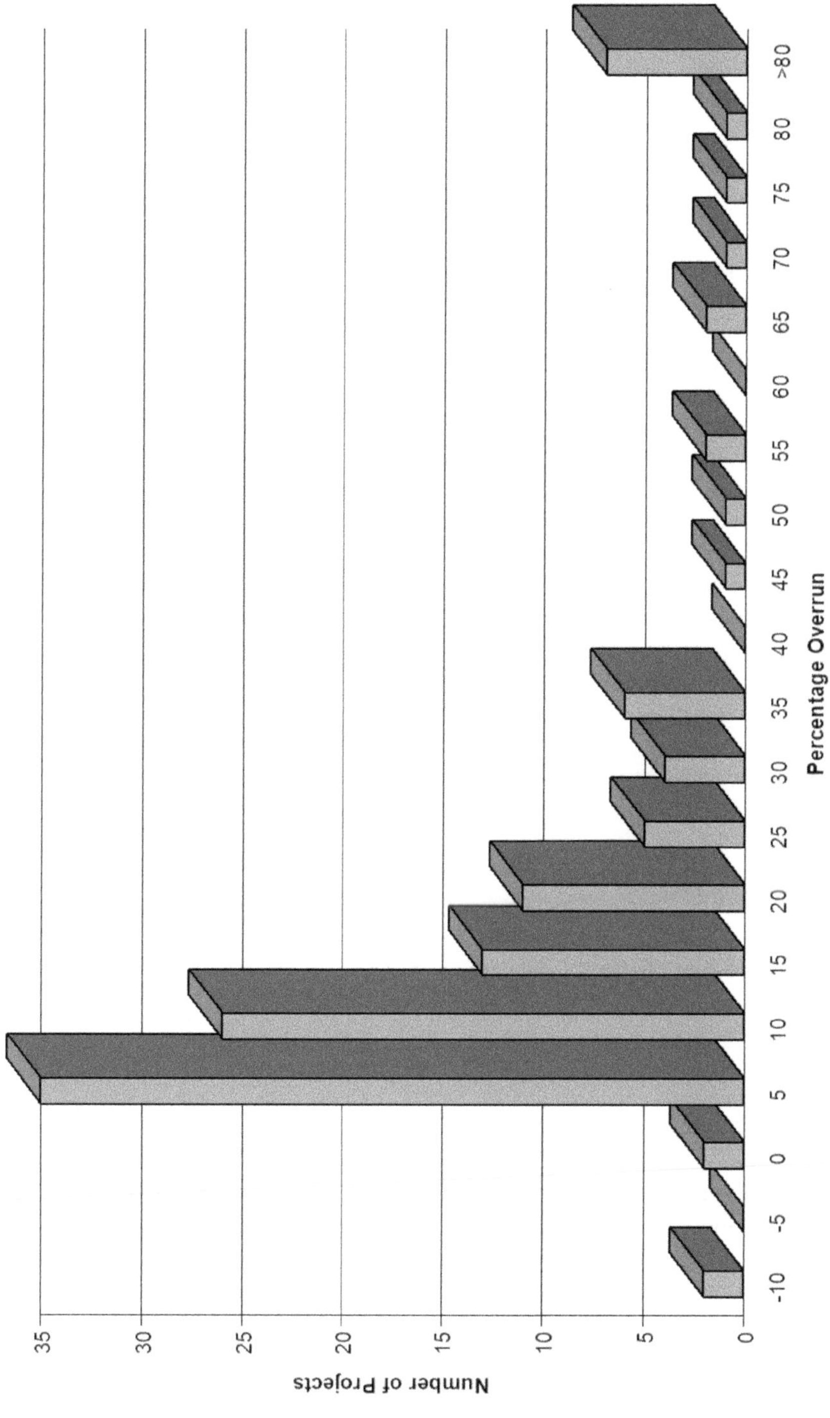

Figure 7.8 Distribution of capital cost overrun for 120 Australian public sector projects (O'Donnell 2002b)

Earlier in the design phase it is common to also allow a design contingency of about 5 percent to cover scope changes (scope creep) as the design is developed. An important part of design management is to manage costs within the budget and this is where cost planning is important. In the absence of good design management and cost planning, cost increases will likely occur due to the design process.

It is important to understand that the contingency allowance is to cover the value of the risk of a cost overrun. In the hypothetical situation shown earlier in Figure 7.6, the 25 percent contingency does not eliminate the risk as there is still about a 1 in 3 probability that the cost will be higher still than the estimate plus contingency. There is, of course, a 2 in 3 probability that the cost will be under the estimate plus 25 percent contract price. This means there is bias in the contract price; however, in competitive tendering, adding too much upside in contingency pricing simply means you do not win the job.

Market forces ultimately mean that contractors win more jobs that have been under-priced and no jobs that have been overpriced. Ideally a portfolio approach is taken to contingency allowances where what contractors will lose on some projects is balanced out by other projects where they earn extra profit. A portfolio approach requires that risk be distributed over multiple projects with a resultant net return over the year that is sufficient to reward the builder for the effort made and risk taken on. A small number of large projects make it difficult to spread the risk in a portfolio approach. For large infrastructure projects, that carry high levels of risk, often there are consortiums and joint venture contracts to spread the risk and reward across multiple entities.

Some key sources of risk on construction projects include the following:

- economic factors – inflation, recession and finance availability
- labour issues – productivity, availability, industrial action and resource shortages
- safety and labour management – accidents, cultural issues and protection requirements
- supply chain delays, cost increases and quality problems
- engineering challenges – structural failures, changes in requirements and plant breakdowns
- latent conditions, contamination, archaeological finds, geotechnical and groundwater issues
- scope creep.

All the material risks should be identified during the tender and a risk matrix prepared along the lines of the simplified example shown in Table 7.2. The risk matrix, covering all aspects of risk, may contain more than a hundred items and require a lot of expertise. For each risk identified, a management strategy is worked out to reduce or eliminate the risk. The probability of things going wrong for each risk item, after implementing the management methods, can be assessed using simulations and by using data from previous projects.

The cost outcome for each risk that occurs is provided by the estimator and/or assessed from historical data from previous projects. The cost is estimated for the residual risk after implementing the management methods. Risk assessment software is commercially available to help prepare the risk assessment and these tend to be more complicated than the example in the following, with more columns and probability distributions to choose from to help value the residual risks. In the example shown in Table 7.2, risk is assessed at 6.7 percent of the estimate value. Executive management may take a view on what to include for risk depending on market conditions and their view of the complexity and riskiness of the project. If they really want to win the project, they may reduce the risk allowance. If they have concerns about the client or the project, they are likely to increase the risk allowance.

Opportunities to save

In addition to the risk value process there are usually opportunities to save that will partly offset the risks. Some examples of the types of things that can be included in an opportunity analysis are listed as follows:

- market pressure on subcontractor pricing
- large project purchasing power
- optimising the design
- enhanced coordination through BIM
- utilising offsite prefabrication including offshore suppliers
- robotics
- new technology
- rationalising the allocation of risk on the project

Before applying a risk contingency, it is always useful to look at the potential up-side. Data is available from previous projects of actual cost savings that were achieved for things like: discounts that were achieved during tender negotiations with subcontractors; material supply discounts achieved on similar large projects; and savings through design refinements.

Large building contractors can usually get better pricing from their subcontractors for example, once the job has been won and they put pressure on the subcontractors to lower their prices. The same applies to purchasing materials and equipment; large contractors can get discounts using their purchasing power, and alternate sources, such as overseas suppliers, can be investigated. An allowance of 2 percent for savings on subcontractor pricing is not unusual and material supply discounts of 5 percent or more for a large project are often achieved.

The probability of achieving the savings can be assessed from experience on previous similar projects and often involves the chief estimator opining on what is the probability of achieving each nominated saving. The savings are based on

Table 7.2 Example: risk matrix and valuation

Risk/opportunity	*Management methods*	*Outcomes*	*Residual probability*	*Cost*	*Risk value*
Project Estimate Risk Matrix				1,000,000	
Building price increases	Escalation clause in contract covering more than 2%. Allow contingency for likely cost increases under 2%. Lock in subcontract prices early	Project may face average cost increases of 1.25%	75%	12,500	9,375
Labour issues	Strong HR Management	Lost production for 4 days due to industrial action	90%	16,000	14,400
Serious accident	Strong safety management procedures and systems	Risk of a Class 1 injury	20%	50,000	10,000
Delayed procurement	Advanced purchase orders. Alternate transport strategies	Critical component delays stopping work for 5 days	35%	20,000	7,000
Engineering failures	Collapse of formwork during pour	Engineer certification and peer review	5%	150,000	7,500
Geotechnical issues	Additional ground investigations. Provisional allowance for rock	Additional rock excavation covered by provisional sum	50%	–	–
Contamination found	Additional ground investigations. Exclude contamination from contract	Excluded under contract	25%		
Ground water issues	Additional ground investigations	Some pumping required	60%	15,000	9,000
Design scope creep	Strong design management	Minor additional scope	25%	40,000	10,000
Total Risk Valuation				**6.7%**	**67,275**

Table 7.3 Example: opportunity matrix and valuation

Risk/opportunity	*Management Methods*	*Outcomes*	*Residual probability*	*Saving*	*Opportunity value*
Project Estimate Opportunity Matrix					
Subcontractor pricing	Use industry strength to discount prices	Reduction in subcontract prices by 2%	90%	1,000,000 12,000	10,800
Supply chain management	Obtain volume discounts	Reduction in material and plant pricing by 5%	90%	15,000	13,500
Value management for design	Hold value engineering workshops	More efficient design with 2% savings	75%	15,000	11,250
Total Opportunities Valuation				**3.6%**	**35,550**
Net Risk and Opportunities				**3.2%**	**31,725**

the discounts that can be obtained and the value of each component of the work. In this simplified example it is assumed that the subcontract work is 60 percent of the estimate sum and the materials and equipment represent 30 percent of the estimate. A simplified example of an Opportunity Matrix is shown in the following example.

The assessed savings are used to offset the risk value calculated earlier (see Table 7.3) resulting in the residual contingency requirement being only 3.2 percent.

Understanding the market

Executive management will choose to adjust the contingency up or down based on their assessment of the market. Large construction projects take years to complete and economic conditions can change rapidly as they did with the Global Financial Crisis in 2007–2008 and more recently with the COVID-19 pandemic in 2020.

The scientific processes employed in estimating and then assessing risks and opportunities can still lead to financial disaster through major economic downturns and catastrophic events such as an earthquake or flood. Insurance is available to cover some risks but not all.

Strangely enough a sudden upturn in the market can lead to serious problems for builders. During depressed market conditions like those of 2020–2021 due to the COVID-19 pandemic, builders and subcontractors have tended to chase work by offering lower prices to keep the business going. Supplier and subcontractor discounts have been readily available and the opportunity assessments will tend to outweigh the risk assessments. In 2021 everyone, including developers, cost consultants and government agencies, became used to lower prices. The lower costs encouraged more development to come to the market, meaning contractors with low pricing could win multiple new projects in a short space of time, but then found that, because of the upturn, the discounted prices had disappeared. Instead of supply chain discounts, builders faced price escalation and a shortage of resources.

When assessing what margin to put on top of the tender pricing, builders need to consider the business as a whole. Project-specific costs are built into the estimate, so if a tender is not won, these costs do not occur. Operating costs for running the business, i.e. overheads such as the estimating department, accounting, senior management, human resources, office rental, and the like have to be covered from the profit on projects won.

Margins on large construction projects are rarely more than about 8 percent and a significant proportion of this goes to pay for overheads. The overhead proportion varies between companies but by reviewing net profit margins for large, listed construction contractors, 3 to 4 percent is typical and this means that around half of the margin on construction projects won goes to pay overheads. This is cost recovery rather than profit. The actual profit from a portfolio of projects is typically less than 5 percent.

CIMIC for example declared a net profit of AUD 208.0 million for the 6 months to June 2021 on revenue from continuing operations of AUD 7,127.4 million which represents a net profit margin of 2.9 percent. A year earlier to June 2020, the profit margin was better at 5.7 percent (CIMIC 2021). Balfour Beatty had underling revenue for 2020 of GBP 8,587 million and a net profit of GBP 51m representing a profit margin of only 0.6 percent no doubt impacted by COVID-19. In 2019 their profit margin was a more typical 2.6 percent (Balfour Beatty 2021).

Conclusion

Organisations need to know in advance what a construction project will cost. Owners need to organise finance and, in addition to the construction cost, also need to budget for ongoing costs in the operations phase. Often there is a trade-off between capital cost up front and ongoing life cycle costs. Optimising the balance between Capex (capital expenditure) and Opex (operational expenditure) is particularly important with BOOT, PFI and PPP projects where the private consortium bidding to win the project has to estimate the full costs of designing, constructing and maintaining the facility for several decades.

Traditional estimating processes comprise measuring the quantities of work to be done and putting price rates against each item. The construction work is a combination of materials, labour and plant with fixed and variable cost components. Many variables exist that impact the costs and the actual cost is not known until the work is completed. In the case of operational phase costs, accuracy issues may not be discovered for many years.

The construction industry is conservative and resistant to change despite there being ample evidence of accuracy issues with estimating, especially with large complex projects. Transport infrastructure projects such as road and rail, for example, have an appalling record of cost overruns in many jurisdictions. Businesses relying on estimates tend to allow contingencies to try and cover the risk of cost overruns, however competitive pressures mean that the contractors that allow for large contingency amounts, don't win the job.

Proponents for new infrastructure projects making their business case tend to ignore the history of cost overruns on previous projects and assume they can manage the risks better, demonstrating optimism bias.

To a large extent estimators rely on expertise and judgement with so many variables to cover while under time pressures often in a short tender period. Digital technologies to aid the traditional process have been adopted, particularly with the use of CAD and BIM to provide the quantities of work.

There is little evidence of estimators adopting machine learning (ML) and data science techniques as disrupters to the traditional estimating procedures. In the related field of estimating for asset management costs, some evidence exists for using advanced digital technologies, however these technologies require a great deal of statistical and mathematical expertise that most estimators do not possess; most data scientists have a PhD for example.

Any paradigm change in estimating is likely to flow from education with new tools developed initially by academics working closely with construction industry leaders and key professional groups. It is unlikely that many estimators will become proficient in the complex statistical procedures required for data science but understanding what is possible and being able to communicate and work with data scientists should be the aim. This will facilitate development of ML applications that estimators can use to enhance both their productivity and accuracy.

A parallel field is the development of BIM systems for designers. The architects did not become computer programmers but now routinely work using BIM systems to enhance their work. Advanced ML applications for estimating that interface seamlessly with BIM systems could be at the heart of a new paradigm in estimating.

Notes

1 The terms Building Information Modelling and Building Information Management are both used widely in the literature. Modelling has been used in this chapter but the terms are interchangeable.

2 There are many estimating textbooks that will provide details of the various commonly used forms of estimates, see, for example, Greenhalgh (2013), Schmid (2011), Sierra (2013), Toenjes (2019).

3 Building price indices (BPI) are available in many countries but, as is the case with similar indices such as consumer price indices, they are averages of averages, usually based on some sort of basket of items, weighted according to 'typical' utilisation of the various components. As such they vary in reliability depending on availability of data, and level

of effort (and detail) which agencies such as national statistical offices apply to their production.' The same is true of locational price indices.

4 Text Mining is the process of deriving meaningful information from natural language text.

5 There is a subtle difference between FM and AM with engineering infrastructure people using the term Asset Management and building property people using the term Facility Management. Really the terms are interchangeable.

6 PFI and PPP projects are also often cited as providing poor value for money (eds).

7 The definition is often attributed to Albert Einstein but there is little or no evidence to suggest that he ever said it (History.com) www.history.com/news/here-are-6-things-albert-einstein-never-said

References and further reading

AIQS (2017) *Australian Cost Management Manual (ACMM) Volume 1* (Sydney: Australian Institute of Quantity Surveyors).

AIQS (2018) *Australian and New Zealand Standard Method of Measurement (ANZSMM)* (Sydney: Australian Institute of Quantity Surveyors).

Arthur Andersen and Enterprise LSE (2000) *Value for Money Drivers in the Private Finance Initiative* (London: Report Commissioned by the UK Treasury Taskforce).

AusHFG (2021) *Australian Health Facility Guidelines Home | AusHFG*. https://healthfacilityguidelines.com.au.

Australian Transport and Infrastructure Council (2018) *Australian Transport Assessment and Planning Guidelines – O2 – Optimism Bias* https://o2-optimsim-bias.pdf (atap.gov.au).

Balfour Beatty (2021) https://balfour-beatty-plc-annual-report-and-accounts-2020.pdf (balfour beatty.com).

Bilal, M., Oyedele, L.O., Qadir, J., Munir, K., Ajayi, S.O., Akinade, O.O., Owolabi, H.A., Alaka, H.A. and Pasha, M. (2016) Big Data in the Construction Industry: A Review of Present Status, Opportunities, and Future Trends. *Advanced Engineering Informatics*, **30** (3), 500–521. https://doi.org/10.1016/j.aei.2016.07.001.

Bishop, C. (2006) *Pattern Recognition and Machine Learning* (New York: Springer).

Bowen, P.A. and Edwards, P.J. (1985) Cost Modelling and Price Forecasting: Practice and Theory in Perspective. *Construction Management and Economics*, **3** (3), 199–215.

Chawla, N., Bowyer, K., Hall, L. and Kegelmeyer, W. (2002) SMOTE: Synthetic Minority Over-Sampling Technique. *Journal of Artificial Intelligence Research*, **16**, 321–357. www.jair.org/media/953/live-953-2037-jair.pdf.

Chen, Y. and Jupp, J. (2019) BIM and Through-Life Information Management: A Systems Engineering Perspective. In: Mutis, I. and Hartmann, T. (eds.) *Advances in Informatics and Computing in Civil and Construction Engineering*. Proceedings of the 35th CIB W78 2018 Conference: IT in Design, Construction, and Management (pp. 137–146) (Cham, Switzerland: Springer). https://doi.org/10.1007/978-3-030-00220-6_17.

CIMIC (2021) *Half Year Report – Financial Statements*. CIMIC Group Limited. 2924-02397365-2A1310908. https://markitdigital.com.

Currie, D. (2000) *Regulation Initiative – Funding the London Underground* (London: Business School).

Dictionary.com (2021) www.dictionary.com/browse/estimate.

Flyvbjerg, B. (2017) *The Oxford Handbook of Megaproject Management* (Oxford: Oxford Handbooks). http://doi.org/10.1093/oxfordhb/9780198732242.001.0001.

Friedman, J. (2002) Stochastic Gradient Boosting. *Computational Statistics & Data Analysis*, **38** (4), 367–378.

Gartner (2018) *Gartner Survey Shows Organizations Are Slow to Advance in Data and Analytics*. Gartner Press Release, 5 February. www.gartner.com/en/newsroom/press-releases/2018-02-05-gartner-survey-shows-organizations-are-slow-to-advance-in-data-and-analytics.html.

Greenhalgh, B. (2013) *Introduction to Estimating for Construction* (Abingdon: Routledge).

Hamilton, H. (2012) *Cumulative Gains and Lift Charts*. www2.cs.uregina.ca; http://www2.cs.uregina.ca/~dbd/cs831/notes/lift_chart/lift_chart.html.

Kifokerisa, D., Roupéa, M., Johanssona, M. and Koch, C. (2019) Building Information Models' data for machine learning systems in construction management. *Proceedings of the Creative Construction Conference 2019, Budapest*. https://doi.org/10.3311/CCC2019-112.

Kolyshkina, I. and O'Donnell, P.J. (2018) *Lifecycle Predictive Modelling using Data Science*. Poster presentation at ICEC-PAQS-2018 Conference, Sydney International Convention Centre (unpublished).

Kolyshkina, I. and Simoff, S. (2019) Interpretability of Machine Learning Solutions in Industrial Decision Engineering. *Data Mining: Proceedings of the 17th Australasian Conference (Ausdm 2019), Adelaide*, 156–170. https://doi.org/10.1007/978-981-15-1699-3_13.

Kumar, V. and Reinartz, W. (2018) Data Mining. In: Kumar, V. and Reinartz, W. (eds.) *Customer Relationship Management* (Berlin: Springer).

Lee, T. and Chen, I. (2005) A Two-Stage Hybrid Credit Scoring Model Using Artificial Neural Networks and Multivariate Adaptive Regression Splines. *Expert Systems with Applications*, **28** (4), 743–752.

Lee, T., Chiu, C., Chou, Y. and Lu, C. (2006) Mining the Customer Credit Using Classification and Regression Tree and Multivariate Adaptive Regression Splines. *Computational Statistics & Data Analysis*, **50** (4), 1113–1130.

Lemmens, A. and Croux, C. (2006) Bagging and Boosting Classification Trees to Predict Churn. *Journal of Marketing Research*, **43** (2), 276–286.

Li, Y. and English, J. (1997) Economic Cost Modeling of Environmental-Stress-Screening and Burn-In. *IEEE Transactions on Reliability*, **46** (2), 275–282.

Liaw, A. and Wiener, M. (2002) Classification and Regression by Random Forest. *R News*, **2** (3), 18–22. http://CRAN.R-project.org/doc/Rnews/.

Logan, A. (2004) The New Scottish Parliament Building. *British Journal of General Practice*, **54** (507), 804.

Lowe, D., Emsley, M. and Harding, A. (2006) Predicting Construction Cost Using Multiple Regression Techniques. *Journal of Construction Engineering and Management*, **132** (7). https://doi.org/10.1061/(ASCE)0733-9364(2006)132:7(750).

Menardi, G. and Torelli, N. (2012) Training and Assessing Classification Rules With Imbalanced Data. *Data Mining and Knowledge Discovery*, **28** (1), 92–122.

Miettinen, R. and Paavola, S. (2018) Reconceptualizing Object Construction: The Dynamics of Building Information Modelling in Construction Design. *Information Systems Journal*, **28** (3), 516–531. https://doi.org/10.1111/isj.12125.

Morrison, N. (1984) The Accuracy of Quantity Surveyors' Cost Estimating. *Construction Management and Economics*, **2** (1), 57–75.

Newton, S. (1992) Methods of Analysing Risk Exposure in the Cost Estimates of High Quality Offices. *Construction Management and Economics*, **10** (5), 431–449.

NewVantage Partners (2019) *Big Data and AI Executive Survey 2019*. Executive Summary of Findings. https:// newvantage.com/wp-content/uploads/2018/12/Big-Data-Executive-Survey-2019-Findings.pdf.

O'Donnell, P.J. (2001) *Functional Area Cost Evaluation Technique Using Regression Modelling*. Doctoral Thesis (Sydney: University of Technology).

O'Donnell, P.J. (2002a) Cost Modelling at the Front End. In: *Proceedings of ICEC Melbourne 2002* (Melbourne: 3rd World Congress on Cost Engineering, Project Management and Quantity Surveying incorporating the 6th Pacific Association of Quantity Surveyors Congress (PAQS)).

O'Donnell, P.J. (2002b) PPP and Risk Management. In: *Proceedings of ICEC Melbourne 2002* (Melbourne: 3rd World Congress on Cost Engineering, Project Management and Quantity Surveying incorporating the 6th Pacific Association of Quantity Surveyors Congress (PAQS)).

Parker, D.E. (1994) Project Budgeting for Buildings. *Cost Engineering*, **36** (12), 13–18.

RIB (n.d.) *5D Modelling in Structural and Civil Engineering*. www.rib-software.com/en/solutions/5d-modeling.

Rogers, D. (2017) In the Digital Era, Growth Is about More Than Technology. In: *Ideas and Insights* (Faculty of Executive Education, Columbia Business School). https://www8.gsb.columbia.edu/articles/ideas-work/digital-era-growth-about-more-technology.

Rolling, C. (2017) *Package 'Gains'*. Version 1.2 [ebook]. www.gains.pdf (r-project.org).

Schmid, K.F. (2011) *Construction Estimating, a Step-by-Step Guide to a Successful Estimate*. [e-book] (Momentum Press). https://www.worldcat.org/title/construction-estimating-a-step-by-step-guide-to-a-successful-estimate/oclc/784886061/editions?referer=di&editionsView=true.

Shepherd, J. (2004) *Social and Economic Transformation in the Digital Era* (IGI Global). www.igi-global.com/chapter/digital-era/29024.

Sierra, J. (2013) *Estimating Practice Techniques and Procedure* (Sydney: Australian Institute of Quantity Surveyors).

Smart Vision (2021) *What Is the CRISP-DM Methodology?* www.sv-europe.com/crisp-dm-methodology/

Strobl, C., Malley, J. and Tutz, G. (2009) An Introduction to Recursive Partitioning: Rationale, Application, and Characteristics of Classification and Regression Trees, Bagging, and Random Forests. *Psychological Methods*, **14** (4), 323–348.

Tella, A. (2019) *Handbook of Research on Digital Devices for Inclusivity and Engagement in Libraries* (Hershey: IGI Global).

Terrill, M. (2016) *Grattan Institute Report on Cost Overruns in Transport Infrastructure*. https://878-Cost-overruns-on-transport-infrastructure.pdf (grattan.edu.au).

Thomas, N. and Thomas, A. (2016) Regression Modelling for Prediction of Construction Cost and Duration. *Applied Mechanics and Materials*, **857**, 195–199.

Tibshirani, R. (1996) Regression Shrinkage and Selection via the Lasso. *Journal of the Royal Statistical Society, Series B (Methodological)*, **58** (1), 267–288.

Toenjes, L.P. (2019) *Construction Estimating* (3rd edition) (Orland Park: American Technical Publishers).

Victorian Auditor-General's Office (1997) *Major Civic Projects: Work in Progress* (Melbourne: Performance Audit Reports to Parliament).

Wikipedia (2021) General Linear Model. *Regression Analysis*. https://en.wikipedia.org/wiki/Regression_analysis#:~:text=In%20statistical%20modeling%2C%20regression%20analysis%20is%20a%20set,independent%20variables%20%28often%20called%20%27predictors%27%2C%20%27covariates%27%2C%20or%20%27features%27%29.

Witten, I., Frank, E., Hall, M. and Pal, C. (2017) *Data Mining: Practical Machine Learning Tools and Techniques* (4th edition) (Cambridge, MA: Elsevier).

Wu, X., Kumar, V., Ross Quinlan, J., Ghosh, J., Yang, Q., Motoda, H., McLachlan, G., Ng, A., Liu, B., Yu, P., Zhou, Z., Steinbach, M., Hand, D. and Steinberg, D. (2007) Top 10 Algorithms in Data Mining. *Knowledge and Information Systems*, **14** (1), 1–37.

Xie, M. and Lai, C. (1996) Reliability Analysis Using an Additive Weibull Model With Bathtub-Shaped Failure Rate Function. *Reliability Engineering & System Safety*, **52** (1), 87–93.

Zhang, T. (2004) Solving Large Scale Linear Prediction Problems Using Stochastic Gradient Descent Algorithms. In: *ICML'04: Proceedings of the Twenty-First International Conference on Machine Learning* (New York: Association for Computing Machinery).

Editorial Comment

Two themes that recur through the three books in this series are (i) the great diversity that is characteristic of so many facets of construction and, (ii) the difficulty of defining exactly what the terms *construction* and *construction industry* mean. In the following chapter, these questions are explored in the context of developing countries. A number of questions are posed, asking, for example, whether the term *construction industry* has the same meaning in every country or group of countries, and then whether it is appropriate, or necessary, to have a more targeted and/or precise definition of the term that is specific to developing countries.

In the 21st century the diversity in construction, in terms of building type, design, scale, materials, procurement and many other attributes has never been greater than it is now and that is particularly so in developing countries (e.g. India, Zimbabwe), where large modern buildings such as up-market hotels operated by well-known international players including Hilton, Marriott, Sheraton and Sofitel, sit cheek by jowl with all manner of less sophisticated structures ranging from tenements to low rise commercial building to houses of many types, as well as informal dwellings and other small structures (sheds, animal shelters and so forth) constructed from simple materials including mud bricks, sheet metal and scrap timber.

International contracting firms use modern materials, methods and equipment to construct high-rise buildings in developing countries that are often largely indistinguishable from buildings with similar function that are found in the major industrialised countries. In contrast, the rest of the building work carried out in those countries, which is likely to represent a very large part of their total volume of construction output (volume, that is, in terms of total floor area or total enclosed space rather than monetary value), will be carried out by all sorts of firms (of all sizes) and people, including substantial amounts of building being done by households and communities. Labour is often largely unskilled, and the use of mechanical equipment is limited. It is reasonable to ask whether it is appropriate to even consider that all of this construction activity should, or can be, considered as a single industry.

In the following chapter, the author, a researcher with many years of investigation into construction in the developing world, addresses a broad range of questions relating to construction in developing countries and puts forward a number of suggestions aimed at improving performance, and increasing the capacity and capability of those who make up the construction sector (firms and individuals) in order that the populations of developing countries can enjoy a better quality of life.

8 Defining and describing construction in developing countries

George Ofori

Introduction

> *The year was 1998. We were about to travel on the short domestic flight from Dar es Salaam to Arusha in Tanzania to attend the first meeting of the CIB Task Group 29 on Construction in Developing Countries. One of the people I was travelling with, a very experienced researcher and practitioner from South Africa, looked around the landscape near the airport and asked: 'Does this country have a construction industry?' I assured him that this was the reason why we were going to attend the conference in Arusha.*

What makes a researcher and practitioner and administrator who knows so much about the construction industry ask such a question? Part of the answer lies in the lack of consensus on exactly what 'the construction industry' is; the term can have different definitions and conceptualisations in different works in the literature. In particular, definitions and conceptions of the industry in developed or industrialised economies may not be appropriate when discussing construction in developing countries where work often occurs in contexts and ways that are different from what pertains in the industrialised countries.

The aim here is to formulate an appropriate definition and description of the construction industry, from the perspective of the developing countries. In doing that the nature of the industry in developing countries is examined and principles established for the creation of appropriate definitions and descriptors. In addition, the need for action to improve the industries in these countries is considered, the magnitude of the task of realising these improvements is described, and actions that can lead to the development of the construction industries in developing countries are proposed.

The following questions in relation to construction in developing countries are addressed:

- What does the term 'the construction industry' refer to?
- Why is it important to get the definition right?
- What is the current situation of the construction industries in the developing countries?

DOI: 10.1201/9781003102403-8

- In what ways has the lack of a clear definition of the construction industry led to the poor performance of the industries in these countries?
- How will having a clear definition help in the transformation of the economies mainly from domestic resources as many governments now aim to do?
- How likely is it that agreement on a common definition of the construction industry will be reached? Who is to lead efforts to attain this definition?
- What is the future of construction in the developing countries?

These questions provide a starting point for a discussion of construction in developing countries which highlights the impact of the continuing lack of agreement on efforts to improve performance in the construction industries in these countries. It is pertinent to note that for the developing countries, defining the construction industry in the most appropriate manner is not a mere academic exercise; adopting inappropriate definitions is at least part of the reason for the lack of progress in addressing the problems facing the industries in those countries; it is sufficient to note here that the definition has real-life implications, and possible impacts on the efforts to improve the quality of life of the citizenry.

Defining a developing country

The common definition of a 'developing country' is based on gross national income (GNI) per capita. The World Bank classifies countries based on per capita GNI (Serajuddin and Hamadeh 2020) as shown in Table 8.1.

Low-income and lower-middle income nations are classified as developing countries. Many authors disagree with the classification and its use to measure development. Meikle (2019) suggests that developing countries differ significantly, and their situations can change. Indeed, the classification is undertaken every year, and countries can be reclassified; in 2020, seven countries moved to higher categories, and three countries to lower categories. It is also pertinent to note that developing countries are dissimilar in many regards, including their prospects for development, including their natural and human capital, and infrastructure base. Finally, some countries with high GNI per capita have poverty, malnutrition and other human and infrastructure indicators similar to those of a typical developing country.

There is a large and growing volume of literature in the field of Development Economics on the subject of measuring development (see, for example, Sen 1999; Stiglitz *et al.* 2010) which has led to the development of various alternative

Table 8.1 World Bank classification by level of GNI (World Bank 2021)

Income level	*Range – USD per capita*
Low Income	$1,035 or less
Lower-middle Income	$1,036 to $4,045
Upper-middle Income	$4,046 to $12,535
High Income	$12,536 or more

indicators of development, such as the United Nations Human Development Index (HDI) and the Global Multidimensional Poverty Index. The HDI was created to stress that people and their capabilities should be the criteria for assessing the development of a country, and not economic growth alone. For example, the HDI can be used to ask how a country with a high GNI per capita can have a relatively low human development outcome (UNDP 2020).

The HDI is 'a summary measure of average achievement in key dimensions of human development: a long and healthy life, being knowledgeable and have [*sic*] a decent standard of living. The HDI is the geometric mean of normalised indices for each of the three dimensions' (UNDP 2020) while the GMPI '. . . complements traditional monetary poverty measures by capturing the acute deprivations in health, education, and living standards that a person faces simultaneously (OPHI 2020, 2021).

While there is debate on the appropriate definition of development, and about the choice of the criteria for determining the categorisation of countries, the GNI per capita definition is widely used, while others, such as HDI and GMPI are used to moderate and refine the results of the economic measures. With respect to construction, Ofori (2012) noted differences between developed and developing countries by considering: features of the construction industry, projects and products, performance of the construction industries, and aspects of the driving forces of construction in the two groups of countries including environmental issues and climate change, construction safety and health, poverty alleviation, globalisation, technology development and innovation, information and communications technology (ICT), quality and productivity, and disaster prevention and reconstruction. The analysis showed that there is a difference between the construction industry in the developing and developed countries; Ofori (2012) called this *developing country exceptionalism*.

Defining the construction industry

It can be said that, until relatively recently, practitioners and researchers in construction focused almost completely on the project level and, to some extent, on organisations. Work on the macro-level issues of construction were pioneered by the University College Economics Research Group at University College London (Ofori 2019). The group completed many studies on the role of construction in the economy and national development, following its first major study by Turin (1969). In the group's work it was necessary for a definition of the construction industry to be formulated. This influenced the conceptualisation and definition adopted by Hillebrandt (1985) in *Economic Theory and the Construction Industry*, the first book which applied the principles of economics to aspects of construction. Other authors have adopted or proposed different definitions and typologies, as discussed in the following (see, for example, de Valence 2019).

There have been many attempts to define and describe the construction industry and delineate its boundaries; however, there remains a debate on key aspects of the definition. Researchers, administrators and even practitioners do not agree on a definition and common conceptualisation of the construction industry. Groak (1994) was one to ask whether construction is an industry or a sector, and even whether there is such a thing as the construction industry. It seems that the question frequently arises as to whether the construction industry exists, and if it does, in what way it does.

In the debate over the definition of the construction industry, some authors consider it to be a distinct part of the economy which accomplishes all the tasks relating to the realisation of built products (Hillebrandt 1985). Others argue that construction is a sector comprising a group of industries, each of which undertakes particular types of work, and therefore, has different needs (Carassus 1998; Ive and Gruneberg 2000). For example, Meikle (2019) argues that construction in each country comprises several separate industries with different products and processes, technology choices and project arrangements, as well as its own problems and necessary actions. Thus, broad sectors should be identified, and their different issues addressed. Groak (1994) suggests that it would be more fruitful to regard construction as organised agglomerations of projects, rather than as a discrete industry or a fixed constellation of firms. He proposes that there should be a greater analytic emphasis on external linkages of construction.

There are also arguments on the boundaries of construction. Some authors, for example, include the manufacturers and suppliers of the materials required by construction. Hillebrandt (2000) suggests that this approach could have some justification. The discussion is extended to include the participants and stakeholders in construction. For example, is the client part of the industry; are the intended beneficiaries of the project, who are legitimate stakeholders also part of the industry? Ive and Gruneberg (2000) analyse the construction industry by considering it as one part of the process of producing and maintaining the built environment; de Valence (2019) suggests that while the construction industry can be depicted in many ways, it is most preferable to do so by emphasising how the built environment is created and maintained. Carassus *et al.* (2006) suggested that the construction industry no longer provides a single item in the form of a building or item of infrastructure but a variety of services and improvements to the human environment. Thus, it is more appropriate to adopt as the framework for analysis 'the entire construction and property sector – the 'built environment cluster' (Carassus *et al.* 2006: 169).

Legal definitions of the construction industry (in Acts and regulations on aspects of construction) actually beg the question as it does not define 'construction', but assumes the word, and then provides a list of activities, and of particular goods and services, that the industry includes or does not include. For example, whereas the CIDB Act of South Africa defines the construction industry as 'the broad conglomeration of industries and sectors which add value in the creation and maintenance of fixed assets within the built environment' (CIDB 2000); the CIDB Act of Mauritius defines it as: 'the industry dealing with construction works and construction services' (CIDB 2008). The Standard Industrial Classification of Industrial Activities (SIC), which guides the collection of data on construction, is considered in other chapters in this book.

There is also a confusion with the terms *building*, *construction* and *built environment*. The confusion is made worse as the terms have different meanings from country to country. For example, Singapore has the Building and Construction Authority, where the 'building' part deals with building control activities, and the 'construction' part is concerned with the development of the construction industry. Notably, the mission of the BCA is: 'to transform the Built Environment sector and shape a liveable and smart built environment for Singapore' (BCA 2021). In South Africa a differentiation is made between building and construction, with the latter being almost equal to civil engineering.

The administrative consideration of construction also adds to the lack of clarity. In most countries there is not a single dedicated government ministry in charge of construction. The responsibilities of two of the few examples (in Myanmar and Vietnam) are set out Table 8.2. Some of the previous dedicated ministries are being 'lost', as some have become absorbed into broader mandate super ministries. For example, the Ministry of Construction in China became the Ministry of Housing and Urban-Rural Development; and that in Japan became part of the Ministry of Land, Infrastructure, Transport and Tourism in 2001. In many cases, the ministries in charge of construction have multiple functions. In South Africa, the title Ministry of Public Works which is actually in charge of all 'works' does not cause the confusion or tension that one could expect. Indonesia's Ministry of Public Works was merged with the Ministry of Housing to create the current ministry which covers more than public construction or housing In Ghana, the Ministry of Works and Housing, which is in charge of construction, has retained its name despite having spawned, over time, the Ministry of Roads and Highways (in 1974) and, more recently, the Ministry of Sanitation and Ministry of Railway Development (in 2016).

Table 8.2 Functions of some ministries responsible for construction

COUNTRY	*RESPONSIBILITIES*
Vietnam	The Ministry of Construction in Vietnam is 'responsible for state administration on construction, building materials, housing and office buildings, architecture, urban and rural construction planning, urban infrastructure, public services; and representing the owner of state capital in state-owned enterprises.' (CTCN 2021)
Myanmar	'The Ministry of Construction oversees the following organisations to plan and carry out the following tasks: (a) carrying out the duties assigned by the State; (b) planning of roads and bridges for the safety of the roads; repairing of existing roads and bridges; (c) airports assigned by other Ministries; Factory Education, Health, etc.; (d) building, maintaining and maintaining state owned buildings; (e) urban and regional planning; (f) expansion of new cities for the improvement of living standard; Building new territories; (g) developing housing plans to solve the problems of the people and staff; (h) the construction and maintenance of state owned buildings; (i) carrying out the construction work in accordance with the guidelines and policies laid down for the development of the State.' (Development Aid 2007)

Possible implications of the lack of a definition

Does it matter how the construction industry is defined? Is it necessary for a correct definition to be found? In what ways has the lack of a clear definition led to poor performance in construction? Ofori (1993) suggested that research on the construction industry in developing countries was at a crossroads. Subsequently, Ofori (1994) also suggested that the practice of construction industry development is at a crossroads. He noted that the policies and programmes which have been suggested for developing countries have not led to much success. He provided these possible reasons:

- the recommendations made by researchers are incorrect
- the recommendations are correct, but they have not been implemented
- the recommendations are correct, but they have been wrongly applied
- the recommendations are correct, they have been appropriately applied, but it will take a long time for the initiatives to yield results.

To these reasons this fundamental one could be added: the analysis behind the formulation of the policies is flawed. A failure to conceptualise the construction industry could also be a factor.

The lack of agreement on the conceptualisation, delineation and definition of the construction industry has implications. First, it is not possible to delineate the industry and determine the components and overall size of the industry and its role in the economy. For developing countries this means that the size of construction is usually underestimated, and its linkage effects not well appreciated. Therefore, the importance of construction is underplayed. Possible linkages are not determined and taken advantage of; this also means relevant segments which are not included in the definition of the industry are not targeted with statutory control, development initiatives (including incentives), or management efforts.

Second, with no cohesive organisation and common structure, it is difficult to argue for governments to take the industry seriously and give it the support and guidance it needs. The main stakeholders of the construction industry in each country are also often unable to find common cause and form an umbrella organisation which can make an effective case to the government on the industry's needs, challenges and aspirations. For example, in the UK, the Construction Industry Leadership Council, the Construction Industry Council, Constructing Excellence, Build Britain and many other umbrella organisations co-exist. Also, the Ghana Chamber of Construction Industry represents the largest number of constituents but there are other groupings, such as the Ghana Institute of Construction and the public-private Construction Industry Development Forum.

Third, fragmentation persists and militates against the formulation of a common strategy for the industry. Fragmentation is often referred to as a major defining feature of construction. The UK, where the members of the Commonwealth draw inspiration, is used here to illustrate the phenomenon. A strategic report (HM Government 2013) noted that the industry in the UK was more fragmented than its counterparts in France, Germany and the USA. BIS (2013) found that for a project costing about GBP 25 million in the UK, the main contractor may engage about 70 subcontractors, with segments of the work worth GBP 50,000 or less. This fragmentation is the result of the high proportion of self-employment and a high number of small and micro businesses. The Infrastructure and Projects Authority (IPA 2016) reported that there are some 956,000 small and medium enterprises in the UK construction industry (99 percent of the total). The Tier One firms manage their finances by withholding payments to those in the lower tiers. ILO (2001) and Wells and Jason (2010) note an increasing international trend to casualise workers, i.e. to lay off permanent workers and reemploy them on a casual basis. Another peculiar aspect related to fragmentation is the adversarialism which the different groups of players show towards each other.

Fourth, without an appropriate conceptualisation and definition of the construction industry, it is not possible to discuss how to examine the challenges of the industry, to restructure it, and to put it on a path towards effective integration and management for productive and profitable operations. The interdependencies are not determined and not explored or used, and the prerequisites for success are not examined. Possible synergies are not investigated and realised.

The developing countries are most in need of a common definition and conceptualisation of the construction industry to guide their efforts in building the capacity and capability of their industries and realise performance improvement.

The importance of a clear definition of construction in the developing countries

It is important to have the right definition and conceptualisation of the construction industry before positions become entrenched. A different way should be found to visualise, delineate and define the industry in the special circumstances of developing countries. For example, a value chain or cluster consideration may help to develop the related segments such as the professional services, materials manufacturing and supply, plant and equipment supply, and relevant segments of the banking and information and communication technology industries.

The developing countries need practical and purposeful definitions and descriptions of the construction industries which will help them to identify their needs, strengths, weaknesses and challenges, and their future. Progress in industry development and improvement does not mean modernisation in terms of leaving behind the good elements of the industry, which may be unique, such as practice, procedure, material or skill, technique or, indeed, a built item, in the name of decolonising the industry. Neither should it mean maintaining and protecting the obsolete and the inefficient in the name of patriotism or heritage. How to make the traditional better should be a guiding principle.

It is time for the construction industries in the developing countries to shed their colonial features and appendages, and the related schedule of functions and responsibilities, together with the resultant power structures which have been constraining the countries they originate from, and develop their own identities. Of these features, the understanding of what the construction industry is, and what it constitutes, is a basic one and a major determinant of progress in developing the industry and improving its performance.

Current state of the construction industries in developing countries

Ofori (2007) notes that work on the construction industries of developing countries has its genesis in the work of Turin's group at University College London which started in the late 1960s. The field underwent significant development in the 1980s and 1990s (as outlined in Giang and Low 2010). Some multi-lateral organisations such as the World Bank (1984), the International Labour Organisation (see, for example, Edmonds and Miles 1984) and the then United Nations Centre for Human Settlements (UNCHS 1992) and United Nations Industrial Development Organisation (UNIDO 1969) as well as bilateral development organisations directly contributed to, and or commissioned studies on the subject. A Task Group on Construction in Developing Countries (TG29) was set up in 1997; this was upgraded to a working commission (W107) in 2001.

Recent significant works on construction in developing countries include the *Construction Industry Capacity Framework*, a diagnostic tool for analysing any construction industry (Arup 2018), the efficacy of which has been established in Uganda, and a major study on the development of the construction industry in India (Loganathan *et al.* 2017).

The literature on construction in developing countries provides the framework for formulating industry development policies and works from it are cited in some policy documents. Examples include the construction industry development strategies for Sri Lanka (NACC 2014) and Ethiopia (MUDC 2012).

The nature of the construction industries in developing countries

What is the current nature of the construction industries in developing countries? Mir *et al.* (2007) reviewed literature on these industries and found that the challenges facing construction industries in developing countries include:

- insufficient education and training
- absence of government commitment
- lack of long-term vision and planning for the industry
- fluctuations in workload
- ineffective budgeting
- defective contract documents
- corrupt procedures
- payment-related delays
- problems of bonding and insurance
- lack of adequate financial resources
- foreign exchange constraints
- non-availability of equipment and spare parts
- poor information.

Other studies which are on the challenges in particular countries include: Ssegawa and Ngowi (2009) on the construction industry in Botswana; Ofori-Kuragu *et al.* (2016) on the industry in Ghana; and Loganathan *et al.* (2017) on the industry in India.

Poor performance by the construction industry in any country can have major national impacts. In their study of the industry in South Africa, Watermeyer and Philips (2020) noted that time and cost overruns on two coal power plants in South Africa had huge consequences, as they as failed to resolve the energy shortfall they were planned to address, caused damage to the economy due to rolling blackouts, left the national utility company reliant on a government bailout and put the sovereign balance sheet at risk. In the light of these outcomes they stressed the importance of improving industry performance.

Nearly three decades earlier, Ofori (1993) suggested that it is necessary to consider the root causes of these difficulties in order to be able to propose suitable courses of action. More recently, Barrett (2007), in presenting an international study on revaluing construction, proposed 'an over-arching "infinity" model' which is broadly conceptualised towards root causes, and highlights the dependence of each area on others. These questions may be considered:

- What are the contributing factors to these issues?
- What factors really underlie them?
- How can the challenges be resolved?

Frimpong *et al.* (2020) considered how the characteristics of the construction industry affect the nature, practices and performance of the industry in developing countries, with a focus on health and safety.

It is generally acknowledged that there is much to be done to improve the performance of the construction industries in the developing countries. CIB W107 on *Construction in Developing Countries* (Rwelamila and Ogunlana 2015) suggests a range of topics in its research roadmap including:

- A new model of development
- Urbanisation and rural development
- Sustainability in housing
- Human resource development
- Modernising the traditional
- Gender equity
- Financing and procurement
- Governance and management
- Project management
- Industry development.

This indicates that action is required in all segments of the industry.

There has been less work on the industries in developing countries during the past two decades. It would appear that a new era of work at a different level of sophistication is required. Ofori (2016) explored ways to improve the performance of construction industries in developing nations, considering the possibility of leapfrogging. He suggested that these industries can offer lessons to their counterparts in industrialised countries, including:

- construction as a contributor to value and wealth creation
- effective, culture-sensitive and contextually relevant project team selection and leadership
- innovative community involvement in projects
- innovative context-relevant value chain formation and management.

Usually, 'construction in developing countries' is seen in negative terms owing to the many problems which studies show that they have. It is important to remember that the construction industry in every country has issues. For example, studies have shown that labour productivity and production processes in construction have not shown the improvements that have been realised in other sectors. A range of internal and external barriers to digital innovation have been identified; they include fragmentation of the industry, inadequate collaboration with suppliers and contractors, a lack of vertical and horizontal integration, lack of an appropriately skilled and trained workforce, and insufficient knowledge transfer from project to project (Shojaei 2019).

For example, in the UK, as noted by the Construction Leadership Council (CLC 2018), various strategic reviews of the construction industry have not led to much. Finding effective solutions in the developing countries will be beneficial to all countries. It is also necessary to ask the question: Are there any positive points to the construction industries of the developing countries?

Why action to develop the construction industries in developing countries is necessary

The need to improve the performance of the construction industries in developing countries can be established by framing the tasks and expectations of those industries through many possible lenses; some of these are:

- construction is required to contribute to the economy as it grows, as a segment in its own right (Hillebrandt 1985; Lopes 2012), providing jobs (Mella and Savage 2018; ILO 2019), and stimulating activities in other segments (Yousaf *et al.* 2019)
- construction should provide the built items needed for production and services (EIU 2017), and for social activity to realise national development (Turin 1969; Kafandaris 1980), and thereby increase incomes and improve the quality of life and wellbeing of the citizenry
- construction is expected to contribute to the efforts towards sustainability and reduced climate impact through its inputs and methods (Du Plessis 2001; Goubran 2019; Opoku 2019)
- construction should help to prevent disasters where possible, and to address the impact of disasters when they occur (Keraminiyage *et al.* 2007; Bosher and Dainty 2010; Chang *et al.* 2012; Wang 2021).

The Post-2015 Development Agenda (Bates-Eamer *et al.* 2012) for addressing the socio-economic development of all countries is based on attaining 17 Sustainable Development Goals (SDGs) by 2030 (UN 2015). The Economist Intelligence Unit (EIU 2017) notes that infrastructure is at the heart of efforts to meet the SDGs; most of the SDGs imply improvements in infrastructure. Ofori (2016) classifies the 17 SDGs from the perspective of construction:

- Basic human and national needs
- What construction must do
- Some of construction's results
- Inputs and methods of the construction industry.

This approach to the SDGs will cover the first three lenses mentioned earlier.

Much is expected of the built environment sector in the development effort. In 2016 Ban Ki-moon, the then Secretary-General of the United Nations, said at the Habitat III Conference in Quito (UN 2017):

> *The New Urban Agenda should be participatory, action-oriented, and implementable. We are committed to build cities that offer opportunities for all, stimulate structural and productive urban transformation based on equity and shared prosperity, and aim to eradicate poverty and advocate social inclusion.*

The disaster prevention and response lens (discussed earlier) is adopted here in considering the subject in detail. It is evident that the need for action in this regard is greatest in the developing countries. This conclusion recognises that many of these countries are more exposed to disasters and are less well equipped to address them. For example, direct economic losses from disasters were estimated to be USD3 trillion between 1998 and 2017 (UNDRR 2018); climate-related disasters formed 77 percent of the total. Climate-related and geophysical disasters claimed some 1.3 million lives. Studies show that poverty is a driver of disaster risk. It is estimated that disasters kill 130 people for every one million people in low-income countries, compared to 18 per one million in high-income countries (UNDRR 2018).

Of the many types of disasters, those relating to the built environment are relevant here. IFRC (2020) reports that, since 1960, there have been over 11,000 disasters resulting from natural hazards. The periodic number has also been increasing. In 1960, there were 33 disasters; these rose to an annual peak of 441 disasters in 2000. Aon (2020: 11) defines a natural disaster as an event meeting at least one of these criteria:

- economic loss of USD50 million
- insured loss of USD25 million
- ten fatalities
- 50 injured
- 2,000 homes and structures damaged, or claims filed.

CRED and UNISDR (2016) considers megadisasters as those claiming more than 100,000 fatalities. Aon (2020) considers 2010–2019 the world's costliest decade. In that period total direct economic damage and losses totalled USD2.98 trillion; there were 2,850 disasters. About 83 percent were caused by climate- and weather-related events, mainly floods (1,298), followed by storms (589). IFRC (2020) notes that these disasters affected 1.8 billion people who were injured, rendered homeless or deprived of their livelihoods, thus setting back the progress made in sustainable development; 97 percent of them were affected by extreme weather and climate events including catastrophic earthquakes, tsunami, inland floods, tropical cyclone events, hurricane landfalls and severe convective storm outbreaks. Thus, the frequency and intensity of climatological events are increasing substantially. Loss of natural resources, food insecurity, direct and indirect health impacts and displacement are also increasing. CRED and UNISDR (2016: 6) observe that the three megadisasters which marked the 20-year period 1996–2015: the Indian Ocean Tsunami, Cyclone Nargis and the Haitian earthquake 'demonstrate the truth of the statement that the worst disasters which could happen have not happened yet.' They suggest that this underlines the importance of preparing for worst-case scenarios where the evidence demonstrates that such events are predictable, and require strong disaster risk governance at the local, national, regional and global levels.

Describing construction in developing countries

The analysis of the construction industry in developing countries can be facilitated by an understanding of the structure of the industry and its various sectors.

A matrix for construction

Turin (1973) developed a matrix to facilitate such analysis. Considering ways in which the capacity of the construction industry might be expanded, he delineated four sectors, identified the major resources they utilised, and outlined the factors most likely to limit the expansion of each sector. The four sectors are:

- *International-modern*, which builds major civil engineering works and large buildings, utilising imported materials and components, plant and equipment, and large numbers of qualified professional and managerial personnel. Expatriate contractors dominate the sector and the major constraint is the availability of foreign exchange
- *National-modern*, embracing the delivery of mainly middle-sized urban buildings, which requires some imported materials and plant, and a high proportion of skilled personnel. The contractors are usually local, and the constraints are the availability of foreign exchange, technicians and managers
- *National-conventional*, characterised by a mixture of traditional materials and techniques and modern inputs, and involving small contractors and self-employed local artisans. It handles most private houses in urban and

semi-urban centres and rural infrastructure, and its constraints are similar to those of the national-modern sector
- *Traditional*, prevalent in the rural areas, and mostly occurring outside the sphere of the monetary economy.

It is evident that the matrix focuses on the building segment of the construction industry, and also on contractors. It does not consider the design and administration professionals. Moreover, the informal segment of the industry is not singled out.

Ofori (1989) suggested another matrix of the construction industry in developing countries which also identifies different sectors, each with its own determinants of demand and operating constraints, and the relationships between them. This matrix consists, first, of two main divisions: formal and informal. The difference between the two parts of the industry is generally regulatory. For projects in the formal part of the industry, designs are prepared, and planning permission obtained. Projects in the informal part are not designed, nor are permits sought before they are undertaken. and this sector is not constrained by laws, codes and levels of professional and administrative capacity. Though more geographically extensive than the formal sector, this part is largely ignored by writers, governments and planners, and its products inadequately covered by national statistics. The construction industry may then be further divided into six sectors, roughly forming a continuum of decreasing technological sophistication:

- formal sectors: international, conventional-large, conventional-small/medium, self-help projects
- informal sectors: monetary-traditional, subsistence.

Both of these categories embrace buildings and works as well as new work and repair and maintenance.

The main features of these sectors are summarised in Table 8.3.

The *International* segment in each country forms part of the global construction industry (Han 2010). It is the domain of the large global players; the size and technical complexity of the projects makes them viable for only a few of such firms (see a list of such international firms in the annual survey of Top 250 firms in *Engineering News-Record* (ENR 2021). Here, the 'build now or wait to develop the capacity and capabilities' question arises; the developing countries face criticism for awarding the projects to international players, thereby foregoing the opportunity to enable the indigenous firms to use these projects to develop their capabilities, or to undertake them after they have developed the necessary expertise (Hillebrandt 1999; Rwelamila and Ogunlana 2015). Countries have to choose between developing their own construction industries or having the items they require built now, even if it is with foreign inputs and expertise.

The companies in the *Conventional-Large* segment would be the largest indigenous firms in a country. However, many construction firms from overseas, especially from China, are increasingly competing for such projects. Governments have

Table 8.3 A matrix for the construction industries of developing countries (Ofori 1989)

Type	*Resources*	*Limiting factors*
International (dams, motorways, special factories, conference complexes, luxury hotels)	Foreign-owned contracting firms Some foreign consultancy Imported materials and equipment, professional and managerial personnel	Foreign exchange (S/MT) Executive capacity of client organisation (L/MT) Professional and managerial skills (MT)
Conventional-large (offices, highways, factories, blocks of flats)	Large local contractors Some imported materials Plant and equipment Skilled personnel	Foreign exchange (S/MT) Professional skills (MT) Intermediate skills (MT) Skilled operatives (MT)
Conventional-medium or small (modern, medium or small: schools, feeder roads, houses)	Small local contractors Imported/local materials Few skilled managerial or technical staff Skilled operatives	Intermediate skills (MT) Efficiency of contractors (LT) Government's extension services (LT)
Self-help (social communal facilities, roads)	Government technical aid Local/imported materials Voluntary labour by community	Local initiative and enthusiasm (Co-operative habit) (L/MT)
Monetary-traditional (houses, shops and similar small buildings)	Traditional materials and tradesmen	Attitudes of clients and users (LT)
Subsistence (houses, barns, communal facilities)	Traditional materials Owner/co-operative labour	Attitudes of clients and users (LT) Clients' initiative and leisure time (ST)

ST: short term; S/MT: short and medium terms; MT: medium term; L/MT: long and medium terms; LT: long term.

to consider policies that include setting floor limits by project size, and formulating and implementing local content regulations.

Companies in the *Conventional-Medium or Small* segment are the most numerous in the industries in developing countries, as they are elsewhere (as conceptualisations of construction industry present it as having a pyramidal structure with a large majority of firms being small). These companies form the foundation on which the development of the industries can be based. They require support from their business partners and a conducive operating environment to develop. For many decades, it has been suggested that conscious efforts should be made to develop construction firms. The seminal work by Relf (1987) provided a template for contractor development programs. UNCHS (1996) suggested the key factors which determine the success of such programs and provided guidelines for action. The governments of several countries, such as Ghana and South Africa, have had contractor development programmes over many years, and the latter has the most comprehensive ongoing program (see, for example, Department of Public Works 2011).

The *Self-help* segment recognises the projects undertaken by the community, with the citizenry making financial contributions and volunteering their labour. Houses and small projects such as basic schools and health centres are typical of this segment.

The *Monetary-Traditional* segment involves the building of houses, agricultural storage facilities and community centres in rural areas using mainly traditional materials, techniques and tradespersons.

In the *Subsistence* segment, citizens in rural areas build their homes and other structures by themselves or in community groups. It combines, in part, the resources and personnel in the self-help and monetary-traditional segments.

There have been significant developments in construction at both the national and international levels since Ofori presented this matrix in 1989. The size, and technological, managerial and contractual complexity of construction items has increased. Competition has increased, and the traditional players which used to dominate the market now face stiff competition from Chinese and Korean firms. This has made the *International* segment even more exclusive. The *Conventional-Large* segment now includes many foreign companies wishing to remain and operate in the developing countries. Again, this has increased competition locally. Countries have introduced, or are being urged to introduce, local content requirements in construction. In spite of these developments, Ofori's (1989) matrix is still relevant as a framework of analysis for the construction in developing countries, if not the whole world.

The traditional segment of the construction industry needs most development, in order to preserve nations' cultural assets and build on existing technical capital such as the traditional trades and the related apprenticeship systems where they exist. It is also potentially useful to find ways to formalise (not replace) the traditional and derive synergies from combining the modern with the traditional. Attention should also be paid to the informal segment of the industry, in order to harness the entrepreneurship, innovation and desire to succeed which is evident in this segment.

Defining and describing construction in developing countries

From the foregoing discussion, the following definition of construction can be offered:

> *Construction is the part of the economy which plans, designs, builds, operates, and repairs, maintains, alters and refurbishes as necessary, and eventually demolishes buildings and items of infrastructure. It includes the professionals undertaking the physical planning and design, and facilities management, as well as the client. It also embraces materials and plant manufacturing and supply where these are devoted to construction.*

This value chain consideration in the definition enables all relevant aspects of construction to be taken into account, synergies to be realised, and comprehensive development of construction to be achieved. The implications in the context of a developing country can be derived. For example, Lizzaralde *et al.* (2013) are among authors who show that the nature of the industry and operating environment in developing countries call for a different approach to the management of projects. As another example, among the project performance parameters in these countries would include affordability; resilience and future proofing; and legacy from the project in terms of feeding back into other projects in future (Ofori 2016). Realisation of the range of stakeholders helps in determining strategies for addressing their needs and aspirations; the literature on community engagement on construction projects would be helpful here (see Close and Loosemore 2014). For example, considerations of health, safety and wellbeing should include the residents in the environs of the project. The construction industries in developing countries reveal the importance of the operating environment of firms, and the potential impact it can have on their operations and performance on projects.

Possible topics for research on construction in developing countries

The discussion so far shows that the construction industries in developing countries require improvement. Some possible current topics on which research could be undertaken are presented in Table 8.4.

The future

Most of the developing countries use the planning and building control regulations, policies and their administrative and enforcing systems and mechanisms which they inherited from the former colonial administrators. These have not been effective and have resulted in building collapses (Windapo and Rotimi 2012), flooding (IFRC 2020), and delays on projects (Mir *et al.* 2007; Watermeyer and Philips 2020). These countries need development planning and building control statutes, policies and systems which will work in the context of the particular

Table 8.4 Some key research topics

Project level	*Corporate level*	*Industry level*
• Productivity – appropriate measures which involve also job creation and the right jobs • Project organisation structures featuring local traditional structures • Project performance including wider perspectives such as safety and health concepts and procedures which meet the perspective of danger and personal security in these countries • Technology applications with search for leapfrogging possibilities • Stakeholder management, including social systems in project management • Supply chain management on the project	• Joint ventures and strategic alliances: local: local and local: foreign • Corporate development (both consultants and contractors) which takes the context of the industry into account • Appropriate public-private relationships for business at local, district, regional and national levels • Value chain management based on the culture	• Preventing and addressing fragmentation • Industry development • Exploring the informal sector • Institution building and synergisation

country, taking into account the dispersed nature of construction projects and the perceived further hazards from natural disasters in most of these countries. The annual *Doing Business* survey of the World Bank (2020) which many countries are studying closely with the view to improving their ranking, has a major segment, *Dealing with Construction Permits*, which can provide a basis for reform.

As noted earlier, technology development in construction has been a major research topic for decades (see UNIDO 1969). More recently, authors have considered the application of new technologies such as Building Information Modelling (Olanrewaju *et al.* 2020) and the elements of other smart practices (Olawumi and Chan 2020) in developing countries. Ofori (2016) asserted that the technological development of the construction industries in developing countries should be based on the advances in developed countries as well as a constant search for the potential of leapfrogging technologies and practices. The concept of pushing the envelope, rather than automatically considering these countries not to be suitable for the application of advanced technologies, such as advanced digital applications, new materials, and off-site construction applications. Indeed, there has been research on, and application of, off-site applications in developing countries for many decades (Richard 2005). Malaysia's *Industrial Building Systems* was formally launched in 1999, and has been implemented in a series of roadmaps (CIDB 2008; Fateh and Mohammad 2017).

The developing countries have a history of developing alternative appropriate local technologies (materials, techniques and their production and application systems) (UNIDO and International Centre for Science and High Technology 2008). These include the development of modernised traditional technologies and materials, e.g. in Ghana, the modified *Atakpame* building (mud walls reinforced with a wooden frame) and landcrete blocks (which have a low cement content). Such materials are often sustainable, however, most of them have failed to gain traction or have been discarded. They could be revisited and the reasons underlying their lack of success identified (van Egmond-deWilde de Ligny and Erkelens 2008). Incentives for developing such items, and initiatives for their application through demonstration effects in their usage, could be studied. There should be more research on the labour-intensive road construction program developed and applied by the ILO (Tembo and Blokhuis 2004) with the view to upgrading and upscaling it.

Human resource development programs should be given a focus reflecting the situation in the developing countries (Debrah and Ofori 2005). For example, with regard to training, there is scope in combining the conventional and the modern with traditional apprenticeship schemes, as is being done in some counties, such as Ghana (GFA Consulting Group 2019). The development of the appropriate orientation among the educated and trained professionals to seek out solutions in their routine practice which suit the local context should also be given priority. Studies can also be undertaken on ways of balancing employment approaches which enable jobs to be created in construction with optimised transaction costs of construction firms.

The informal segment of the construction industry is important in all countries, both developing and industrialised; it is no longer a fringe or even an undesirable economic activity. This segment has been of interest to many authors, such as ILO (2010) and Wells (2007). Mlinga and Wells (2002) explored the potential of obtaining synergies from collaboration between formal and informal construction firms. As noted earlier, Ofori (1989) highlighted the under-utilised potential of the informal sector and emphasised the need for appropriate policy. This should be a subject of interest in the developing countries

Ofori's (2016) proposal that there should be post-occupancy evaluation of built items in developing countries could be explored. In the UK, the Construction Leadership Council (CLC 2018) also suggests that post-occupancy evaluation should be integrated into procurement. In developing countries, it should be routine to undertake a post-completion technical audit, followed by periodic (five years and ten years) performance assessment of major projects, especially public projects, and those used by large numbers of people. There is a need for research to develop templates for these procedures which would be suitable for application in the context of these countries.

The government in each developing country should recognise the role of the construction industry in national development and prepare sector strategies for it in national plans and policy initiatives and directions. For example, the focus of the governments of many developing countries on district and rural industrialisation should be a driving force of appropriate business decision making among construction enterprises. The development of allied industries such as the manufacturing and effective distribution of materials, components, equipment and tools should be given attention. The role of the government in developing such an integrated industry also requires study. Considering the role of the government in construction, the question is how important government considers the industry and the position of the ministry in charge of it within the community of government organisations. This has an impact on the effectiveness of the ministry. The suggestion by some industry players that the agencies which had been established to manage the strategic development of the construction agencies such as Singapore's Building and Construction Authority (BCA) and Malaysia's Construction Industry Development Board (CIDB) should be under the Ministry of Finance is instructive in showing the lack of influence of the ministries in charge of construction. This argument has most recently been made in Ghana with respect to the proposed Construction Industry Development Agency (CIDA) (see Ofori-Kuragu *et al.* 2016).

IFRC (2020) notes that the resources needed to adapt to current and imminent climate-driven disaster risks are within reach. As an example, it would take an estimated USD50 billion annually to meet the adaptation requirements set out by 50 developing countries for the coming decade. IFRC (2020) also notes that this amount is dwarfed by the global response to the economic impact of COVID-19 which passed USD10 trillion in just about one year from 2020 to early 2021. Built environment researchers should develop techniques for formulating proposals for all aspects of the rehabilitation of large areas devastated by natural disasters which are occurring with greater frequency and potency, destroying huge areas as exemplified by the devastation wrought by Hurricane Irma in the Caribbean and the USA in 2017, followed by Cyclone Idai in Mozambique in March 2019.

Conclusion

The construction industry needs to be defined and described appropriately from the perspective of the developing countries to enable it to be analysed with the right framework and approach so that all its segments and the linkages among them are considered, and to realise sustainable long-term improvement in its capacity, capability and performance. For example, if this is the era of increased natural disasters, especially those which are climate-related, then there is a need for an effective and efficient construction industry in each country, building good quality, durable and sustainable built items, and contributing to the resilience of the societies. This era of frequent climate-induced events which can cause significant damage to the infrastructure shows the danger of the capacity, capability, and performance deficits in developing countries.

The need to understand the nature of construction in developing countries is as important as that of finding out what it can contribute to the study of construction in general. The two-way flow of knowledge is another legitimate subject for study. It is pertinent to end by pointing out that describing construction appropriately in the developing countries has the potential to improve lives and livelihoods in many countries and is a matter of life and death for the majority of the population of the world.

After gaining a better understanding of construction in developing countries, are there lessons for the developed countries, and thus, for mainstream construction management and economics, from the consideration of construction in developing countries? Some of the possible lessons include: the importance of the role of the government and strategic consideration of construction; and the merits of construction extending welfare and safeguarding to the community, meaning stakeholder management is even more important. The latter lesson means that stakeholder development should be a consideration in construction, implying awareness building and education of the beneficiaries to prepare them to play their roles in the effective delivery of projects. The third possible lesson is corporate development involving contractors and consultants; this involves applying extension functions in construction, as in agriculture, to build up the capability of construction firms. Finally, it is necessary to consider the matrix of the construction industry in developing countries and explore the possibility of deriving lessons for the mainstream. For example, the construction industry everywhere has the involvement of foreign enterprises in various forms and types of projects.

References and further reading

Aon plc (2020) *Weather, Climate and Catastrophe Insight: 2019 Annual Report* (Chicago: Aon plc).

Arup (2018) *Construction Industry Capacity Framework* (London: Department for International Development).

Barrett, P. (2007) Revaluing Construction: A Holistic Model. *Building Research & Information*, **35** (3), 268–286. https://doi.org/10.1080/09613210601068286.

Bates-Eamer, N., Carin, B., Lee, M.H., Lim, W. and Kapila, M. (2012) *Post-2015 Development Agenda: Goals, Targets and Indicators*. Special Report (The Centre for International Governance Innovation and the Korea Development Institute). https://sustainabledevelopment.un.org/index.php?page=view&type=400&nr=775&menu=1515.

BCA (2021) *About BCA* (Singapore: Building and Construction Authority). https://www1.bca.gov.sg/about-us/about-bca.

BIS (2013) *Supply Chain Analysis Into the UK Construction Sector: A Report for the Construction Industrial Strategy*. BIS Research Paper No. 145 (London: Department for Business Innovation & Skills).

Bosher, L. and Dainty, A. (2010) Disaster Risk Reduction and 'Built-in' Resilience: Towards Overarching Principles for Construction Practice. *Disasters*, **35** (1), 1–18. https://doi.org/10.1111/j.1467-7717.2010.01189.x.

Carassus, J. (1998) *Production and Management in Construction: An Economic Approach*. Les Cahiers du CSTB (Paris: Centre Scientifique et Technique du Batiment. Livraison 395).

Carassus, J., Andersson, N., Kaklauskas, A., Lopes, J., Manseau, A., Ruddock, L. and de Valence, G. (2006) Moving from Production to Services: A Built Environment Cluster Framework. *International Journal of Strategic Property Management*, **19**, 169–184.

Chang, Y., Wilkinson, S., Potangaroa, R. and Seville, E. (2012) Managing Resources in Disaster Recovery Projects. *Engineering, Construction and Architectural Management*, **19** (5), 557–580.

CIDB (2000) *Construction Industry Development Act, No. 38 of 2000* (Government of Republic of South Africa). www.cidb.org.za/publications/Documents/leg_regs_gg21755_cidb_act_38_2000.pdf.

CIDB (2008) *IBS Implementation in Malaysia*. Construction Industry Development Board, Malaysia (Kuala Lumpur: CIDB).

CLC (2018) *Procuring for Value: Outcome Based, Transparent and Efficient* (London: Construction Leadership Council).

Close, R. and Loosemore, M. (2014) Breaking Down the Site Hoardings: Attitudes and Approaches to Community Consultation During Construction. *Construction Management and Economics*, **32** (7), 816–828.

Construction Industry Development Board (2008) *Construction Industry Development Act 2008*. www.cidb.mu/wp-content/uploads/2018/12/Construction-Industry-Development-Board-Act-2008.pdf.

CRED and UNISDR (2016) *Poverty & Death: Disaster Mortality, 1996–2015* (Geneva: Centre for Research on the Epidemiology of Disasters and UN Office for Disaster Risk Reduction).

CTCN (2021) *Ministry of Construction* (Copenhagen: Climate Technology Centre & Network). www.ctc-n.org/about-ctcn/organisations/ministry-construction.

Debrah, Y.A. and Ofori, G. (2005) Human Resource Development of Professionals in an Emerging Economy: The Case of Tanzanian Construction Industry. *International Journal of Human Resource Management*, **17** (3), 440–463.

Department of Public Works (2011) *Framework: National Contractor Development Programme* (Pretoria: Department of Public Works and Construction Industry Development Board).

de Valence, G. (2019) Accounting for the Built Environment. In: Best, R. and Meikle, J. (eds.) *Accounting for Construction: Frameworks, Productivity, Cost and Performance* (Abingdon: Routledge).

Development Aid (2007) *Ministry of Construction of Myanmar*. www.developmentaid.org/#!/donors/view/145607/ministry-of-construction-of-myanmar.

du Plessis, C. (2001) *Agenda 21 For Sustainable Construction in Developing Countries*. CIB Report Publication 237 (Rotterdam: CIB).

Edmonds, G.A. and Miles, D.W.J. (1984) *Foundations for Change: Aspects of the Construction Industry in Developing Countries* (London: Intermediate Technology Publications Ltd). https://doi.org/10.3362/9781780442242.000.

EIU (2017) *The Critical Role of Infrastructure for the Sustainable Development Goals* (London: The Economist Intelligence Unit).

ENR (2021) ENR's 2020 Top 250 International Contractors. *Engineering News-Record*. www.enr.com/toplists/2020-Top-250-International-Contractors-Preview.

Fateh, M.A.M. and Mohammad, M.F. (2017) Industrialized Building System (IBS) Provision in Local and International Standard Form of Contracts. *Journal of Construction in Developing Countries*, **22** (2), 67–80. https://doi.org/10.21315/jcdc2017.22.2.5.

Frimpong, B.E., Wang, C.C. and Sunindijo, R.Y. (2020) Characteristics of the Construction Industry in Developing Countries and Its Implications for Health and Safety: An Exploratory Study in Ghana. *International Journal of Environmental Research in Public Health*, **17** (11), 4110. https://doi.org/10.3390/ijerph17114110.

GFA Consulting Group (2019) *Ghana Skills Development Initiative (GSDI III)*. www.gfa-group.de/projects/Ghana_Skills_Development_Initiative_GSDI_III__3884439.

Giang, D.T.H. and Low, S.P. (2010) Role of Construction in Economic Development: Review of Key Concepts in the Past 40 Years. *Habitat International*, **35** (1), 118–125. https://doi.org/10.1016/j.habitatint.2010.06.003.

Goubran, S. (2019) On the Role of Construction in Achieving the SDGs. *Journal of Sustainability Research*, **1** (2). https://spectrum.library.concordia.ca/986048/2/JSR_1126.pdf
Groak, S. (1994) Is Construction an Industry? *Construction Management and Economics*, **12** (4), 287–293. https://doi.org/10.1080/01446199400000038.
Han, S.H., Kim, D.Y., Jang, H.S. and Choi, S. (2010) Strategies for Contractors to Sustain Growth in the Global Construction Market. *Habitat International*, **34** (1), 1–10, January.
Hillebrandt, P.M. (1985) *Economic Theory and the Construction Industry* (London: Macmillan).
Hillebrandt, P.M. (1999) Problems of Larger Local Contractors: Causes and Possible Remedies. In: Tindiwensi, D. and Bagampadde, U. (eds.) *Second Meeting of the CIB Task Group 29 (TG 29): Construction in Developing Countries (Africa Region) Contractor Development* (Kampala, Uganda: CIB Publication 244), 25–26 June.
Hillebrandt, P.M. (2000) *Economic Theory and the Construction Industry*. 2nd Edition, Macmillan, Basingstoke. https://doi.org/10.1057/9780230372481.
HM Government (2013) *Construction 2025: Joint Strategy From Government and Industry for the Future of the UK Construction Industry* (London: UK Government).
HM Government (2018) *Industrial Strategy: Construction Sector Deal* (London: UK Government).
IFRC (2020) *World Disasters Report 2020 – Come Heat or High Water* (Geneva: International Federation of Red Cross and Red Crescent Societies).
ILO (2001) *The Construction Industry in the Twenty-first Century: Its Image, Employment Prospects and Skill Requirements* (Geneva: International Labour Organization).
ILO (2019) *Developing the Construction Industry for Employment-Intensive Infrastructure Investments* (Geneva: International Labour Organization).
IPA (2016) *Government Construction Strategy 2016–20. Infrastructure Projects Authority, UK* (London: UK Government).
Ive, G. and Gruneberg, S.L. (2000) *The Economics of the Modern Construction Sector* (London: Palgrave-Macmillan).
Kafandaris, S. (1980) The Building Industry in the Context of Development. *Habitat International*, **5** (3–4), 289–322.
Keraminiyage, K., Amaratunga, D. and Haigh, R. (2007) Role of Construction in Managing Disasters in Developing Economies. In: *Annual Bank Conference on Development Economics* (Slovenia: The World Bank, Bled), 17–18 May. http://eprints.hud.ac.uk/id/eprint/22642/.
Lizzaralde, G., Tomiyoshi, S., Bourgault, M., Malo, J. and Cardosi, G. (2013) Understanding Differences in Construction Project Governance Between Developed and Developing Countries. *Construction Management and Economics*, **31** (7), 711–730.
Loganathan, S., Srinath, P., Mohan Kumaraswamy, M., Kalidindi, S. and Varghese, K. (2017) Identifying and Addressing Critical Issues in the Indian Construction Industry: Perspectives of Large Building Construction Clients. *Journal of Construction in Developing Countries*, **22**, 121–144. https://doi.org/10.21315/ jcdc2017.22.supp1.7.
Lopes, J. (2012) Construction in the Economy and Its Role in Socio-Economic Development. In: Ofori, G. (ed.) *New Perspectives on Construction in Developing Countries* (Abingdon: Routledge).
Meikle, J. (2019) A Response to George Ofori's Special Note. *Journal of Construction in Developing Countries*, **24** (2), 207–208. https://doi.org/10.21315/jcdc2019.24.1.10.
Mella, A. and Savage, M. (2018) *Construction Sector Employment in Low Income Countries* (London: Department for International Development).
MUDC (2012) *Construction Industry Policy* (Addis Ababa: Ministry of Urban Development and Construction).

Mir, A.H., Tanvir, M. and Durrani, A.Z. (2007) *Development of Construction Industry – A Literature Review*. Pakistan Infrastructure Implementation Capacity Assessment, Report No. 43185 (Washington, DC: World Bank).

Mlinga, R.S. and Wells, J. (2002) Collaboration Between Formal and Informal Enterprises in the Construction Sector in Tanzania. *Habitat International*, **26** (2), 269–280.

NACC (2014) *National Policy on Construction*. National Advisory Council on Construction (Colombo: Ministry of Housing and Construction).

Ofori, G. (1989) A Matrix for the Construction Industries of Developing Countries. *Habitat International*, **13** (3), 111–123.

Ofori, G. (1993) Research on Construction Industry Development at the Crossroads. *Construction Management and Economics*, **11** (3), 175–185. https://doi.org/10.1080/01446199300000017.

Ofori, G. (1994) Practice of Construction Industry Development at the Crossroads. *Habitat International*, **18** (2), 41–56. https://doi.org/10.1016/0197-3975(94)90049-3.

Ofori, G. (2007) Construction in Developing Countries. *Construction Management and Economics*, **25** (1), 1–6.

Ofori, G. (2012) Reflections on the Great Divide: Strategic Review of the Book. In: Ofori, G. (ed.) *Contemporary Issues in Construction in Developing Countries* (pp. 1–19) (Abingdon: Spon Press). https://doi.org/10.4324/9780203847350

Ofori, G. (2016) Construction in Developing Countries: Current Imperatives and Potential. In: Kähkönen, K. and Keinänen, M. (eds.) *Proceedings of the CIB World Building Congress 2016* (Vol. 1, pp. 39-52) (Tampere, Finland: Tampere University of Technology).

Ofori, G. (2019) Construction Industries in Developing Countries: Need for New Concepts. *Journal of Construction in Developing Countries*, **23** (2), 1–6.

Ofori-Kuragu, J.K., Owusu-Manu, D.-G. and Ayarkwa, J. (2016) The Case for a Construction Industry Council in Ghana. *Journal of Construction in Developing Countries*, **21** (2), 131–149. https://doi.org/10.21315/jcdc2016.21.2.7.

Olanrewaju, O.I., Chileshe, N., Babarinde, S.A. and Sandanayake, M. (2020) Investigating the Barriers to Building Information Modeling (BIM) Implementation Within the Nigerian Construction Industry. *Engineering, Construction and Architectural Management*, **27**(10), 2931–2958.

Olawumi, T.O. and Chan, D.W. (2020) Key Drivers for Smart and Sustainable Practices in the Built Environment. *Engineering, Construction and Architectural Management*, **27**(6), 1257–1281.

Opoku, A. (2019) Biodiversity and the Built Environment: Implications for the Sustainable Development Goals (SDGs). *Resources, Recycling and Conservation*, **141**, 17.

OPHI (2020) *Charting Pathways Out of Multidimensional Poverty: Achieving the SDGs* (Oxford: Oxford Poverty and Human Development Initiative and United Nations Development Programme). https://ophi.org.uk/wp-content/uploads/G-MPI_Report_2020_Charting_Pathways.pdf.

OPHI (2021) *Policy – A Multidimensional Approach*. Oxford Poverty and Human Development Initiative (Oxford: Department of International Development). https://ophi.org.uk/policy/multidimensional-poverty-index/.

Relf, C. (1987) *Guidelines for the Development of Small-Scale Construction Enterprises* (Geneva: International Labor Organization).

Richard, R.-B. (2005) Industrialised Building Systems: Reproduction Before Automation and Robotics. *Automation in Construction*, **14** (4), 442–451. https://doi.org/10.1016/j.autcon.2004.09.009.

Rwelamila, P.D. and Ogunlana, S. (2015) *Construction in Developing Countries Research Roadmap: Report for Consultation*. CIB Working Commission 107 on Construction in Developing Countries (Rotterdam: CIB).

Sen, A. (1999) *Development as Freedom* (Oxford: Oxford University Press).

Serajuddin, U. and Hamadeh, N. (2020) *New World Bank Country Classifications by Income Level: 2020–2021*. https://blogs.worldbank.org/opendata/new-world-bank-country-classifications-income-level-2020-2021.

Sheffield, P. (2020) The Future Is Already Here – Civil Engineering at the Cutting Edge. *Civil Engineering*, **173** (1), 3–9.

Shojaei, R. (2019) *Understanding the Socio-Economic Inhibitors to the Take Up of Digital Innovation in Construction – State of the Nation Sector Report* (Cambridge: Cambridge Centre for Housing & Planning Research).

Ssegawa, J.K. and Ngowi, A.B. (2009) *Challenges in Delivering Public Construction Projects: A Case of Botswana* (Saarbrücken: Lambert Academic Publishing).

Stiglitz, J.E., Sen, A. and Fitoussi, J.-P. (2010) *Mismeasuring Our Lives: Why GDP Doesn't Add Up* (New York: New Press).

Tembo, S. and Blokhuis, F. (2004) *Manual for Supervision of Labour Based Road Rehabilitation Works* (Harare: ILO ASIST).

Turin, D.A. (1969) *The Construction Industry: Its Economic Significance and Its Role in Development* (London: Building Economics Research Unit, University College Environmental Research Group).

Turin, D.A. (1973) *Construction and Development: Building Economics Research Unit* (London: University College Research Group, University College London).

UNCHS (1992) *Technology in Human Settlements: The Role of Construction* (Nairobi: United Nations Centre for Human Settlements).

UNCHS (1996) *Policies and Measures for Small Contractor Development in the Construction Industry* (Nairobi: United Nations Centre for Human Settlements).

UNDP (2020) *Human Development Report 2020: The Next Frontier – Human Development and the Anthropocene* (New York: United Nations Development Programme).

UNDP (2021) *Human Development Reports – Human Development Index (HDI)* (United Nations Development Programme). http://hdr.undp.org/en/content/human-development-index-hdi.

UNDRR (2018) *Economic Losses, Poverty and Disasters, 1998–2017* (Geneva: United Nations Office for Disaster Risk Reduction).

UNIDO and International Centre for Science and High Technology (2008) *Available Technologies for Local Building Materials* (Trieste: International Centre for Science and High Technology).

United Nations (2015) *Transforming Our World: The 2030 Agenda for Sustainable Development*. A/RES/70/1 (New York: UN).

United Nations (2017) *New Urban Agenda* (New York: UN).

United Nations (2019) *The Sustainable Development Goals Report 2019* (New York: UN).

United Nations Industrial Development Organisation (UNIDO) (1969) *Industrialization of Developing Countries: Problems and prospects – Construction Industry*. Monograph No. 2 (New York: UNIDO).

United Nations Task Team on Habitat III (2017) *One United Nations for Habitat III* (New York: United Nations).

van Egmond-deWilde de Ligny, E. and Erkelens, P. (2008) Construction Technology Diffusion in Developing Countries: Limitations of Prevailing Innovation Systems. *Journal of Construction in Developing Countries*, **13** (2), 43–63.

Wang, J. (2021) Vision of China's Future Urban Construction Reform: In the Perspective of Comprehensive Prevention and Control for Multi Disasters. *Sustainable Cities and Society*, **64**, 102511, January. https://doi.org/10.1016/j.scs.2020.102511.

Watermeyer, R. and Philips, S. (2020) *Public Infrastructure Delivery and Construction Sector Dynamism in the South African Economy* (Pretoria: National Planning Commission and GIZ).

Wells, J. (2007) Informality in the Construction Sector in Developing Countries. *Construction Management and Economics*, **25** (1), 87–93. https://doi.org/10.1080/01446190600601339.

Wells, J. and Jason, A. (2010) Employment Relationships and Organizing Strategies in the Informal Construction Sector. *African Studies Quarterly*, **11** (2–3), 107–124, Spring.

Windapo, A.P. and Rotimi, J.O. (2012) Contemporary Issues in Building Collapse and Its Implication for Sustainable Developments. *Buildings*, **2** (3), 283–299. https://doi.org/10.3390/buildings2030283.

World Bank (1984) *The Construction Industry: Issues and Strategies in Developing Countries* (Washington, DC: World Bank).

World Bank (2020) *Doing Business 2020: Comparing Business Regulation in 190 Economies* (Washington, DC: World Bank).

World Bank (2021) *World Bank Country and Lending Groups – Country Classification*. https://datahelpdesk.worldbank.org/knowledgebase/articles/906519-world-bank-country-and-lending-groups.

Yousaf, A., Sabir, M. and Muhammad, N. (2019) A Comparative Input-Output Analysis of the Construction Sector in Three Developing Economies of South Asia. *Construction Management and Economics*, **37** (11), 643–658. https://doi.org/10.1080/01446193.2019.1571214.

Editorial Comment

'The only constant in life is change'.[1]

'When you are finished changing, you are finished'.[2]

'Even if you're on the right track, you'll get run over if you just sit there'.[3]

These quotes span around 2500 years of our history; those in charge of any business today should remember these observations; and those in charge of large businesses should remember that they apply to all businesses regardless of how big and/or successful they are. History is littered with hugely successful corporations that have ended in bankruptcy, often because they failed to react to changes in the business environment in which they were operating, Kodak, the photographic giant, is a prime example. Founded when photography was in its infancy, in the late 19th century, Kodak grew to have 90% of film sales and 85% of camera sales in the US in the 1970s as well as large markets in other countries all around the world. In 1975 the company developed the first handheld digital camera but dropped the idea because it was thought that it would damage their market for the film and other consumables that were the mainstay of the business. As digital photography grew in popularity and digital devices began to take over so many aspects of daily life it was inevitable that analogue photography was unlikely to survive. By 2011 the company had filed for bankruptcy. There are numerous other examples of highly successful companies that traded very profitably for long periods that eventually failed; examples include Pan-American Airways (1927–1991) and Toys R Us (1948–2017). Failure has often been a result of failure to move with the times or through poor management of other aspects of the business, such as over-ambitious plans for expansion.

Construction firms operate in an environment that presents them with additional challenges: firms compete for work on a project-by-project basis, risk levels are high, completion times are tight, and as they bid for contracts, they must predict price levels for materials and labour up to several years in advance. It is little wonder that contracting firms often fail, and when they do the effects are felt by many apart from those in the firms who are directly affected when their jobs disappear, often with little warning.

In this chapter the author addresses questions related to the business models that currently underpin many construction firms and asks why recurring failures are not better understood and thus why business models have not been developed and adapted so that construction firms can be more resilient.

9 Transforming the construction firm?

Hedley Smyth

Introduction

There has over the years been considerable attention given across most developed countries to improving construction project performance. This is not because the sector, and main contractors in particular, have been and are at the forefront of performance improvement, rather it is the reverse, where calls for improvement have yielded minimal results. While incremental steps have been taken, a transformation of performance has not occurred. Repeatedly, outcomes have fallen short of expectations, for example from the internationally influential Egan Report (Egan 1998) in the UK (e.g. Green 2011; Smyth 2010). There are multiple reasons, yet attention has not been given to one of the prime causes residing with the firm. A recent report (Smyth 2018) has pointed towards the transactional business model of the main contractor, and the failure to reform management to embrace transformation. It points out that this is much needed because the large main contractors are no longer serving the needs of clients and society, nor serving the long-term interests of the shareholders or indeed firms themselves.

The prime focus of that report was the biggest main contractors in the UK. This chapter updates the situation in the UK and places the argument in the wider international context because the issues around managing the construction firm are a problem across the globe, although the expression of the problem is at different stages of unfolding in different countries and regions. The modern construction company in the UK has faced repeated calls to improve project performance since the Second World War (e.g. Murray and Langford 2003; Ive 1995). The Simon Report (Simon 1944) started this and a report on the state of play has appeared in the UK in most decades since. The Latham (1994) and Egan (1998) reports were the last major ones in print form after which a plethora of online reports has been published in the UK (e.g. Wolstenholme 2009). In the UK, both the roadmap developed by the UK's Construction Leadership Council for the post-COVID era, through its message of *restart, reset and reinvent*, and the Government *Construction Playbook* report (HM Government 2020) have all the hallmarks of repeating past patterns. For example, the last phase of 'reinvent' envisages transforming the industry, delivering better value through collaboration

DOI: 10.1201/9781003102403-9

and partnership over a 12–24-month period, which is strategically unrealistic and fails to address the management and business models of the firms.

Why have the responses across industry and results fallen below expectations? This requires some unpacking. There are some good well-rehearsed project-related reasons around complexity, uncertainty and the temporary organisation (e.g. Lundin and Söderholm 1995; Packendorff 1995; Bakker *et al.* 2016), yet one of the main reasons is that those wishing to see improvement have simply aimed at the wrong target – the project – or overlooked a primary target. Perhaps the target should be something more stable, sure and constant, or at least relatively permanent. That more enduring target is the firm that manages the project, where strategy, systems, procedures, capabilities and routines might be expected to be enduring. The construction firm, in particular its business model and management, is an important part of the target. It is neglected in both research and practice where the project lens dominates and the firm is frequently viewed as a bundle of projects. Yet the firm faces entropy long term and without management reform permanence is threatened.

These problems are global among tier one or main contractors. These are the large contractors that are defined as *systems integrators* (Davies *et al.* 2007). They are the organisations appointed by clients to deliver the projects, although many do not execute projects directly, the work being carried out on-site by subcontractors, with off-site fabricators producing components, many of which are quite large, for example, bathroom pods, and complex (e.g. air conditioning modules). The situation is worst among the medium and large companies that are listed on the stock market, because the short-term horizon of investors, sometimes referred to as representing *impatient capital* (Narayanan and DeFillippi 2012), demands high dividends and they are unconcerned about long-term perspectives and management approaches. They shun stocks where management pursue low-cost change and incremental investment at the expense of dividends. Directors are incentivised to follow suit as bonuses are frequently paid on the basis of annual shareholder returns. The UK is at the forefront of this process. The USA is not far behind. But it is not the stock markets that are the problem; it is the business model. Low cost, low investment and cash flow management encourage main contractors to live off trade credit from suppliers, delay of payment to subcontractors and investing surplus working capital to earn an interest. In a good market this may yield a return on capital employed (ROCE) of 20–25 percent. The business model needs to be managed, especially to facilitate changes in the model in markets that challenge the viability of the model.

A 'good market' is defined as one with reasonable rates of interest. In the prevailing environment, viability is severely challenged when squeezed by margins on projects and when interest rates are low or even negative. Low productivity exacerbates the situation. Across Europe, construction productivity has fallen over five or more decades. In France, for example, according to *The Economist* (2017) construction productivity fell by around 16 percent over the last 20 years. In the USA it has halved since the 1960s. This can partly be explained by improvements made in offsite construction and even more by traditional products being reconfigured off-site to make assembly of more sophisticated elements easy, leaving the difficult tasks that can less easily be substituted by components or improved through standardised methods in the hands of contractors and subcontractors. This is exacerbated by contracting becoming less capital intensive due to access to low-cost labour, often foreign or migrant labour, imported from overseas. The result is that workers 'replace' the potential for productive machinery and technologies, rather than machinery replacing labour. This is especially the case for SMEs, but of course many SMEs are subcontractors to the large main contractors, and so all tiers and projects in the market are affected.

SMEs are more agile in being able to shrink in size in downturns and expand again in upturns, where the consequences can be absorbed internally. The same applies to privately owned firms, including the larger ones. This chapter largely focuses upon the large main contractors undertaking large complex projects, many of which are operating in international markets.

The construction market continues to become more international, but not global, in the sense that a few dominant players span most regional markets in their operations and presence on the ground. Contractors from different countries and regions have adopted different practices. US firms tend to follow their clients and therefore fail to establish a permanent presence on the ground. This is a strategic approach yet it works against establishing a global presence. When British contractors enter regional markets, they tend to set up branch offices but typically do not give them adequate management support. The common view is that the presence may be temporary, although typically this is not the case in North America

This is a pragmatic approach. Mainland European contractors adopt a more strategic approach, tending to establish branch offices in key markets. A few Australasian contractors have tried to do the same, albeit with mixed success. A strategic approach generally helps support geographic expansion. Chinese contractors are following government policy where access and ownership overseas is being facilitated by infrastructure and commercial development. However, the management practices tend to follow the patterns from either the host country or the approach in the domestic market of origin.

In recently developed and developing countries, especially in China and South East Asia, the industry may look quite healthy; for example, the Chinese market has been buoyant. Main contractors have insulated themselves from uneven workloads in the growth period by not paying subcontractors to protect their revenue and similarly subcontractors have not paid their hired workforce. That system began to break down as the economy matured. New capabilities have been developed on the back of rapid urbanisation and mega-infrastructure projects. However, the current focus is more tactical in the search for different means of management, typically new models and new software to provide solutions. Strategic management receives less attention. Issues remain hidden as the slowing domestic market is compensated for by the growing export market for the larger contractors as China tries to dominate economies more and more. This exactly mirrors the failure of British contractors to address the growing problems with their productivity and profitability in the late 1950s through to the early 1970s when their domestic decline was compensated for by going overseas, especially to the Middle Eastern oil nations, to sustain profits.

Falling fast

Chinese firms are in the ascendency still, largely because they operate in a highly protected domestic market, which is sustainable as long as the economy continues to grow. Yet they embed the same problems that have led to the demise of contractors spearheading development in the post-Second World War era, namely from Western firms, especially the US firms, which headed the ENR rankings by turnover for decades. These post-war firms are now dwarfed by the Chinese firms – see Table 9.1.

The 2019 ranking shows ACS, Hochtief and Vinci moving into the top four (ENR 2019), but the data, and hence the validity of the rankings, may be partly distorted by China's response to growing protectionism through under-reporting and may also partly be an accurate reflection of slowdown in the Chinese economy prior to the COVID outbreak.

British contractors were second only to US contractors on the world stage in the 1950s and 1960s. By the end of 1993–96, in the aftermath of the recession during which leading British contractors lost over 18 percent market share to their European counterparts, only one British contractor (in terms of where ownership resides) appeared in the top ten European contractors by turnover (Smyth 2018). European contractors strategically secured a leading market position and began to dominate the upper end of the market. The same occurred in the wake of the 2008 financial crisis. For example, Vinci consolidated its position by acquiring Taylor Woodrow Construction in 2008 and Bouygues also made acquisitions in the UK. In 2011 *Building* suggested that:

> *Bigger firms are likely to dominate over the coming years as they have the scale and expertise to take advantage of the larger contracts . . . . The bigger companies*

Table 9.1 The top ten international contractors in 2018 (ENR 2018)

2018 World Ranking	*Contractor Organisation*	*Country of Origin*
1	China State Construction Engineering Corporation	China
2	China Railway Group	China
3	China Railway Construction Corporation	China
4	China Communications Construction Group	China
5	Vinci	France
6	Power Construction Corporation of China	China
7	ACS (Actividades de Construccion Y Servicios)	Spain
8	Bouygues	France
9	Shanghai Construction Group	China
10	China Metallurgical Group	China

> *should emerge stronger than their smaller rivals when the sector moves back into its growth phase.*
>
> (*Building* 2011: 38)

And this is what happened, the overseas owned contractors became dominant as the smaller rivals shrank in terms of market share.

Therefore, the history has been that the shortcomings of management are most evident in economic downturns and recessions. The shortcomings of UK management were most evident during the downturn of the early 1990s and, to a degree, in the financial crisis of 2008–10 (Smyth 2018). The entry of European majors into the UK had taken place in the 1990s, from which they gradually consolidated their position. There was a retreat among major British contractors in the wake of increased competition. The millennium saw an upturn in the form of PFI and PPP procurement and some respite from the growing competition, coupled with diversification into facilities management and maintenance which broadened into support services.

In 2003, the top five British contractors by turnover were, in descending order, Amec, Balfour Beatty, Wimpey, Taylor Woodrow, and Carillion (*Building* 2003). Amec subsequently focused on the oil and gas part of its business and left construction. Wimpey decided to focus on housebuilding. By 2011, the major European contractors were now the dominant Tier 1 contractors in the UK. The top ten contractors across Europe no longer had a British owned contractor in their number, with Balfour Beatty slipping to 11th position (*Building* 2011; Smyth 2018). The largest were still able to bid for the large contracts and megaprojects, however, the capacity and capability was lacking to undertake substantial parallel and serial contracts at the top end of the market. Yet this was neither fully understood by management according to a report by accountants Grant Thornton (Hartnell *et al.* 2010, cited in Smyth 2018) nor acted upon, even though it had been pointed out to the industry:

> *UK contractors would be foolish to ignore this trend and there are concerns about big European firms bidding against them for work.*
>
> (*Building* 2011: 37)

This is neither a covert nor overt point about protectionism, it is a comment about the need for effective management to keep pace with market change.

By 2019, the financial reporting for the previous year showed that only one British contractor, Balfour Beatty, was of a considerable size; see Table 9.2. It tends to be ranked on the cusp of the top ten by turnover among European contractors for their global operations. Balfour Beatty suffered management issues, having acquired the US professional services organisation Parsons Brinckerhoff in 2009, subsequently failing to integrate it and then being forced to dispose of it in 2014. They remain a very transactional firm, trying to operate in a sophisticated market of complex large projects, programs and megaprojects, where size and capabilities that are transformational are needed, although they are not alone in failing to

Table 9.2 The top ten British contractors (*The Construction Index* 2019)

2019 UK ranking	*Contractor organisation*	*Turnover (GBP million)*
1	Balfour Beatty	7,802.0
2	Kier	4,512.8
3	Interserve	3,225.7
4	Galliford Try	3,132.3
5	Morgan Sindall	2,971.5
6	Amey	2,667.8
7	Mace	2,350.0
8	ISG	2,237.62
9	Keller	2,224.5
10	Laing O'Rourke	1,935.4

develop such capacity. Arguably their acquisition of Parsons Brinckerhoff can be seen as an attempt to make substantial change.

The second placed main contractor in Table 9.2 is Kier, which is about half the size by turnover, and has experienced difficulties over several years. It is, in effect, a medium-sized contractor on a European scale and quite insignificant on the global stage. The second biggest contractor was Carillion until January 2018, when it went into liquidation.

The large British contractors have been divided into two types. One type is the dedicated main contractor, for which the core business is managing construction whereby they largely manage the process and subcontract all work. The other type comprises bi-sectoral or diversified groups. Historically, diversification into related areas, such as housing development, property development, aggregates and mining, were a means to smooth profit earned from contracting (Ive 1994). Balfour Beatty was a diversified group as was Tarmac, which later gave rise to Carillion.

Over the last two decades UK contractors have increasingly become specialised, focusing upon horizontal diversification through specialist subcontracting and vertical diversification, largely via PFI and PPP contracts, into facilities management (FM). Diversification into FM originally increased competitiveness to win PPP-type contracts, and smoothed profits from contracting, as did some PPP contracts in general, and increased profitability as FM carried higher margins. However, that has been eroded as PPP-type contracts have become less popular. Government has pursued austerity policies and driven down public sector FM and support services margins via its funding regimes of ministries and departments with the result that support services have become a problem area in regard to payment and margins since 2010. This has had some effect on troubled groups such as Carillion, Interserve and Kier.

Before examining what needs to change and how, a detailed case is presented, which is an extreme example yet helps to illustrate the current trajectory.

The case of Carillion

Carillion emerged from the Tarmac group of companies. Tarmac had rapidly expanded in the late1970s. It acquired a range of construction businesses, including Holland Hannen and Cubitts in 1976, which initially posed problems along with other operations acquired overseas (*Financial Times* 1979), and which carried considerable fixed assets and thus reduced financial agility. Then, in the 1980s, Tarmac acquired a series of construction-related businesses, including the high-profile US management contractor, Schal. Tarmac became a diversified construction group that led the way in minimising its working capital (Ive 1994), comprising mining, aggregates, building materials (particularly concrete, brick and block production), construction and housing development. Tarmac was trading well by the end of the 1980s. Although interest rates reached 15 percent in 1989, providing there was already a healthy forward workload and insubstantial loans, contractors had good opportunities to earn a good rate of return from investing surplus working capital not required short term. This also provided opportunity for takeovers, such as Tarmac's acquisition of the mechanical engineer, Crown House (Tarmac 1989).

Tarmac faltered in the 1990s. The recession at the end of the 1980s hit the aggregates and minerals business first. The Chairman stated:

> *The United Kingdom construction industry endured an even more difficult year in 1990 than was predicted.*
>
> (Pountain 1990: 2)

Construction followed as projects came to an end without being replaced. In retrospect, the first seeds of divestment were probably sown at this point. Restructuring, including some asset sales to reduce debt, and cost reduction plans were implemented. In the upturn in 1995, Tarmac announced the exchange of its housing development business with Wimpey for its heavy building materials industry (Tarmac 1995). Wimpey were beginning to increasingly specialise in housing development, while the building materials were beginning a good recovery. Tarmac's Group Chief Executive stated how the problem alluded to by Pountain was addressed:

> *The disposal of the private sector housing business through the asset exchange has reduced the volatility of our cash flows and earnings and has also removed a long term drain on our cash in the form of land bank renewal and growth.*
>
> (Simms 1996: 7)

This was the first step in divestment and the refocus on its core business, Tarmac, in essence, returning to its roots. It later demerged its construction business to form Carillion in 1999. The demerger costs covering management, legal and other expenditure amounted to GBP 28.4 million (Tarmac 1999). Carillion's first full year of trading was 2000. The Private Finance Initiative (PFI) had provided a stream of work for construction and the demerged firm quickly focused on PFI contracts including FM and maintenance work alongside construction. The turnover was GBP1.9 billion and the pre-tax profit was just under GBP42 million, yielding a 2.2 percent return (Carillion 2000).

PFI projects, and subsequently PPP work, typically involving private finance and management in various forms, was a staple of the business from the outset.

PFI contracts gave access to complementary income streams. Contracts had largely relied on stage payments from capital projects to replenish working capital and opportunities to secure ROCE. PFI projects were funded through debt and repaid through client expenditure during operations. This was one income stream, essentially a remodelling of the traditional payment method for capital projects. It opened up a second income stream based upon diversification into maintenance and facilities management. At the time of diversification, these support services secured a regular income flow at higher margins. This was providing a smoothing effect to traditional revenue flow for contractors and higher rates of profit.

It also provided an opportunity for the support services business to be expanded to include other clients and thus broaden the customer base and then add other support services. Carillion followed this path and while it broadened its customer base, many of the clients were essentially public sector organisations and subject to public sector policies and procurement regimes. In the wake of the financial crisis of 2008–10, the austerity policies of government permeated every area of work. Support services contract prices were driven down upon renewal to lower margins and tighter contract terms, placing the support services business under pressure for many organisations, including Carillion, which had 38 percent of its revenue in this sector (Rogers 2018).

PFI and PPP work became a problem area, disguised by Carillion's opaque accounting practices. The debt was growing after 2012. It was generally assumed to be part of its capital projects, especially PPP-type contracts. However, there were support services PPP contracts too. Debts were also accumulated but hidden as 'trade payables', which was part of what is known as *reverse factoring* (S&P 2018; Rogers 2018). The consequence was that the long-term debt burden was larger than was generally appreciated and it continued to grow. Among other things, it allowed the company to pay out more in dividends to shareholders than was sustainable, hence feeding what in hindsight looks like at least optimism bias, or a lack of critical assessment, but the courts will determine whether either of those perceptions are realistic. There were other accounting problems, namely the use of 'rosy assumptions' about the degree of completion of projects (Rogers 2018; S&P 2018), which had the dual effect of disguising the demand on working capital and suggesting an imminent flow of revenue. When this did not materialise, it fuelled debt under trade payables in order to continue the illusion of a viable business. The third accounting issue was the disdain for the pension scheme which was allowed to run a deficit to bolster the working capital and service the debts of the company, adding to the false portrayal of the financial position.

While much has been made in the press about the accounting practices, these are symptoms of deeper issues concerning passive management and a transactional business model that became acute in the case of Carillion. This deeper malaise became apparent soon after the formation of Carillion, during the first five years or so of the new millennium. Over successive years, the annual reports and accounts for the firm, whose main audience is the investment markets, portrayed a dynamic organisation, announcing successive reorganisations and adjustments of the structure to take account of the changing markets. This presented a picture of proactive management, yet the reverse was largely the case. The company tended to rebadge existing activities under new headings, although the structures and management largely remained the same. The City analysts either chose not to draw attention to the implications, especially long term, or ignored the issue, as did investors. Carillion management therefore got into a habit of portraying Carillon as a dynamic organisation while it remained passive in practice. It was not developing the capabilities to manage change. This became embedded in the organisation, so that key personnel changes did not fundamentally change this position (Parliamentary Report 2018). The result was that when the market became constrained following the financial crisis and government austerity policies drove margins down the management did not have the strategic creativity, management capabilities and underpinning routines to make the necessary changes. Financial managers were arguably driven into implementing reactive measures that merely staved off failure of the firm. The external auditors for Carillion did not pick up the issues, despite their severity, and the accounts were sufficiently questionable to warrant retrospective investigation following the demise of the company.

The share price collapsed in 2017. Debt had risen to an estimated GBP 1.7 billion. Carillion went into liquidation in January 2018. It was the UK's second-largest construction company, and the largest UK trading organisation to go into liquidation in modern times. The failed business model resulted in company failure. Morrell (2019) called the business model for Carillion 'insane'. The failure proved costly for government that needed to manage the situation by maintaining Carillion staff to provide continuity. The BBC (2018) estimated the cost to the taxpayer as GBP 148 million. The risk of driving down prices does not pay long-term. In this case, government did not undertake sufficient risk assessment or management oversight across the multitude of ministries, departments and organisational agencies that employed the firm. Contractors were managed at the overall level of government as if they were standard product suppliers in many ways, rather than asset-specific providers.

Carillion had developed a bad reputation as an untimely payer of subcontractors. This helped cash flow management but only helped mask the underlying debt burden. This is not uncommon among contractors. Kier is another example, but many other contractors are and have not been exactly timely payers. Carillon was not the only British contractor to face difficulties; Interserve was to follow and it, too, failed to survive, with accountants taking over and endeavouring to salvage resources by collecting payments and selling parts of the business. The support services part of the business was also a problem area. Diversification had become part of a broad problem with management and the business model. The value of diversification was eliminated when austerity hit FM and support services hard too, for it only worked as long as it provided a compensatory source of revenue and profit. Kier was another large company forced into asset sales, selling its housebuilding, recycling and waste management businesses in an effort to raise GBP 55 million per annum between 2019 and 2021 (BBC 2019).

The Carillion case has some unique features, yet the overall position it presents is far from unique. Indeed, it is representative of the state of play among many contractors in the UK, although its decline was far more dramatic than that of other UK contractors, and main contractors in many other countries.

Creating fertile ground – towards new business models

The current trajectory is largely the same the world over, but there is nothing deterministic about the trajectory. It can be changed and this needs to be examined. The changes need not be uniform, indeed there is market advantage in differentiation with investment in capabilities serving various needs and value outcomes. Indeed, diverse business models are healthy for both suppliers and clients. To explore this, the British context is used as the point of departure once again.

Although the UK government claimed prior to Brexit, British businesses will be able to bid for global public sector contracts worth an annual GBP 1.3 trillion after leaving the EU, and the World Trade Organization (WTO) confirmed that 46 nations voted to let Britain remain in the Government Procurement Agreement, of which the EU is part (Department of International Trade 2019; see also WTO 2014), the post-Brexit era, now coupled with a post-Covid era has yet to unfold. There are no current signs that the UK government is committed to enabling British contractors to transform. The UK Government *Construction Playbook* report (HM Government 2020), although couched as industrial strategy, merely takes forward the rhetoric of change at the project level (Green 2021).

The opportunity to bid is not the same as having the capacity to prequalify and bid; it is the capacity to prequalify and bid that is largely in doubt for most British contractors over the long term unless transformative strategies and new business models emerge. This can be broken down further. Essentially, recasting the business model strategically can be divided into the following options or potential combinations that are set out as follows:

- The emergence of agile and transformative main contractors from the existing stock of contractors based upon investment in management and technical capabilities, with acceptable profit margins, but where profit growth largely comes from growing the firm rather than profit margins *per se*. The ROCE-based transactional model is superseded in this option; investment in management capabilities will develop incrementally. And more critically, specialist subcontract work, such as M&E or other specialisms depending on the strategy for differentiation, will be selectively taken in-house (Smyth 2018).
- The emergence of the professional consultant, especially the former independent quantity surveyors in the UK, cost consultants and project managers, to occupy the systems integrator role, based on professionalism and their knowledge base. It will also include some of the very large multidisciplinary global consultancy firms acting in this role. This will be supplemented by BIM

and digital technologies as professional consultants will have adopted this technology more effectively, ahead of the existing contractors (Smyth 2018).

- The emergence of new agile contractor entrants to the top of the national league tables within a short time span as a result of having developed their strategies and business models to create competitive advantage over the incumbents. They will be flexible and vertically integrated and not wedded to a discipline or technology, but will use whatever works to configure and reconfigure agile supply chains, and apply certain in-house specialist expertise as appropriate and available.
- The emergence of new agile entrants to the top of the national league tables within a short time span due to their strategies and business models, whose origins have less to do with construction and more to do with service and digitisation. There are already players emerging, especially in the USA and China that are integrating certain supply chains. They are, at present, as much if not more embedded in property development and e-procurement, using their capabilities to manage or project manage construction (Hall *et al.* 2019), yet it is only a matter of time before the prime focus for some entrants is mainstream construction.

The real challenge for the existing main contractors, particularly in relation to the first option aforementioned, can be unpacked. One dimension is that there are no signs of any of the incumbents making the move. Short termism among shareholders, transactional finance managers and the incentive structures of the main boards simply militate against facing up to the challenge. Being the chief executive and on the senior management teams is sufficiently challenging without undertaking a transformation which will not realise the benefit for the current management, that is, during their tenure. In this sense the passivity of the management of Carillion in its early days represents the sectoral problems for current main contractors in the UK.

Second, there is an awareness and mindset problem. British-owned contractors did not see or understand the consequences of losing over 18 percent market share to mainland European counterparts in the recessive conditions between 1993 and 1996. In 2021 there is no emergent sign of management addressing matters within their control nor of government driving transformative change at sector and firm levels. Essentially the sophistication of clients, and complexity of their project and service demands now outstrip the capabilities of firms to meet these demands. Even though there are improvements, they are insufficient. The development gap between the demand in terms of satisficing client expectations, meeting economic and policy needs with the associated project complexity and size where relevant, is increasing annually, as depicted in Figure 9.1.

Firms are not thinking in transformative ways; when met with a challenge, they address it within their current frame of reference. Even when the intention is to make successive step changes, the current industry and policy mindset frames the responses as small scale, which amount to no more than a single step, typically at the project level. It has been evident over successive generations of management

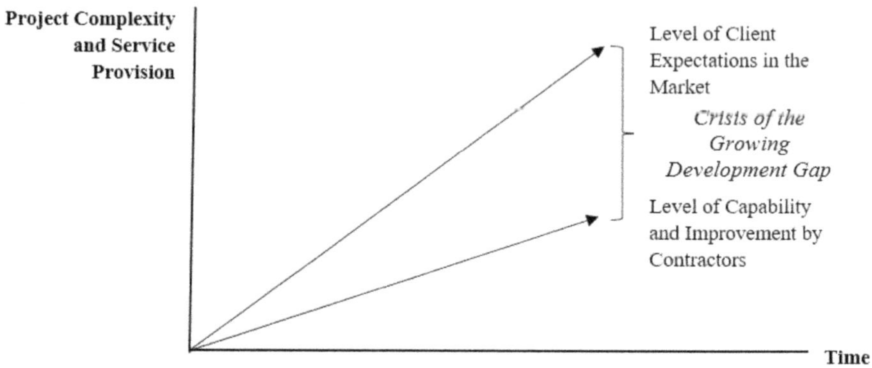

Figure 9.1 The development gap between client demands and contractor provision

and government reports (e.g. Green 2011; Smyth 2018; Murray and Langford 2003; Ive 1995).

What are some of the main actions incumbent management need to take to transform their organisations? There is not a single recipe, and part of the aim of transformation is to differentiate the firms; from this the range of value propositions in the marketplace will change. However, there are patterns of action which offer some guidance, and these are outlined in the following.

From focus on the management of the project towards a prime and proactive focus on the management of the firm

Contractors comprise more than a bundle of projects, and firm level management will need to become more strategic in their approach towards investment, developing capabilities and creating structures, systems and procedures that enable internal integration in the firm, and systems integration at the project level. This will include portfolio management and effective program management as well as project management. In sum, improvement needs support from the senior management team, and that includes support for portfolio and program management. Such improvements provide the basis for transforming the firm, yet also potentially support improvements at the operational level.

From short-term strategies of survival towards long-term strategies for growth

This represents a shift in corporate responsibility: recognising clients, supply chain members and employees as stakeholders having greater importance than hitherto, rather than giving preference to shareholders and the self-interest of top management. Financial management and business criteria will remain critical, but with a different focus that relates to investment and management for a broader range of priorities. Each firm will need to select their areas of focus for investment, based upon transformative strategic objectives and a business case in support of these objectives. Profitability flows from serving these broader business objectives, based upon a return on investment. In sum, a more service and client-oriented approach that is also employee-centred to acknowledge the strategic sources of revenue and the importance of social capital respectively is required.

From a transactional business model towards a transformational business model

Flowing from the previous point, this places priority upon performance improvement across a broader range of fronts and on productivity, which can only be improved where contractors selectively invest, including directly employing for critical specialisms and specialist services. The emphasis shifts from the current focus on the 'exchange', or more accurately on administering the series of exchanges or stage payments within the terms of the contracts, which absorbs the largest resources in project and construction management. It shifts towards value in service delivery and construction content in ways that yield benefits, impact and valuable outcomes in use for clients and users. This represents a shift from the dominant preoccupation with risk and contract administration. In sum, an investment-led approach is incrementally introduced to enhance the social capital with a greater emphasis upon technology to improve delivery and value in the use of construction output.

From cash flow management and the Return On Capital Employed (ROCE) towards being reliable payers and securing an adequate Return On Investment (ROI)

Moving from a reliance on trade credit on purchases and through delayed payments, coupled with investing surplus working capital outside the main business, to yielding a return from investments made in people, technical capabilities and knowledge, and in equipment and technologies to improve performance. The benefit to clients has to first become apparent in order to justify higher charges in bidding and at the final account stages. In sum, profit margins become a prime source for profit, coupled with improved reputation to drive up growth, hence market share and an increased level of profitability for the business as a whole.

Investment to differentiate service provision

Investment allocation needs to be focused on differentiated improvement; it is likely to be concentrated on areas of strength (and correcting some serious shortcomings on occasions), typically in incremental ways that successively build upon related areas that can be mutually reinforcing. The result will be differentiation between contractors, incorporating general contracting with specialisation, which helps manage the market and the ability to raise bid prices, while improving the value delivered to clients and end users. In sum, resource scarcity concentrates investment in areas of strength with the result of differentiating service provision from the competition and delivering valuable outcomes to the selected client base. Investment selection is part of differentiation, but the types of investment that can be undertaken include:

- robust knowledge transfer capability at program level that goes beyond the provision of an IT platform and includes management learning and transfer
- health and safety capabilities that extend beyond compliance to care for the workforce
- linking safety management systems based upon information to knowledge management systems for effective knowledge transfer
- client management capabilities and key account management functions at portfolio and program management levels
- specialist technical capabilities to reduce the carbon footprint or improve service provision
- concerted adoption of digital technologies along the lines of aviation and aerospace where the systems are (re)designed to optimise the technology application rather than grafted onto existing systems at suboptimal performance levels.

Strategic and fixed overarching management templates at an international level, cascading down to more agile and contextual operational strategies at the local level

High levels of autonomy and silo working help hold down transaction costs, whether between international offices, different divisions or business units as well as between firm and project, yet constrain improvement. Integrated strategies are needed for a consistent service and implementation of coherent differentiation. Overarching management templates help ensure consistency, while giving flexibility to adjust to regional cultures, where appropriate, operational contexts and client needs. In sum, flexibility and discretion to take the initiative is therefore nested under a standardised and consistent service design, and indeed helps frame action that at times will give rise to new lessons learned while maintaining a consistent service and construction quality at all scales of operation.

Investment to develop firm capabilities, especially at program management level

This is weak among contractors; repeat business clients tend to have more developed program management than main contractors. Contractor program management is essential for coordinating capabilities and activities to improve performance across projects; important areas include key account management and effective knowledge management. These require resourcing outside project budgets and require coordination to embed the capability and maintain engagement from the project level. Program management requires support from above, for example, human resource policies that hold personnel and line managers to account for all policies, which can comprehensively engage with knowledge management. Engagement between human resource policies and knowledge management might include guidance from training and coverage in personal development plans and annual reviews to ensure the capabilities are valued. In sum, capabilities need configuration and articulation in the firm and particularly through effective program management at the interface with projects.

Management capabilities can be introduced incrementally over successive years, each one potentially mutually reinforcing those introduced previously

Management capabilities hold the greatest opportunities for transformation over the short and medium term. This is especially the case for main contractors. Care needs to be taken as management capabilities are more challenging to initiate and maintain, yet are more easily introduced incrementally than high cost technologies. Mutually reinforcing improvement also carries the opportunity for iterative learning and readjustment to fine-tune the capabilities. The main conceptual benefit of management capabilities is that they are social capital, which accrues in value with use, whereas technologies and most fixed capital depreciates with use and over time. In sum, management capabilities are an obvious way forward because they fit the systems integrator role of the main contractor, support differentiation, and are more difficult for competitors to copy.

Change the basis of pay at senior management level and link to long-term performance of the firm

Incentives and bonus schemes are typically aligned to dividend payments, whereas rates of growth over longer periods are not only ignored but stymied by annual incentives. Shifting to longer term performance, say seven years, would change the situation. Encouraging managers to invest in shares can generally help. Specifically, bonuses can be retained and paid on future performance, or even better, government could introduce a new form of 7-year share, that cannot be sold, traded or licensed on any basis, for such payments across the board. In sum, longer term management horizons that consider the survival and growth of the firm are necessary, and, indeed, highly beneficial for both the firm and the economy.

Create awareness for the need for change across the industry ecosystem

Restructuring, which tends to be dramatic, and change management both carry risk for the firms and in the market and institutional level in the broader ecosystem of industry stakeholders. Therefore, awareness of the need for change needs to be built across institutions and informal networks. This includes government, as industrial policy maker, as well as major clients. Coverage also includes trade and professional bodies as well as clients and key lobby groups. The pace of awareness creation goes hand in hand with the demonstration effect, whereby the benefits of change become evident well after the changes are made. In sum, stakeholders need to be taken along on the journey, understanding both process and benefits.

Develop a program of educating investors on the need to switch from ROCE to ROI

Within the framework set out earlier, investors, especially shareholders, need to be made aware of the changes and, in particular, the potential benefits for long-term investors. This is critical to the initiation of change, which again will need to be supported subsequently by dividend payments flowing from higher margins and growth. This may lead to some short-term turbulence in share prices and thus contractors need to be in a reasonably sound financial position to start making the changes. Transformational change is not an option as a rescue plan for a failing company because any short-term share trading turbulence may prove too damaging. In sum, share investors need long-term outlets in which to invest and many investors carry long-term investments in their portfolios. Construction is becoming an even higher risk investment and this needs to change.

Create awareness among clients, particularly those in the public sector, of the incremental change

Public sector clients are the most important in terms of scale and scope. Procurement is highly decentralised and creating awareness of the needs for, and benefits of, change requires carefully managed campaigns. The public sector does not only need to be aware, but also need to buy into the process. Value has come more to the fore in many countries; in Denmark, for example, the lowest bidder on public contracts cannot be appointed, yet price remains dominant and will still impose a considerable discipline as the bid price gives some perceived certainty in an otherwise uncertain landscape, even though final account outturn costs often bear little relation to the bid price (e.g. Smyth 2015). It may be possible for the demonstration effect to be shown in private sector work if the public sector proves to be late adopters of the message. In sum, public sector clients will receive better value outputs and outcomes in terms of meeting policy and wider societal objectives, yet few will be impressed by the promise of 'benefits tomorrow' without both a thorough understanding of the benefits, and these being demonstrated as a result of changes made.

The first nine of the 12 management actions are largely inward facing, whereas the last three are largely concerned with external facing activity. These three management actions are critical to the success of the others. Should management need templates on which to plot their transition, there are useful tools, such as the business model canvas (Strategyzer 2019). The balanced scorecard (Kaplan and

Norton 1992) provides another potential starting point although it was designed more to monitor and refine existing practices than to facilitate transformational change. However, both the business model canvas and the balanced scorecard are useful for providing a framework rather than guiding management on what to do next. The purpose is to populate the boxes with strategic aims and objectives that set out what to do and how to do it – formulation and implementation. Achieving the aims and objectives will require new systems and procedures, adaptations of old ones, and new practices, so that capabilities are aligned and integrated so as to work effectively.

Discussion

There is not a singular prescription, which is itself an important point, as investment should both start where there is perceived strength and lead to differentiation because there are insufficient available resources to allocate to all potential areas of transformation (Penrose 1959). Let us take a few starting points by way of illustration and explore some ways in which transformation may be initiated so that changes are embedded within, and spread across, the organisation.

Let us first assume the procurement department has strong models for qualifying and selecting suppliers, linked to supply chain management. If the company is typical it will have a static approach to qualification and selection, whereby suppliers and, more importantly, subcontractors have to meet threshold criteria. Clients tend to be the same in this respect. The change comes in moving towards a dynamic set of qualification and selection criteria based upon rates of change. Here, firms first have to demonstrate how they are improving performance over successive projects, which puts paid to the 'cut and paste' practices which are prevalent among many main contractors and subcontractors, where declared improvements are seldom implemented (Smyth and Duryan 2019). Measurement is a challenge because attribution and intangible elements render measures hard to identify or establish. A more evidence-based approach is easier, using any available hard and soft evidence to make qualitative assessments. Any final assessment of the rate of change is therefore likely to be largely qualitative, perhaps using Likert scales that could also be weighted by the significance of recent improvements made among managers and to reflect any polls of customers and stakeholders. Significance could be assessed on a basis of both alignment to the firm's strategy for differentiation and tactical alignment with the specific service and content required by key clients.

Procurement, based upon rates of change, has the effect of not only driving transformation in the firm, but instigating it in the supply chain, especially among subcontractors. Once embedded, the transformation needs to be spread. How does performance link to business development, which typically, like procurement, is siloed? Business development could try to lever improved information about the value wanted by clients to inform tactical procurement decisions; this serves to not only move practice away from seeing value as inputs but seeking opportunities to improve the value delivered and assess the impact upon the organisational and end user solutions that the projects are trying to address (Smyth and Duryan 2019). This potentially spans the procurement-business boundary but can begin to cross other, and integrate over, organisational boundaries, including bid management and estimating as well as project management. The risk register can be extended to include the positive side – non-contractual commitments or promises to add value delivered in both service experience and content. This in turn would be handed to construction and project managers to ensure they deliver against the register. The process of change can also link to service design, and in this way, procurement begins the incremental development of new management capabilities. Indeed, any of these other functions can be used as a starting point for transformation.

A second capability that can act as a starting point is key account management (KAM), which has been partially adopted by many international contractors (e.g. Smyth 2015). Partial, for it has not been adequately resourced, embedded and used as a springboard for transformation. The aim is to manage the client, not just the project. This spans projects and needs resourcing at the program management level, which is typically poorly developed among contractors (e.g. Smyth 2015, 2018). KAM increases in value if linked to client or customer lifetime value (CLV), whereby the firm not only sees itself in terms of a pipeline of projects, but the amount of business that key clients are likely to give to the business. The lifetime may be assessed over a seven- or ten-year period, the estimated value being the CLV, while historic data yields the past CLV. As a guide, 20 percent of the clients provide 80 percent of turnover, hence a few clients can be managed carefully through KAM (Smyth 2015). Clearly the aim then becomes to increase the CLV through serving clients better. This can be achieved through, for example, service design, decisions on projects that are not just taken from a project budget viewpoint. Applying a CLV perspective, and by understanding the value projects provide in meeting the objectives of the client and end users, the value outcomes are potentially improved. This takes thinking and practices away from a sole or predominant focus on inputs *per se*, although the approach can be fed into improvements in tailoring value propositions (Smyth and Duryan 2019).

A third service capability can be focused upon health and safety (H&S), for example. Currently, H&S statistics have plateaued in most developed countries after successive decades of improvement (e.g. Smyth *et al.* 2019b). Therefore, some of the organisational behavioural characteristics evident in current business models need to be changed to move beyond marginal improvements to a substantial reduction in the number of incidents and accidents as well as improve wellbeing on site and in relation to people's quality of life (Smyth *et al.* 2019a, b). For example, most contractors have effective safety management systems (SMS) and they have partial knowledge management systems, which are not considered part of health and safety. SMS are typically robust, with high levels of engagement and effective mechanisms for sharing information, however, being able to convert the information systematically into knowledge for transfer and application is absent. Knowledge management systems are poorly supported at program management level with low levels of engagement (e.g. Duryan *et al.* 2020; Smyth and Duryan 2020). It therefore becomes logical to connect the two, and use the strengths of each respective system to improve H&S (Smyth *et al.* 2019a). More fundamental changes to organisational behaviour to reduce fatigue and stress will have positive H&S effects, challenging the current model of long working hours to yield improvements in wellbeing.

Conclusion

The transactional approach focuses upon inputs and requirements, costs and risks, and returns that use construction as a cash generation activity rather than a profitable activity in its own right. This is the basis of the approach among British contractors, who are a long way down the road towards failure. The trajectory is the same in other countries although the position along the continuum towards failure varies across nations. Demand has shifted and clients expect greater value, which they receive from every other sector. There is a development gap in construction between the shape of demand and what the leading contractors are able to deliver in their respective markets and along the supply chain. A more systematic and integrated approach is needed, which requires transformation and hence a change to the prevailing business models. There is little or no sign that the incumbents are responding and therefore their fate may be to be overtaken by disrupters and new entrants who change the management process and employ new approaches, typically linked to applying digital technologies.

There is no turning back. The evidence to date among main contractors is that there is a series of rigidities in firms. Rigidities are embedded in senior management mindsets, the lack of investment, the resultant lack of management and capabilities to transform. What does this mean for the future? Prediction is a dismal business, but there are indications in place. First, the growth of project management services among the built environment professions is one option that will act as an alternative to the main contractor as consultancies come to occupy the systems integrator function (Smyth 2018). Second is the rapid emergence of digital disruptors who are integrating supply chains that provide options for certain market segments by creating new integrated supply chains (Hall *et al.* 2019). These digitally driven organisations are offering differentiated alternatives but may even grow to configure multiple supply chain options. Third are new entrant construction companies that are agile and vertically integrated which are not wedded to a discipline or technology but use whatever works to configure and reconfigure in-house supply chains. Many of these entrants will have in-house specialisms to differentiate their value proposition and will be able to improve productivity because they employ and manage directly. These companies will also need to control the supply chain tightly in order that the service provision complies with the new business model.

In conclusion, if the tier 1 contractors do not reform themselves, they will become redundant, especially in countries such as the UK and USA where the business model is proven to be broken.

Notes

1 Heraclitus (c. 535–c. 475 BCE) Greek philosopher.
2 Benjamin Franklin (1706–1790) US statesman and polymath.
3 Will Rogers (1879–1935), American humourist, columnist, actor – date unknown.

References and further reading

Bakker, R.M., DeFillippi, R.J., Schwarb, A. and Sydow, J. (2016) Temporary Organizing: Promises Processes and Problems. *Organization Studies*, **37** (12), 1703–1719.

Bartlett, K., Blanco, J.L., Rockhill, D. and Strube, G. (2019) Breaking the Mold: The Construction Players of the Future. In: *Voices* (London: McKinsey & Co.), September.

BBC (2018) *Carillion Collapse to Cost Taxpayers £148m*. www.bbc.co.uk/news/business-44383224. Accessed 21 November 2019.

BBC (2019) Today Program. *BBC Radio 4*, 17 June 2019.

Building (2003) Top 100 Contractors and Housebuilders by Turnover: Industry Rankings. *Building Magazine*, 25 July.

Building (2011) Euro Contractors: Is the UK Market in Their Grip? *Building Magazine*, 21 January.

Carillion (2000) *Annual Report* (Wolverhampton: Carillion).

The Construction Index (2019) *Top 100 Construction Companies*. www.theconstructionindex.co.uk/market-data/top-100-construction-companies/2019. Accessed 12 August 2020.

Davies, A., Brady, T. and Hobday, M. (2007) Organizing for Solutions: Systems Seller vs. Systems Integrator. *Industrial Marketing Management*, **36**, 183–193.

Department of International Trade (2019) *Bidding for Overseas Contracts: What to Expect From 1 January 2021*. www.gov.uk/guidance/bidding-for-overseas-contracts-what-to-expect-if-theres-a-no-deal-brexit. Accessed 21 February 2020.

Duryan M., Smyth H., Roberts A., Rowlinson S., and Sherratt F. (2020) Knowledge transfer for occupational health and safety: Cultivating health and safety learning culture in construction firms. *Accid Anal Prev.* May;139:105496. doi: 10.1016/j.aap.2020.105496. Epub 2020 Mar 18. PMID: 32199157.

The Economist (2017) Least Improved. *The Economist*, 19 August, 55–56.

Egan, J. (1998) *Rethinking Construction* (London: HMSO).

ENR (2018) Top 250 Global Contractors. *Engineering News Record*. www.enr.com/toplists/2018-Top-250-Global-Contractors-3. Accessed 18 July 2019.

ENR (2019) Top 250 Global Contractors. *Engineering News Record*. www.enr.com/gdpr-policy?url=https%3A%2F%2Fwww.enr.com%2Ftoplists%2F2019-Top-250-Global-Contractors-1. Accessed 12 August 2020.

Financial Times (1979) Tarmac Starts to Fill in the Cracks. *Financial Times*, 9 November.

Green, S.D. (2011) *Making Sense of Construction Improvement* (Oxford: Wiley-Blackwell).

Hall, D.M., Whyte, J.K. and Lessing, J. (2019) Mirror-Breaking Strategies to Enable Digital Manufacturing in Silicon Valley Construction Firms: A Comparative Case Study. *Construction Management and Economics*, **34** (4), 322–339.

Hartnell, C., Hiddleston, K. and Westerman, P. (2010) *A Long, Hard Road Ahead: Business Model Change in the UK Property & Construction Sector* (Grant Thornton). www.grant-thornton.co.uk/pdf/property_report.pdf. Accessed 5 October 2015, no longer available online.

HM Government (2020) *The Construction Playbook: Government Guidance on Sourcing and Contracting Public Works Projects and Programs* (London: Cabinet Office, HMSO) https://bit.ly/3rz0kqR. Accessed 1 April 2021.

Ive, G. (1994) A Theory of Ownership Types Applied to the Construction Majors. *Construction Management and Economics*, **12** (4), 349–364.

Ive, G. (1995) The Client and the Construction Process: The Latham Report in Context. In: Gruneberg, S.L. (ed.) *Responding to Latham: The Views of the Construction Team* (Ascot: CIOB).

Kaplan, R.S. and Norton, D.P. (1992) The Balanced Scorecard: Measures That Drive Performance. *Harvard Business Review*, **70** (1), 71–79.

Latham, M. (1994) *Constructing the Team* (London: HMSO).

Lundin, R.A. and Söderholm, A. (1995) A Theory of the Temporary Organization. *Scandinavian Journal of Management*, **11** (4), 437–455.

Morrell, P. (2019) An Insane Business Model. *Construction Manager*, January. https://constructionmanagermagazine.com/carillions-insane-business-model. Accessed 6 April 2021.

Murray, M. and Langford, D. (eds.) (2003) *Construction Reports 1944–98* (Oxford: Blackwell).

Narayanan, V. and DeFillippi, R. (2012) The Influence of Strategic Context on Project Management Systems: A Senior Management Perspective. In: Williams, T. and Samset, K. (eds.) *Project Governance: Getting Investment Right* (Basingstoke: Palgrave Macmillan).

Packendorff, J. (1995) Inquiring into the Temporary Organization: New Directions for Project Management Research. *Scandinavian Journal of Management*, **11** (4), 319–333.

Parliamentary Report (2018) *Carillion Inquiry* (Business, Energy and Industrial Strategy and Work and Pensions Committees). www.parliament.uk/business/committees/committees-a-z/commons-select/work-and-pensions-committee/inquiries/parliament-2017/carillion-inquiry-17–19/publications/. Accessed 12 August 2020.

Penrose, E. (1959) *The Theory of the Growth of the Firm* (Oxford: Basil Blackwell).

Pountain, E. (1990) *Chairman's Statement*. Annual Report (Wolverhampton: Tarmac).

Rogers, R. (2018) Not-so-Sudden Death: How Carillion Disguised Its Ailing Finances Just Enough. *Construction Research and Innovation*, **9** (2), 44–47.

Simms, N. (1996) *Group Chief Executive's Review*. Annual Report (Wolverhampton: Tarmac).

Simon, E. (1944) *The Placing and Management of Building Contracts* (London: Ministry of Works and Planning, HMSO).

Smyth, H.J. (2010) Construction Industry Performance Improvement Programs: The UK Case of Demonstration Projects in the 'Continuous Improvement' Program. *Construction Management and Economics*, **28** (3), 255–270.

Smyth, H.J. (2015) *Market Management and Project Business Development* (Abingdon: Routledge).

Smyth, H.J. (2018) *Castles in the Air? The Evolution of British Main Contractors* (London: The Bartlett School). www.ucl.ac.uk/bartlett/construction/castles-in-the-air.

Smyth, H.J. and Duryan, M. (2019) Service Design and Knowledge Management in the Construction Supply Chain for an Infrastructure Program. *Built Environment and Property Asset Management*, **9** (1), 80–86.

Smyth, H.J. and Duryan, M. (2020) Knowledge Transfer in Supply Chains. In: Pryke, S.D. (ed.) *Successful Construction Supply Chain Management: Concepts and Case Studies* (Chapter 14, e-edition) (Chichester: Wiley Blackwell).

Smyth, H.J., Duryan, M. and Kusuma, I. (2019a) Service Design for Marketing in Construction: Tactical Implementation in Business Development Management. *Built Environment and Property Asset Management*, **9** (1), 87–99.

Smyth, H.J., Roberts, A., Duryan, M., Xu, J., Toli, M., Rowlinson, S. and Sherratt, F. (2019b) The Contrasting Approach of Contractors Operating in International Markets to the

Management of Wellbeing, Occupational Health and Safety. In: *Proceedings of the CIB World Building Congress* (Hong Kong: Hong Kong Polytechnic University), 17–21 June.
S&P (2018) Carillion's Demise: What's at Stake? *S&P Global Market Intelligence*, **23**.
Strategyzer (2019) *The Business Model Canvas*. www.strategyzer.com/canvas/business-model-canvas. Accessed 12 August 2020.
Tarmac (1989) *Annual Report* (Wolverhampton: Tarmac).
Tarmac (1995) *Annual Report* (Wolverhampton: Tarmac).
Tarmac (1999) *Interim Results* (Wolverhampton: Tarmac).
Wolstenholme, A. (2009) *Never Waste a Good Crisis: A Review of Progress on Rethinking Construction and Thoughts for Our Future* (London: Constructing Excellence).
WTO (2014) *Agreement on Government Procurement* (World Trade Organization). www.wto.org/english/tratop_e/gproc_e/gp_gpa_e.htm. Accessed 21 February 2020.

Editorial Comment

In industrialised ('developed') countries most business is conducted by business entities which can be collectively called 'firms'. The Collins English Dictionary defines a firm in this sense as 'an organization which sells or produces something, or which provides a service which people pay for'.[1]

There are a bewildering number of different sorts of firms around the world and while many have different names in different places, they are often very similar in nature. Conversely there are some types that are peculiar to one or perhaps a few countries.[2] These are some of the generic types:

- private company limited by shares
- public limited company
- limited liability partnership
- general partnership
- statutory corporation
- state-owned enterprise
- holding company
- subsidiary company
- sole proprietorship

Construction firms range from the very small, comprising just a single tradesperson (perhaps with a family member assisting with tasks such as invoicing and booking work) to very large multinational corporations which can comprise numerous subsidiary companies and have many thousands of employees under the one umbrella. Even larger are the consortia that are formed for very large projects that see two or more of these very large firms working together and, in turn contracting with a great many smaller firms that provide all manner of inputs to projects including materials, components, labour, supervision as well as design and other professional services.

Economists have put forward various theories in an effort to better understand how firms work. Teece (2016[3]) suggests that the theory of the firm is:

> *a general topic encompassing models that seek to answer a number of questions about firms, including why they exist, what determines their boundaries, how the differing interests of owners and managers can be aligned, how firms should be organized internally for efficiency and why firms differ.*

In the following chapter the author, who has written extensively on the topic of firms in the construction industry, reflects on the complex and diverse nature of such firms and distils much of what has been written to date regarding the nature of the construction industry (in all its complexity) and the firms that make up the industry and operate within and across it.

10 The measurement and characteristics of construction firms in theory and practice

Gerard de Valence

Introduction

There is an old joke that an economist is someone who, finding something that works in practice, wonders if it will work in theory. The joke is so old its origins have been lost in time, and it relies on the stereotype of economics as abstract and theory driven. Like all stereotypes this is partially true, models using the basic tools of economics, like supply and demand or preferences and expectations, are based on theory and tested, mostly successfully, every day. However, while economic theory is comprehensive it is not complete. There are topics with unresolved theoretical issues in macroeconomics, such as monetary policy and productivity, and there are topics, such as firms, where theory is notable for its absence.

It has been said there is no theory of the firm. Archibald (1987: 357) thought 'It is doubtful if there is yet general agreement among economists on the . . . 'theory of the firm' and the purpose of the theory of the firm 'is to investigate the behaviour of firms as it affects allocation and distribution' of resources. He then divided approaches to firms into 'optimizing' mathematical models and profit-seeking 'other' models, in particular Nelson and Winter's (1982) evolutionary theory of firms. He concluded that while there may not be *a* theory of the firm, there are many models that have been developed. Many of those models have been applied to construction firms by construction economists.

Machlup (1967) found a firm can be an organisation, a decision-making system, a collection of assets and liabilities, a legal entity, or a form of business unit, and concluded these concepts are all different and useful. However, the 'theory of the firm' as reviewed by Machlup (1967) is a theory of how production and cost functions interact with demand in the market. Microeconomic theory takes firms as a production function, and the treatment of the firm as a production function shifted the meaning of 'firm' towards the optimizing model of a production process, avoiding questions of position, definition, internal structure and external boundaries. Those questions were addressed by industry economists on a topic of shared interest: how economic activity is organised and studied (Schmalensee 1989). This is where topics such as imperfect competition and oligopolistic markets, auction theory, game theory, and buyer and supplier power are relevant, topics that cannot be discussed without reference to firms.

DOI: 10.1201/9781003102403-10

Industry economics is also concerned with the empirical problem of classifying firms and measuring industrial concentration. Firms are classified as belonging to specific industries on the basis of common characteristics in products, services, production processes and logistics (UN Statistical Division 2008). There are statistical and taxonomic issues with industry data, for example, a firm can have separate plants, or a factory might have a number of different production processes or products. Often the data cannot be disaggregated, is unrepresentative, or uncharacteristic of dynamic and evolving industry structures. In many cases data is not available or infrequent. There are major measurement problems in the treatment of quality changes in the inputs used and the capital intensity of factor usage. National statistical agencies do not use 'firm', but estimate the number and size of business units, enterprises, establishments and entrepreneurs, and the data is presented in different ways (Gruneberg 2019).

The precision of microeconomic models of optimizing firms is not possible in industry economics, partly because of the number of variables involved but mainly because of the lack of quality data. However, industry economics and microeconomic theory are both concerned with explaining why things happen: why are prices lower under one set of conditions than under another, why do some products succeed, why does a firm outsource some functions? Both view the type of market structure that links producers with suppliers and consumers as an important variable, but they differ in the scope of the variables they include (Carlton and Perloff 2005). Theory provides principles, which are the tools for defining problems, analysing issues and interpreting results. The analysis of firms, industries and markets requires the use of such tools, but the particular tools used depend on the theoretical framework chosen. Rodrick (2016) explains the role of the models that economists use. There are many valid models, and the challenge is to know which is applicable to a specific problem or issue. This is the main point of his book: different models have different applications, there are no wrong models, only badly and inappropriately applied models. Rodrick thus regards contemporary economics as a collection of models, not as a single grand theory (or a quest for one). The economist's craft lies in knowing which model is appropriate to the task at hand.

Construction firms present challenges on both sides of these theoretical and empirical issues, and the chapter is broadly divided into two main parts to reflect that. The chapter first surveys the data on construction firms from Europe and the United States (USA), which establishes the broad outline of an industry structure with many small firms and fewer large ones. This is followed by Australian data from five construction industry surveys done between 1986 and 2011, identifying trends and showing how changes in the number and type of firms have affected the structure of the industry, as the number of large firms significantly decreased. Based on the data presented there is a discussion on industry structure and concentration, which is the market share of the largest firms. More recent Australian data provides data by firm size based on employment, with micro firms fewer than 5, small firms 5–19, medium firms 20–199 and large firms more than 200 employees. Industry income and output shares by firm size and per employee are analysed.

The second half of the chapter discusses the theoretical issues, following developments in the treatment of firms in construction economics. Starting in the 1970s, there has been an evolving conception of both construction firms and the construction industry as the characteristics of the industry have been incorporated into economic models of firms and markets. The increasing sophistication of these models over the four decades to 2020 is detailed. The chapter also discusses the alternative approaches to firms found in relational contracting and the concept of hybrid firms. The important point these approaches raise is the boundaries of firms in a project-based industry with extensive subcontracting, an issue also addressed by transaction cost economics (TCE). The development of more complex models of firms since 2000 is then discussed, followed by the conclusion.

Construction is unusual as an industry, with a large number of small firms bidding for work in local and regional competitive markets, with little or no control over prices, coexisting with a small number of firms that, as main contractors and/or project managers, deliver large projects. Then there are the relatively few large national and international corporations in construction. Similar industry structures are found in retail, where there are a few large regional and sub-regional shopping malls and many widely distributed small shops, and health, with large hospitals, clinics and individual practices. In those industries, supermarkets use their buying power to drive down costs, and hospitals are often monopsony buyers in their region. Neither retail nor health, however, are project-based industries. Construction is a geographically distributed industry that brings together many suppliers in many different locations, and is different from other project-based industries like shipbuilding or energy generation which bring together many suppliers at a few locations. In construction, from project to project, temporary teams of clients, contractors, consultants and suppliers interact, bound by contracts between the legal entities that are the firms they represent. This is similar to a theatre or movie production, where a large number of different people and a wide range of different skills are involved at different times during development, production and distribution, as the end credits demonstrate.

Firm size and distribution in construction

The structure of an industry is the number of firms categorised by size, typically the number of employees. Firms are classified as small, medium or large, with the numbers used varying by country and industry, as the following tables show. Generally, small firms employ fewer than 20 people and have many working proprietors. Data on firms (often called enterprises in the statistics) is presented using the *International Standard Industrial Classification* (UN Statistical Division 2008). Section F in ISIC includes the complete construction of buildings (division 41), the complete construction of civil engineering works (division 42), and specialised construction activities or special trades, if carried out only as a part of the construction process (division 43). Also included is repair of buildings and engineering works. Although there are national variants on the Standard Industrial Classification (SIC) format there is also a great deal of commonality (for a detailed discussion of construction firms and the SIC see de Valence 2019). Economic activities are subdivided in a four-level structure. Activities are first divided into 'sections', which are alphabetically coded. These sections divide productive activities into broad groupings such as 'Agriculture, forestry and fishing' (A), 'Manufacturing' (C) and 'Information and communication' (J). The classification is then organised into numerically coded categories, which are two-digit divisions, three-digit subdivisions or groups, and four-digit classes with the greatest level of detail. SIC codes therefore represent industries, and firms are classified (or often self-classify) to industries on the basis of common characteristics in products, services, production processes and logistics.

As the data in the following tables shows, construction statistics include a large number of small firms. These firms include many tradespeople (working proprietors) engaged in the alteration, repair and maintenance of the built environment as well as contractors and subcontractors for new builds. The broad base of small firms is a distinctive feature of the overall construction industry as national statistical agencies define it. There is a long tail of small firms, typically family-owned businesses in construction trades doing repair and maintenance work. Although this is a well-known feature of construction there is considerable variation across countries. The most recent data from Europe, the USA, and Australia illustrates the point.

Eurostat compiles data from the 27 members of the European Union (EU) and publishes an annual *Construction of Buildings Statistics*. The most recent (Eurostat 2020) has firm size data for 2017 for employment, value added and 'apparent labour productivity' (their term for value added per employee), for the EU total (shown in Table 10.1). Eurostat micro firms have fewer than ten, small firms 10–49, medium firms 50–249 and large firms more than 250 employees.

The enterprise size structure of the construction of buildings sector is dominated by micro enterprises (employing fewer than ten persons). Almost 95 percent of all enterprises were categorized as micro enterprises, they employed 46.2 percent

Table 10.1 Key class size indicators, Construction of buildings (NACE Division 41), EU-27, 2017 (Eurostat 2020)

	Number of enterprises	*Number of persons employed*	*Value added*	*Apparent labour productivity*
	(thousands)		*(EUR million)*	*(EUR thousand per head)*
All enterprises	799.8	2 894.2	117 614.3	40.6
All SMEs	799.4	2 584.7	95 982.2	37.1
Micro	753.1	1 337.0	42 600.5	31.9
Small	41.4	794.5	31 842.3	40.1
Medium-sized	4.9	453.2	21 539.4	47.5
Large	0.4	309.5	21 632.1	69.9

Note: The sum of all categories does not equal the total of all enterprises due to estimated values with lower reliability

of the persons employed in the EU-27's construction of buildings sector in 2017 and accounted for 36.2 percent of its value added. While the 400 large enterprises (employing 250 or more persons) contributed 10.7 percent of the employment, they generated 18.4 percent of total value added, and thereby recorded the apparent labour productivity of EUR 69.9 thousand per person employed. (Eurostat 2020: 6)

Eurostat also provides this data for each of the individual countries, table 10.2, which shows significant regional variation:

Micro and small enterprises collectively employed three quarters of the construction of buildings persons in 2017 for EU total and more than half in nearly all

Table 10.2 Number of persons employed by enterprise size class, Construction of buildings (NACE Division 41), EU-27, 2017 (Eurostat 2020)

	Total (thousands)	*SMEs*	*Micro*	*Small (% of total)*	*Medium-sized*	*Large*
EU - 27	2 894.2	89.3	46.2	27.5	15.7	10.7
EU - 28	3 342.9	87.4	45.3	26.7	15.4	12.6
Belgium	83.6	90.5	50.7	19.7	20.1	9.5
Bulgaria	62.2	95.9	20.3	40.0	35.6	4.1
Czechia	89.7	92.8	43.5	31.8	17.4	7.2
Denmark	26.1	75.4	18.0	28.2	29.2	24.6
Germany	301.7	88.2	21.3	43.1	23.8	11.8
Estonia	17.3	56.3	56.3	-	-	-
Ireland	36.3	93.0	56.7	25.7	10.6	7.0
Greece	31.7	85.1	61.7	23.3	-	-
Spain	490.6	97.8	68.6	22.5	6.7	2.2
France	208.7	51.5	23.3	16.5	11.7	48.5
Croatia	37.9	91.8	37.5	33.0	21.3	8.2
Italy	300.5	98.1	64.0	25.9	8.2	1.9
Cyprus	11.4	100.0	43.1	28.7	28.2	0.0
Latvia	23.3	96.2	32.3	39.0	24.9	3.8
Lithuania	41.7	92.4	24.3	41.3	26.8	7.6
Luxembourg	12.9	53.5	13.6	-	39.9	17.7
Hungary	62.3	84.1	48.3	35.7	-	-
Malta	3.5	66.7	35.4	-	31.4	-
Netherlands	139.4	90.4	55.6	17.7	17.0	9.6
Austria	67.6	72.2	14.1	28.6	29.5	27.8
Poland	286.7	91.4	57.7	18.3	15.4	8.6
Portugal	147.1	95.5	50.7	31.1	13.7	4.5
Romania	172.4	94.1	33.4	36.6	24.1	5.9
Slovenia	13.8	77.0	41.6	35.4	-	-
Slovakia	35.6	96.1	59.9	23.7	12.5	3.9
Finland	77.0	84.9	38.8	32.3	13.8	15.1
Sweden	113.2	74.4	31.1	25.9	17.5	25.6
United Kingdom	448.6	75.1	39.2	21.9	14.1	24.9
Iceland	7.1	91.1	44.3	28.9	18.0	8.9
Norway	87.1	85.7	36.0	33.6	16.0	14 3
Switzerland	83.8	76.7	11.6	23.2	42.0	23.3

(-) not available

Table 10.3 US Construction 2012 (US Census Bureau 2012; Table 2)

Enterprise employment size	*Number of enterprises*	*Sales or revenue USD 1,000,000*	*Annual payroll USD 1,000,000*	*Number of paid employees*
All enterprises	581,601	1,349,346	260,606	5,006,131
Fewer than 100 employees	576,272	812,924	154,461	3,336,286
100–499 employees	4,788	226,818	46,899	817,823
500–999 employees	na	82,320	14,787	222,481
1,000–2,499 employees	141	79,475	14,968	211,141
2,500–4,999 employees	45	62,749	10,516	145,875
5,000–9,999 employees	17	38,072	7,497	113,133
10,000 employees or more	9	46,988	11,476	159,392

na = not available due to sampling issues

of the EU Member States, the exceptions were Denmark, France and Austria. The share of the persons employed in large enterprises was less than one-quarter in most Member States, the exceptions being France, Austria and Sweden. In value added terms, the contribution of large enterprises reached 55.2percent in France, but was less than 10.0percent in Bulgaria, Ireland, Spain, Italy, Cyprus, Portugal, Romania and Slovakia. (Eurostat 2020: 7)

In the USA the Census Bureau collects data on industries and enterprises, the latest being 2012. The website has this notice: 'Due to limited resources and competing priorities of critical programs within the Census Bureau, the Enterprise Statistics Program has been suspended.' Reflecting the scale of the American economy, the size range of firms is much greater than the EU and the largest firms much larger, as in Table 10.3. Again, over 95 percent of US firms are small, in this case with fewer than 100 employees, and have on average five or six employees. However, there were 212 firms with 1,000 or more employees that had a total 630,000 employees, of which nearly 160,000 were employed by the nine largest firms.

Australian construction industry surveys

There have been five construction industry surveys (CIS) by the Australian Bureau of Statistics (ABS), the most recent is for 2011–12. All five surveys found the construction industry is overwhelmingly made up of small firms which contribute most of the industry's output and account for almost all of the number of enterprises. Table 10.4 shows the breakup between contractors in *Building and Engineering* and the subcontractors in *Construction services* (which were called trades in the earlier surveys). The 2002–03 survey used different categories of businesses (not establishments) in residential, non-residential and non-building, and trade services and is not comparable with the other surveys. In 2002–03 there were 339,981 businesses of which 269,228 were trade services and 70,753 were residential, non-residential and non-building businesses

How the size of firms is measured in the CIS has changed twice. The three surveys in 1996–97, 1988–89, and 1984–85 divided firms into three sizes: employ

Table 10.4 Construction, number of firms by industry subdivision'000.

	1984–85	*1988–89*	*1996–97*	*2011–12*
Building construction	24.5	19.6	33.1	31.3
Engineering construction	3.4	3.9	3.1	5.7
Construction services	77	74.5	158	173

Source: ABS Construction Industry Survey. Australian Bureau of Statistics, Cat. No. 8772.0.1996–97, 1988–89 and 1984–85 Private Sector Construction Establishments: Number operating during the year. 2011–12 Number of establishments at end June

Table 10.5 Construction firm size and number of operating businesses

Selected indicators	*Employment fewer than 5*	*Employment 5 to 19*	*Employment 20+*
Operating businesses	93.67	5.71	0.62
Employment	68.62	17.74	13.63
Wages and salaries	39.39	28.24	32.38
Total income	48.13	23.55	28.32
Operating profit before tax	74.42	11.28	14.30
Industry gross product	53.50	22.14	24.36

fewer than 5, employ 5–19, and employ 20 or more. The 2011–12 survey divided firms into small 0–19, medium 20–199 and large with over 200 employees. The 2002–03 survey divided firms by income and the data cannot be compared to the other surveys, however, although income was used to classify firms, the 2002–03 survey produced a similar result. The ABS found: 'The construction industry was characterised by a large number of very small businesses. During 2002–03, 64.7% (219,926) of construction businesses earned income less than AUD 100,000 and a further 25.3% (86,035) earned income between AUD 100,000 to less than AUD 500,000' (ABS 2013). Using the different metric of income, it was found 90 percent of firms were small or very small. Here the 1996–97 survey and the 2001–12 survey data is presented. The breakup of firms by size is in Table 10.5.

In the 1996–97 survey businesses with fewer than five employees accounted for 94 percent of all businesses and over two-thirds of all employees. Less than 1 percent of businesses employed 20 or more. Businesses with fewer than five employees accounted for slightly less than half the total income and expenses, whereas businesses with employment of 20 or more accounted for almost one-third of these. The data for 1996–97 is in Table 10.6 converted to percentages, showing the importance of the 0.62 percent of large firms. Their 13.6 percent of employees earned 32.3 percent of salaries and wages, generated over 28 percent of income and nearly 25 percent of gross output.

The survey in 2011–12 classified firms by the number of employees into small 0–19, medium 20–199 and large with over 200. The same data for the 2011–12 survey is in Table 10.7. The changes between 1996 and 2012 are revealing. The total number of firms has increased marginally from 195,000 to 210,000, but the share of small firms has increased from 94 to 98 percent as the number of medium

Table 10.6 Percent of total construction by firm size, 1996–97

	Small 0–19 persons	*Medium 20–199 persons*	*Large 200 or more persons*
Businesses	97.7	2.2	0.1
Employment	62.1	19.3	18.6
Wages and salaries	37.9	26.3	35.8
Income	49.0	23.6	27.3
Operating profit before tax	74.0	17.4	8.6
Industry value added	51.6	22.9	25.5

Source: ABS 2011–12. *Construction Industry Survey*. Australian Bureau of Statistics, Cat. No. 8772.0

Table 10.7 Percent of total construction by firm size 2011–12

Construction	*Perfect competition*	*Monopolistic competition*	*Oligopoly*
Subcontractors	Labour based subcontracting	Mechanical services(HVAC), demolition	Lifts, building automation
Contractors	Many small and medium contractors	Some medium-sized contractors	Large main contractors

and large firms fell from 12,300 to fewer than 5,000. There was a trend with the number of medium-sized firms decreasing to less than half, while slightly increasing their share of industry employment.

In 2011–12, fewer than 0.1 percent of firms were large, employing 18.6 percent of the workforce, paying 32 percent of wages and salaries and generating 27 percent of industry income and 25 percent of output. These figures are remarkably similar to the 1996–97 CIS numbers, however, the 186 large firms in 2011–12 had almost the same share of employment, income and output that 1,200 firms had in 1996–97. This was a significant increase in industry concentration. In the 1996 survey the 1,200 firms employing 20 or more had a total of 66,000 employees and accounted for 13.6 percent of employment and 24.4 percent of industry output. In 2012, there were 186 firms employing 200 or more with 177,000 employees, accounting for 18.6 percent of employment and 25.5 percent of IVA. These long-run changes in industry structure can only be the result of a long wave of mergers and acquisitions reducing the number of firms while increasing concentration and reducing competition, and should be of particular interest to major clients and regulators.

Australian industry data

This data is provided in the Australian Bureau of Statistics annual publication *Australian Industry* (ABS 8155), produced annually using a combination of directly collected data from the annual Economic Activity Survey conducted by the ABS, and Business Activity Statement data provided by businesses to the Australian Taxation Office. The data includes all operating business entities and government owned or controlled Public Non-Financial Corporations. *Australian Industry* excludes the finance industry and public sector, but includes non-profits in industries like health and education and government businesses providing water, sewerage and drainage services. The selected industries included account for around two-thirds of GDP. Excluded are ANZSIC Subdivisions 62 Finance, 63 Insurance and superannuation funds, 64 Auxiliary finance and insurance services, 75 Public administration, and 76 Defence. The most recent issue is for 2018–19.

The analysis is based on industry value added (IVA) and industry employment. IVA is the estimate of an industry's output and its contribution to gross domestic product (GDP) and is broadly the difference between the industry's total income and total expenses. IVA is given in current dollars in *Australian Industry*. The data is presented at varying levels for industry divisions, subdivisions and classes, but unfortunately does not include the number of firms. Micro firms have fewer than 5 employees, small firms 5–19, medium firms 20–199 and large firms more than 200 employees.

Figure 10.1 shows the shares of the indicators used in the comparison of the CIS, with similar results. Large firms have 15 percent of employment, 30 percent of wages and salaries and 23 percent of output, not too dissimilar to the CIS data. Medium firms have 18 percent of employment, 27 percent of wages and salaries and 21 percent of output, again not too dissimilar to the CIS data. Micro and small

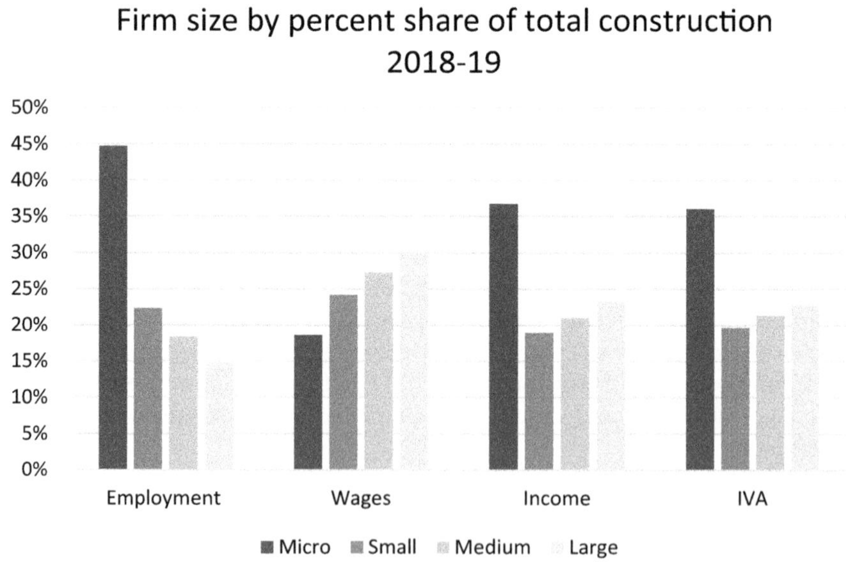

Figure 10.1 Construction firms 2018–19

firms account for approximately 65 percent of employment but only 55 percent of output, thus explaining the problem of the long-run low growth rate of productivity in construction (de Valence and Abbott 2015). Figure 10.2 shows large firms have twice the level of output per employee, measured as IVA per employee, and medium firms nearly 50 percent more. This is an imperfect but useful proxy for productivity. There is no significant difference between micro and small firms, but large firms have twice the added value per employee of micro and small firms. Medium-sized firms are in the middle.

Firms leverage the capital on their balance sheet to maximise revenue and profits. For firms in construction markets annual revenue is the aggregated income from current work, or contracts won but not completed. Construction firms and contracts range widely in duration, size and value, but the amount of work a firm can take on must be related to the capital a firm has available. This relationship between firm size and the annual value of contracts or projects undertaken is based on the assumption that construction firms seek to maximise revenue but are constrained by their working capital. In construction the contract packages reflect the complexity of work, so there is a wide range of contract sizes. Construction contracts can, therefore, be arranged on the basis of contract size and complexity. This is a well-known and widely agreed characteristic of the industry, with the relationship first investigated in the 1980s when Flanagan and Norman (1982) found competing contractors' bids were affected by the type of project and by the value range. Small bidders considered both contract type and size, and large bidders were more successful when bidding for large contracts. Contract size and complexity were also important for Hillebrandt (2000), who argued the wide range of contract sizes in the construction market is the major determinant of the number of firms able to undertake the work. Male (1991) had a 'project-based vertical market defined by project size and complexity', where

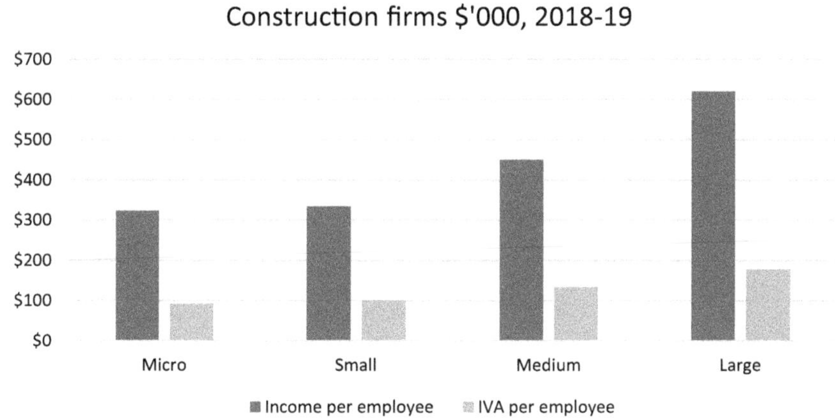

Figure 10.2 Annual income and industry value added per employee

going up the vertical market left fewer companies able to undertake particular types of projects.

Construction has a large number of small firms bidding for work in competitive markets with little or no control over prices. There are a smaller number of firms that can deliver large projects in a given region, and there are a few dozen multinational corporations in construction. Construction economics has a wide range of views on the types of markets these firms operate in and their competitive behaviour (reviewed in de Valence 2011). There is, however, universal agreement that construction is an industry of projects, and firms operate in markets for projects of many different types. Skitmore (1991) found bidders for construction contracts should not be considered as homogeneous or standardised, Drew and Skitmore (1992) concluded consistently bidding for specific types of construction work is a successful strategy, and Drew and Skitmore (1997) found that differences in competitiveness are greater for different contract sizes than for different types of contract. Larger firms have more depth, more resources, and more technical and financial capacity therefore, as Low and Lau concluded (2019: 31). 'The construction industry can also be structured by size of contract and degree of complexity. Large construction firms usually undertake large contracts with a high degree of complexity.'

The relationship between firm size and IVA per employee is not surprising, large firms are typically better managed than small firms (Bloom and van Reenen 2010). Management was identified by Hillebrandt and Cannon (1990) as the most important determinant of the capacity and capability of construction firms, because managerial skills give a contractor greater flexibility. How firms utilise their capabilities differentiates them within a diverse, location-based production system. It is widely recognised there are differences between industries in the way that production is organised and new technology adopted, adapted and applied (Nelson and Winter 1982), but differences within industries generally get less attention (Andrews *et al.* 2015). Important differences are the individual characteristics of

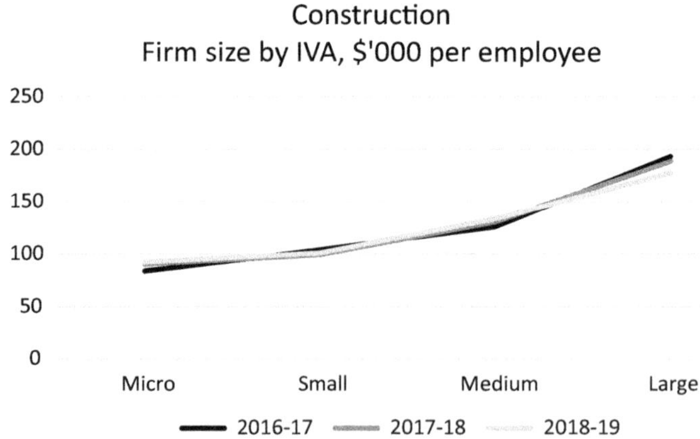

Figure 10.3 Firm size and industry value added per employee

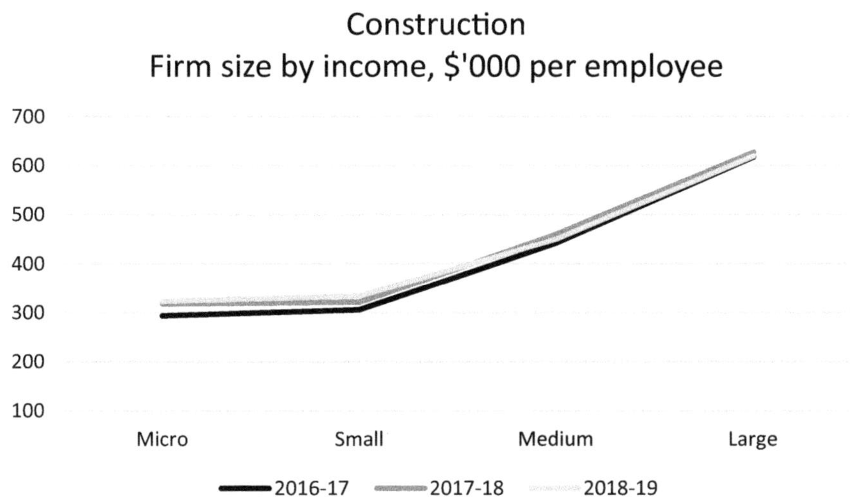

Figure 10.4 Firm size and income per employee

firms such as their size, the effects of competitive dynamics, and how the adoption of new technology by one company in an industry influences the adoption of technology by other companies in that industry (Dosi 1982). For building and construction this is significant, not only because of the number of small and medium-sized firms, but because of the size and reach of the major firms.

The relationship between firm size and contract value is a fundamental reality in construction, and is also the foundation of the relationship between projects and firms. A firm is a legal entity, and the typical reporting period is one year. A firm's income is the cumulative cash flow of their portfolio of projects over a year. The focus on projects and construction management in construction research can miss the role of firms as the ongoing participants in the industry.

Industry structure and concentration

The physical characteristics of location and projects mean construction is organised into local and regional markets around supply and availability of the resources needed to deliver projects for local clients. Cooke (1996: 138) describes the industry as 'dominated by a large number of small firms' and 'geographically fragmented', and Male (2003: 135) thought the level of fragmentation was a direct result of the sophistication and complexity of technology used in commercial buildings, the vagaries and variability of demand and the consequent increasing trend towards specialisation, subcontracting and self-employment. These characteristics were addressed in the earliest research on the organisation of construction. Stinchcombe (1959) contrasted bureaucratic and craft systems of work administration: manufacturing has mass production with economies of scale through standardisation of tasks, but construction uses standardised products and parts. In craft production work administration and control is given to workers and foremen, but they do not make decisions on product type, design and price, which are made by others, variously referred to as administrators, bureaucrats, clients and employers. Stinchcombe argued bureaucratic administration requires long production runs and predictable work-flow, while uncertainty and variability in work-flow will make subcontracting and the craft system more efficient.

Viewing the construction industry as predominantly made up of small firms supports the view of the industry as fragmented with the characteristics of perfect competition (e.g. Runeson 2000). There are parts of the industry that fit the perfect competition model, the small- and medium-sized contractors that rely on low-bid tendering to get work and labour based subcontractors, such as formworkers, steel fixers, bricklayers and concreters. There are few significant barriers to entry to the construction industry for small firms, and such barriers will continue to be low while the industry maintains current practices based on a large number of small, specialised subcontractors. These firms compete on the basis of price, so labour-intensive subcontractors and small contractors can be assumed to operate under perfect competition. There are, however, a limited number of contractors capable of managing large projects, and the barriers to entry at this level in the form of prequalification are significant, based on track record, financial capacity and technical capability (de Valence 2007). The data, which emphasises the number of firms, is deceptive:

> *Because of the very large number of small firms, the entire industry is often characterized as unconcentrated. That description is too broad, however, because not all construction companies do overlapping work and some segments are much less fragmented than others. For example, a limited number of general contractors are capable of managing the very large projects, whereas there are a great many small subcontractors. Competition among large general contractors and among specialty firms seems to be oligopolistic, while rivalry among small contractors who do basic labour tends to be closer to perfect competition . . . when relevant markets are defined, as opposed to considering the whole industry, competition is often limited because many firms are specialized or cannot compete on large projects . . . limited competition and substantial entry barriers can facilitate many different types of anticompetitive conduct, including unilateral and horizontal varieties. In addition, procurement procedures for construction projects are often conducive to collusion.*
>
> (OECD 2008: 9)

Construction is an industry where there is evidence of concentration (McCloughan 2004), and significant barriers to entry exist due to prequalification systems and capability requirements used by clients to select contractors for major projects (Ezulike *et al.* 1997). Oligopolistic competition focuses on competition through product differentiation, or in the case of building and construction through specialisation in particular types of projects (e.g. bridges, high-rise), forms of procurement (e.g. design and build, negotiated work), finance and PFI type projects, or relationships with clients (such as alliancing or partnering), or by region. Suppliers of glass facades, lifts and building automation systems are also oligopolistic because there are few manufacturers of these products. Between these two market structures there are some firms in the industry that are in monopolistic competition. These are medium-sized contractors that have specialised and differentiated their products and services or have developed ongoing relationships with clients and thus get negotiated work (de Valence 2003).

Ofori (1990) recognised there are many heterogenous small firms in the construction industry, and discusses (and provides data for) Singapore, the USA, Canada and Western Europe. He concludes 'whereas the construction industry has a pyramid structure, the distribution of its workload takes the form of an inverted pyramid. In other words, the industry is relatively concentrated. The small firms are generally uncommitted, transient, undercapitalised, have poor access to credit, operate within limited geographical areas, and seldom apply modern management tools' (Ofori 1990: 77). Langford and Male argued 'the construction industry is first, comprised of geographically dispersed and overlapping project-based market structures and second, is hierarchically structured in terms of company size. Fragmentation is high at the smaller end of the industrial structure, for example in repair and maintenance work, whilst for new-build work fragmentation decreases according to project characteristics, including those with an international dimension. Entry and exit barriers to the industry are many but exist often in a subtle form.' (Langford and Male 2001: 26).

The construction industry has its own organisational and institutional features. For Bresnen and Marshall (2001) the industry is a mature, project-based industry with complex professional and organisational contractual and working relationships. The industry is geographically distributed, causing significant horizontal and vertical differentiation within construction firms, with potential for uncoupling between project activities and organisational strategies. Dubois and Gadde (2002) describe a decentralised, project-based structure as a loosely coupled system, where temporary project teams come and go, combining members from a broad range of firms and industries as required. Moreover, the context is one of wider networks containing many small- and medium-sized firms in localised subcontracting markets (Gruneberg and Ive 2000).

Although Bresnen (1990) discusses a range of organisational and institutional relationships where external contracting is common, he sees large contractors delivering major projects as the core of the construction industry's system of production at the end of the twentieth century. By this stage the production system had a clear outline, and a clear structure, for bringing together the products, suppliers and materials needed for building and engineering projects, and had stabilised around particular forms of procuring, financing and managing those projects. Construction materials like cement, concrete and glass, and components like building management systems, interior walls, plumbing fixtures, glass facades, lifts and elevators are all oligopolistic industries in a mature supply chain (Syverson 2008).

The construction industry of today is the outcome of a long development path over the twentieth century. It has a production system that has been developing since the introduction of steam engines over 200 years ago, followed by successive developments in iron, glass, steel and concrete technologies. Many of the industry's global leaders are well-established, Bechtel for example is over 100 years old, and other firms like Hochtief, Skanska, and AECOM can trace their origin stories back over a similar period. Shimizu is over 200 years old. From an industry life-cycle perspective (Hughes 1989), the modern industry is in the late stage of competition and consolidation, where successful firms survive and thrive, and gain both market share and market power over time (Syverson 2019). Consolidation of an industry leads to concentration in a few firms.

A stylised representation of construction industry firms by market type is in Table 10.8, showing how concentrated markets can be the outcome of either firm size or specialisation. Figure 5 relates market type to contract size. As a firm gets larger it takes on bigger projects and competes with fewer other firms. How construction economists sought to reconcile theoretical and conceptual models of

Table 10.8 Construction firms by market type (de Valence 2011)

Construction	*Perfect competition*	*Monopolistic competition*	*Oligopoly*
Subcontractors Contractors	Labour based subcontracting Many small and medium contractors	Mechanical services (HVAC), demolition Some medium-sized contractors	Lifts, building automation Large head contractors

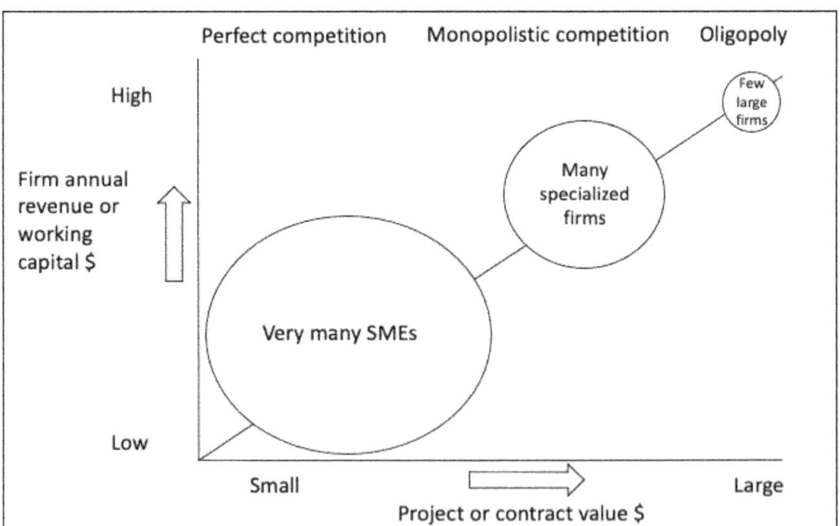

Figure 10.5 Large firms deliver bigger projects and compete with fewer other firms

construction firms with the messy reality of the construction industry is discussed in the next section.

Development of conceptual models of construction firms

In an economics textbook the model of a firm is sparse and lacks detail. Firms are optimisers of scarce resources constrained by technological capacity. Based on cost and demand functions a firm maximises profit as a 'black box' that mysteriously but efficiently turns inputs into outputs as its contribution to total industry output. In introductory texts this neoclassical model is presented graphically as the set of choices a firm faces to minimise marginal cost and maximise profit based on marginal revenue. Firms here are price takers, they are small relative to the market, they do not have market power, and their products are homogeneous (i.e. the same).

Hillebrandt (1974) used this model of a firm in chapters on costs, demand and markets. However, Hillebrandt ranged widely, with alternative views on the objectives of the firm (growth, revenue, managers' incentives), costs for project-based firms, revenue curves and mark-ups, product differentiation, and the effects of different types of markets included. That discussion incorporated the characteristics of construction, based on Hillebrandt's familiarity with the British industry, and led to two key conclusions. The first was perfect competition due to ease of entry. Even with a limited, selected number of tenderers there is 'effective competition', with the same outcome as perfect competition. The short discussion of imperfect competition (pp. 136–38) is conventional and construction is not mentioned. The second conclusion was that marginal analysis is appropriate for project-based firms. This model of the firm can also be found in Briscoe (1988) and Runeson (2000) – they, like Hillebrandt, attempted to reconcile neoclassical economics with construction industry characteristics.

At the end of the 1980s three books taking radically different approaches to the industry and the firm appeared. Ball (1988) gives a Marxist analysis of British construction, focusing on social relations and the contracting system, and claims 'One theoretical avenue which seems of little use in studying the industry is to apply neoclassical economic theories of the firm' (1988: 19). The dual role given to firms as producers (of buildings) and merchants (purchasing inputs) foreshadowed the trade credit/cash farming literature, with Ball arguing the merchanting role predominates at the expense of wages and productivity.

By contrast, Bon's *Building as an Economic Process* (1989) applied Austrian economics to construction. This branch of economics, a polar opposite to Marxism, is relevant to construction because it emphasises capital and the capital stock, the pivotal role of investment in capital formation, the explicit role of the time taken for investment decisions to be fulfilled, the cyclic nature of economic and building activity, and the possibility of production plans or projects failing. Firms here are vehicles for investment, preparing plans and sourcing the capital required, and the building process is a series of decisions on the use of capital. The book strongly linked construction (supply) to the property market (demand) through capital flows and the need for ongoing repair and maintenance, and Bon suggested this relationship should be the basis of a research agenda in construction economics (CE). However, much of that research is now in the atheoretical area of life-cycle costs, and Bon's book remains the only use of Austrian economics in CE.

Hillebrandt and Cannon (1989) argue in their book that it is 'inherently difficult to relate the economic structure, behaviour and performance of contracting firms to theoretical models' (1989: 6). Their alternative is managerial economics (on firm decision-making) and management theory (on business strategy, organisation theory and human resources), explored in seven of the eight other contributions to the book. This illustrates the large grey area between construction management and construction economics, where topics like these are of mutual interest and cannot be considered from an economic viewpoint without reference to industry custom and practice.

Thus, at the end of the 1980s, there were four distinctly different concepts of construction firms, supported by detailed analysis of the industry in Britain, where these researchers worked. These were: Neoclassical firms in competitive contracting markets; Marxist firms acting as producers and merchants; Austrian firms circulating capital between construction and property; and Managerial firms organised for construction.

Alternatives to firms

Coase (1937: 388–390) argued that the firm is an organisation, rather than just a production function. He separated the market from the firm with the 'price mechanism' on one hand and its 'supersession' on the other. For Coase, the alternative to the firm was the coordination of self-employed individual producers by the market, each being his or her 'own master'. In construction and other project-based industries the coordination of individual producers is known as subcontracting and is the responsibility of the main contractor and their project manager.

However, subcontractors are often not engaged in a single transaction, as in the market-based trades of instant exchange and settlement envisaged in economics textbooks. Richardson (1972) identified relationships between firms as important, Hughes (1983) saw production as an organised network of firms, and Goldberg (1980) and Dore (1983) argued the relationship between a large corporation and its subcontractors is typically more durable and intensive than a market relationship. The idea of 'relational contracting' had firms develop long-term ties with subcontractors, often with mutual understanding and trust that are not typical of markets. Instead of using the market, the firm will rely on a trusted supplier, especially when their relationship involves shared knowledge and learning (see Gil 2010 for these issues in a major construction project).

There are also 'hybrid' concepts such as the 'quasifirm' as developed by Eccles (1981) as a response to Stinchcombe (1959), who considered cases in the construction industry where 'relations between the general contractor and his subcontractors are stable and continuous over fairly long periods of time and only infrequently established through competitive bidding. This type of 'quasi-integration' results in what I call the 'quasifirm'.' (Eccles 1981: 339–340). He argued this 'stable and continuous' contracting relationship is not just a form of relational exchange along the lines of Richardson, Goldberg and Dore aforementioned, subcontracting is an 'interface' between the market and the firm. Both relational exchange or quasifirms can be used to describe aspects of subcontracting, Eccles applied it to construction, and Cheung (1983) also used building work as one of his three examples of transactions. Lai (2000) analysed whether a construction project is a market or hierarchy in some detail, in the context of subcontracting as the interaction of firms in a market. His view is that this interaction is largely contractual, following Masten (1991) who forcefully argued for the importance of the firm as a legal entity.

A contract defines the boundary of a firm because it creates a legal person with unambiguous identity. Hodgson (2002: 26) concludes, 'A firm is defined as an integrated and durable organisation involving two or more people, acting openly or tacitly as a 'legal person', capable of owning assets, set up for the purpose of producing goods or services'. He then asks 'why do so many economists evade the obvious, everyday, legally grounded, definition of the firm?' and argues that it is because economists frame the analysis of the firm and the market in universal, ahistorical and relatively de- institutionalised terms. If the boundaries of the firm are indistinct, alternatives like 'internal markets' within firms, the 'quasifirm', 'hybrid firms' and firms as 'quasi-markets' are possible. However, as Lai (2000) and Hodgson (2002) argue, there is no good reason to abandon the formal, legal conception of firms because of relational contracting, networking, subcontracting or other developments. Firms can act strategically, and thus will sometimes form coalitions, and some relationships will be more durable than others.

Ive and Gruneberg argued that subcontractor markets exist in a separate and distinct way from contractor markets: 'firms engaged in each stage of the process of production of the built environment . . . compete directly (actually or potentially) with one another, and thus constitute an industry. The firms of other stages in the process stand not as competitors but as suppliers or buyers from that industry' (Ive and Gruneberg 2000: 7–8). Repeated interaction does not a relationship make. In de Valence (2015) this argument was extended, suggesting a market is created by the main contractor as they go through the subcontracting process. The paper introduced the idea that procurement of subcontractors for a project creates an identifiable, though temporary, market for goods and services. Such a market has distinctive characteristics that make it different from other markets, because subcontractors are a hierarchically organised network of autonomous producers. This is a short-lived market characterised by an imbalance in bargaining power between subcontractors bidding for work and contractors acting as auctioneers, to the potential detriment of both subcontractors and clients.

Development of new models of firms

Issues with the model of firms with perfect information, constant returns to scale, and no market power created a broad research agenda in economics. Two topics in that agenda crossed over and were applied to construction firms in the 1990s. The first was on the boundaries of the firm and transaction cost economics (TCE) based on the work of Coase (1937) and Williamson (1975), who described firms as a 'nexus of contracts' and claimed 'Any issue that arises as or can be reformulated as a contracting issue can be examined to advantage in transaction cost economizing terms' (Williamson 2000: 599). First applied to construction by Winch (1989), TCE offered a mechanism to analyse construction procurement methods (Ive and Chang 2007), conflicts of interests among contracting parties (Li *et al.* 2013), hidden costs associated with pre- and post-contract work (Walker and Wing 1999), and uncertainty and risk (Chang 2013).

TCE became an active research stream (reviewed by Aziz 2021). Firms minimise transaction costs by choosing either internal production or external supply, the make-buy decision investigated by Murray and Kulakov (2019) for international construction firms. This added another explanation for subcontracting to the flexibility and minimizing fixed costs explanation already established by Stinchcombe (1959), where specialisation by subcontractors results in lower cost of supply under specific contract conditions. The extension of TCE to construction introduced issues like the hold-up problem on required investment (Winch 2006), incentives and contracts (Ive and Chang 2007), and reframed information asymmetry between participants as a principal-agent problem (Cerić 2014).

In their book on construction firms Gruneberg and Ive (2000) included a chapter on TCE. They also added to the Hillebrandt and Cannon (1989) managerial model of the firm different types of markets, industry capacity, productivity, decision-making under uncertainty and models of pricing, cost and investment. Construction firms were differentiated by specialisation, size and growth rates. Firms manage portfolios of projects in markets that have barriers to entry and can become concentrated. This is a considerably more complex model of construction firms. Their Prologue concludes 'The aim of this book is to give a clear understanding of some of the economic issues directly confronting construction firms in their operations and provide the economic basis for planning and decision making' (2000: xvii). However, while they follow Hillebrandt and Cannon (1989) on management decisions, they also include discussion on capital circuits and the social structure of accumulation, which builds on Ball (1988).

By including a range of firm-specific and industry-specific factors in pricing decisions, and the subsequent profit margin, Gruneberg and Ive (2000) focused on how average costs at the firm level are marked up, constrained by factors such as union power, sales and promotion activities, the degree of concentration in the market, the level of overheads to be covered and the prices of competing products. Another important distinction introduced by Gruneberg and Ive is the difference in pricing behaviour between large construction firms and small firms, which they describe as near-firms or micro-firms. The large firms are oligopolistic, an important point, but small firms operate under conditions that closely resemble those of perfect competition and are 'constrained by a chronic shortage of operating financial capital' (Gruneberg and Ive 2000: 225).

Research on firms and industry trends became more focused as topics like small and medium enterprises (SMEs) in construction, innovation, and megaprojects became active research streams in the 2000s. That research broadly showed industry trends such as use of BIM, integrated design and manufacture, and framework procurement agreements are driven by the largest firms, and that separation between local, national and international firms and markets was growing ever wider as globalisation gathered momentum in the new millennium (Runeson and de Valence 2009). Over the 1990s there had been increased interest in international contracting and contractors (e.g. Ive 1990; Low 1991; Crosthwaite 1998). Alliances in international construction became a focus of research: Badger and Mulligan (1995) and Bing *et al.* (1999) addressed why alliances are formed, types and benefits of international alliances, trends in global construction and risks in international markets. Kangari and Sillars (1997) looked at Japanese construction alliances. Norwood and Mansfield (1999) studied construction companies in international joint ventures in the Asian market. Ofori (2003) recognised an 'international construction system' where firms chose markets based on firm and national competitive advantages. Firm advantages include name and size, national advantages are related to distance to the market and historical, social and economic relationships. Brockman (2009) differentiated international, multinational and global markets and firms, with international firms competing outside their domestic market. Over these years the perspective on the industry shifted from a fragmented system of production to one where a significant proportion of the value of work done is coordinated by the small number of large firms that deliver large projects.

That new view of the industry underpinned Gruneberg and Francis' *The Economics of Construction* (2019), which provided 'a game theory account of the behaviour of firms', the second crossover from economics. In the 1990s game theory became the dominant approach in industry economics (see Tirole 1988; Carlton and Perloff 2005). They discuss aspects of firms' business models, financing, contractual disputes and power relations at greater length than Hillebrandt, building on the research on construction firms done over the previous three decades. There is also the use of case studies of the collapse of UK contractor Carillion in 2018, Grenfell Tower, construction for the London Olympics and manufactured housing. These illustrate how the business environment a construction firm faces has become significantly more complex over the decades. Hillebrandt's turnover and profit maximizing firm in the 1970s has evolved into one primarily concerned with growth and survival in the 2020s. While that may be a matter of degree, it is not insignificant. Gruneberg and Francis argue contracting markets compete profits down to the point where firms cannot invest in productivity improvements. According to Hillebrandt, prices, costs and profits for a project were determined by a conventional marginal analysis, producing an equilibrium result. In Gruneberg and Francis the last two chapters point to an emerging field of research on the economics of construction projects, combining project financial and feasibility studies with procurement strategies. Construction firms operate in an industry Gruneberg and Francis describe as 'a highly fragmented project-based industry, with very low profit margins and a high risk of failure for the many firms operating in a very complex supply chain'.

From the fragmented perfectly competitive industry of the 1970s, by 2020 the model of firms in the construction industry is a hierarchical supply chain managed by main contractors and project managers, reflected in the system of tiered suppliers, from oligopolistic tier 1 contractors down through layers of monopolistic competition where firms compete among themselves, to a deep layer of small firms in perfectly competitive markets. The industry maintains key elements of hierarchical governance through contracts between tiers of subcontractors and suppliers, optimizing supply chains through standardisation of parts and products. Some firms act variously as contractors, subcontractors, designers or consultants, depending on circumstance. As this shows, there is no clear boundary where discussion of firms crosses over to discussion of an industry, reflecting the grey areas in the relationship between microeconomics and industry economics.

How an industry organises production fundamentally determines the number and nature of firms. Topics like construction innovation and productivity are industry characteristics that cannot be sensibly discussed without reference to the number and size of firms and the structure of the industry. What the idea of a fragmented industry of mainly small firms misses is the importance of the large firms in the industry. From an industry life-cycle perspective (Hughes 1989), the industry is in the late stages of competition and consolidation, where successful firms survive and thrive, and gain both market share and market power over time (Syverson 2019).

Conclusion

The chapter started with data on firm size and industry structure from the EU and USA, showing the general pattern of many small firms and few large ones. Australian data provided more detail, tracking firm numbers through five surveys showed a significant decrease in the number of large firms while their share of industry income was more or less unchanged. This reveals the increasing concentration of the industry between 1996 and 2012. The relationship between firm size and contract value is a fundamental reality in construction and is also the foundation of the relationship between projects and firms. Some aspects of that were explored using industry data from the Australian Bureau of Statistics, showing a clear relationship between firm size and income. Large firms also have twice the level of IVA per employee compared to small and micro firms.

The number, fragmentation and diversity of construction firms make generalizing about their financing, management, behaviour and strategy challenging. To deal with this, researchers investigating aspects of construction often divide the industry into categories such as Tier 1, 2, or 3 firms, subcontractors or consultants, or into market segments like residential building or engineering construction. Alternatively, firms can be assumed to be homogeneous in character as contractors in a project-based industry, while differing in scale and size. Both these approaches have strengths and weaknesses, as the theoretical debates about the nature of construction firms demonstrate. The chapter also finds classifying construction firms by size instead of industry sectors resolves some of the issues raised in relation to the structure and conduct of the construction industry, in a way that division into sectors based on the type of project does not.

This chapter argues that firms are economic agents, whose behaviour is determined by the type of market they compete in. The relationship between firm size and the value of contracts or projects undertaken is used to determine the number of competitors and market type the firm competes in, based on the assumption that construction firms seek to maximise revenue but are constrained by their working capital. The characteristics of large firms in an oligopolistic market, the characteristics of a medium-sized construction firm in monopolistic competition with similar firms, and of a small firm in a perfectly competitive market are identified. The chapter presents a stylised view of firms and places them in the context of contracting markets for projects. Data supporting the empirical relationship between firm size, revenue and value added is presented and analysed.

As this chapter shows, there is no clear boundary where discussion of firms crosses over to discussion of an industry, reflecting the grey areas in the relationship between microeconomics and industry economics. How an industry organises production fundamentally determines the number and nature of firms. Topics like construction pricing, innovation and productivity are industry characteristics that cannot be sensibly discussed without reference to the number and size of firms and the structure of the industry. What the idea of a fragmented industry misses is the importance of large firms in an industry where they typically might be fewer than 1 percent of the number of firms. Construction statistics include a large number of small firms, typically family-owned subcontractors in the trades or engaged in the alteration, repair and maintenance of the built environment. The broad base of small firms is a distinctive feature of the overall construction industry as national statistical agencies define it. There is a long tail of small firms, a few large ones, and even fewer international contractors. However, as Edgerton (2007) notes, old technologies survive long after innovations that were claimed to replace them arrived, such as the telegraph, fax machine and vinyl records with telephones, email and CDs respectively. Stone, brick and wood have been widely used materials for millennia, and industrialised materials like corrugated iron, glass and concrete are ubiquitous. Because a large part of construction work is maintaining and repairing the existing stock of buildings and structures, current skills, technologies and materials will continue to be used by these small firms and there will be continuing demand for their services.

The fact that the construction industry of today is the outcome of a long development path is an important characteristic of the industry. A small group of international contractors have become incumbents in global markets for large projects. Many local and regional markets also have incumbents that are large relative to their subcontractors. For construction this is significant, because these major firms have the management and financial resources required to invest in new products and processes, and, in particular, technological capability, that small firms do not. The issue may be the ability of firms to capture knowledge externalities, adopt new techniques, and adapt to the impact of both emerging technologies and new entrants (Brynjolfsson *et al.* 2019).

Bridging the divide between the empirical data available on firms and theoretical models of firms is difficult. Microeconomic models of firms optimizing along a production function are theoretically strong and work well, other models are less precise. Bröchner (2011: 20) argued 'current theories of industrial organization should be fruitful' when applied by construction economists. These are the models of firms industry economics employs to explain their characteristics and operation, the structure of supplier and client markets, the forms of competition and cooperation between firms, and the effects of economies of scale and scope on industries. Each individual industry has a distribution in the number and size of firms that is specific to that industry, determined by its particular system of production, so a generalised production function may not explain the characteristics and operation of those firms. This becomes a more significant issue if firms are not in a perfectly competitive market, as assumed by microeconomic theory, such as concentrated and oligopolistic markets.

Crossing the divide between data on construction firms and theoretical models of those firms took several decades. The first model was Hillebrandt (1974), which was firmly based on microeconomic theory and applied the production function method to construction firms, applying a standard economic model that proved to be very durable, and appeared to be suitable to an industry made up of small firms in perfectly competitive markets. However, the data that revealed the fragmentation of the construction industry also showed the existence of a relatively small number of large firms. To explain their characteristics required other models from outside microeconomic theory. As discussed earlier, by the end of the 1980s managerial, Marxist and Austrian models of construction firms had been developed, and the concepts of quasi-firms and hybrids had been proposed. Then in the 1990s, TCE was applied to construction and the key theoretical issue of the boundaries of firms in contracting markets. It became increasingly apparent research on important industry topics cannot be done without addressing the issue of the role of large firms: examples are small firms and supply chains, international contractors and megaprojects, auctions and procurement, and innovation and productivity. Through investigating the empirical and theoretical issues these topics raise, a better understanding of the nature of these complex interactions can be found.

An optimizing firm is a theoretical mathematical construct, often described as a 'black box' of some unknown set of resources, routines and knowledge inside a firm required for production. Ideally, such a firm is a profit maximiser (setting output) competing in efficient markets (setting prices), and from that perspective how the firm maximises profit is not that important. Reality is more complex. Many firms go bust or get taken over, successful firms grow but with limits, industries have life cycles of decades and products life cycles of much less. Patterns of demand and supply ebb and flow over time, as the economy goes through short-run business cycles and long-run structural change. Most firms have finite lives, but some survive and thrive. How and why that is the case has been the subject of extensive research in industry economics in general and in construction economics in particular. There may not be much theory, but there is no shortage of topics and issues to investigate.

Notes

1 https://www.collinsdictionary.com/dictionary/english/firm
2 Wikipedia (2021) *List of Legal Entities by Country.* https://en.wikipedia.org/wiki/List_of_legal_entity_types_by_country.
3 Teece, D.J. (2016) Theory of the Firm. In: Augier, M. and Teece, D. (eds.) *The Palgrave Encyclopedia of Strategic Management* (London: Palgrave Macmillan). https://doi.org/10.1057/978-1-349-94848-2_568-1.

References and further reading

ABS (2013) *8772.0 – Private Sector Construction Industry, Australia, 2002–03* (Canberra: Australian Bureau of Statistics). www.abs.gov.au/AUSSTATS/abs@.nsf/DetailsPage/8772.02002-03?OpenDocument.

Andrews, D., Criscuolo, C. and Gal, P.N. (2015) *Frontier Firms, Technology Diffusion and Public Policy: Micro Evidence From OECD Countries*. OECD Productivity Working Papers, 2015–02 (Paris: OECD Publishing).

Archibald, G.C. (1987) Firm, Theory of. In: Eatwell, J., Millgate, M. and Newman, P. (eds.) *Palgrave Dictionary of Economics* (pp. 357–362) (London: Macmillan).

Aziz, A. (2021) Applications of Mainstream Economic Theories to the Construction Industry: Transaction Costs. In: Ofori, G. (ed.) *Research Companion on Construction Economics* (Cheltenham: Edward Elgar).

Badger, W.W. and Mulligan, D.E. (1995) Rationale and Benefits Associated With Alliances. *Journal of Construction Engineering and Management*, **121** (2), 100–111.

Ball, M. (1988) *Rebuilding Construction: Economic Change and the British Construction Industry* (London: Routledge).

Bing, L., Chew, D.A.S., Fan, W.W. and Tiong, R.L.K. (1999) Risk Management in International Construction Joint Ventures. *Journal of Construction Engineering and Management*, **125** (4), 277–284.

Bloom, N. and Van Reenen, J. (2010) Why Do Management Practices Differ Across Firms and Countries? *Journal of Economic Perspectives*, **24** (1), 203–224.

Bon, R. (1989) *Building as an Economic Process: An Introduction to Building Economics* (Englewood Cliffs, NJ: Prentice Hall).

Bresnen, M. (1990) *Organizing Construction: Project Organization and Matrix Management* (London: Routledge).

Bresnen, M. and Marshall, N. (2001) Understanding the Diffusion and Application of New Management Ideas in Construction. *Engineering Construction and Architectural Management*, **8** (6), 335–345.

Briscoe, G. (1988) *The Economics of the Construction Industry* (London: Mitchell).

Bröchner, J. (2011) Developing Construction Economics as Industry Economics. In: de Valence, G. (ed.) *Modern Construction Economics* (Abingdon: Spon Press).

Brockman, C. (2009) Global Construction Markets and Contractors. In: Ruddock, L. (ed.) *Economics for the Modern Built Environment* (Abingdon: Routledge).

Brynjolfsson, E., Rock, D. and Syverson, C. (2019) Artificial Intelligence and the Modern Productivity Paradox: A Clash of Expectations and Statistics. In: Agrawal, A., Gans, J. and Goldfarb, A. (eds.) *The Economics of Artificial Intelligence: An Agenda* (London: The University of Chicago Press).

Carlton, D. and Perloff, J. (2005) *Modern Industrial Organisation* (4th edition) (New York: Harper Collins).

Cerić, A. (2014) Strategies for Minimizing Information Asymmetries in Construction Projects: Project Managers' Perceptions. *Journal of Business Economics and Management*, **15** (3), 424–440.

Chang, C.-Y. (2013) A Critical Review of the Application of TCE in the Interpretation of Risk Allocation in PPP Contracts. *Construction Management and Economics*, **31** (2), 99–103.

Cheung, S.N.S. (1983) The Contractual Nature of the Firm. *Journal of Law and Economics*, **26** (2), 1–21.

Coase, R.H. (1937) The Nature of the Firm. *Economica*, **4**, 386–405.

Cooke, A.J. (1996) *Economics and Construction* (London: Macmillan).

Crosthwaite, D. (1998) The Internationalisation of British Construction Companies 1990–96: An Empirical Analysis. *Construction Management and Economics*, **16** (4), 389–395.

de Valence, G. (2003) *Market Structure, Barriers to Entry and Competition in Construction Markets*. CIB W55/W65/W107 Conference, Singapore, 22–24 October, 819–828.

de Valence, G. (2007) The Significance of Barriers to Entry in the Construction Industry. *Australian Journal of Construction Economics and Building*, **7** (1), 29–37.

de Valence, G. (2010) Innovation, Procurement and Construction Industry Development. *Australasian Journal of Construction Economics and Building*, **10** (4), 50–59.

de Valence, G. (2011) Market Types and Construction Markets. In: de Valence, G. (ed.) *Modern Construction Economics* (London: Taylor & Francis).

de Valence, G. (2015) *Characteristics of a Project as a Temporary Micro-Market*, COBRA *Conference*, Sydney, 8–10 July.

de Valence, G. (2019) Comparing Construction in National Industrial Classification Systems. In: Best, R. and Meikle, J. (eds.) *Accounting for Construction: Frameworks, Productivity, Cost and Performance* (Abingdon: Routledge).

de Valence, G. and Abbott, M. (2015) A Review of the Theory and Measurement Techniques of Productivity in the Construction Industry. In: Best, R. and Meikle, J. (eds.) *Measuring Construction: Prices, Output and Productivity* (London: Taylor & Francis).

Dore, R. (1983) Goodwill and the Spirit of Market Capitalism. *British Journal of Sociology*, **34** (4), 459–482.

Dosi, G. (1982) Technological Paradigms and Technological Trajectories: A Suggested Interpretation of the Determinants and Directions of Technical Change. *Research Policy*, **11** (3), 147–162.

Drew, D.S. (2011) Competing in Construction Auctions: A Theoretical Perspective. In: de Valence, G. (ed.) *Modern Construction Economics: Theory and Application* (Abingdon: Spon).

Drew, D.S. and Skitmore, M. (1992) Competitiveness in Bidding: A Consultant's Perspective. *Construction Management & Economics*, **10** (3), 227–247.

Drew, D.S. and Skitmore, M. (1997) The Effect of Contract Type and Size on Competitiveness in Bidding. *Construction Management & Economics*, **15** (5), 469–489.

Dubois, A. and Gadde, L.-E. (2002) The Construction Industry as a Loosely Coupled System: Implications for Productivity and Innovation. *Construction Management and Economics*, **20** (7), 621–631.

Dyer, D. and Kagel, J.H. (1996) Bidding in Common Value Auctions: How the Commercial Construction Industry Corrects for the Winner's Curse. *Management Science*, **42**, 1463–1465.

Eccles, R.G. (1981) The Quasifirm in the Construction Industry. *Journal of Economic Behavior & Organization*, **2** (4), 335–357.

Edgerton, D. (2007) *The Shock of the Old: Technology and Global History Since 1900* (Oxford: Oxford University Press).

Eurostat (2020) *Construction of Buildings Statistics – NACE Rev 2*. https://ec.europa.eu/eurostat/statistics-explained/pdfscache/16257.pdf.

Ezulike, E., Perry, J. and Hawwash, K. (1997) The Barriers to Entry Into the PFI Market. *Engineering, Construction and Architectural Management*, **4** (3), 179–193.

Flanagan, R. and Norman, G. (1982) An Examination of the Tendering Pattern of Individual Building Contractors. *Building Technology and Management*, **28**, 25–28, April.

Gil, N. (2010) Developing Cooperative Project Client-Supplier Relationships: How Much to Expect from Relational Contracts. *California Management Review*, 144–169, Winter.

Goldberg, V.P. (1980) Relational Exchange: Economics and Complex Contracts. *American Behavioral Scientist*, **23** (3), 337–352.

Gruneberg, S.L. (ed.) (2019) *Global Construction Data* (Abingdon: Routledge).

Gruneberg, S.L. and Francis, N. (2019) *The Economics of Construction* (London: Agenda Publishing).

Gruneberg, S.L. and Ive, G.J. (2000) *The Economics of the Modern Construction Firm* (London: Macmillan).

Hillebrandt, P. (2000) *Economic Theory in the Construction Industry* (3rd edition) (Basingstoke: MacMillan).

Hillebrandt, P.M. (1974) *Economic Theory and the Construction Industry* (Basingstoke: Macmillan).

Hillebrandt, P.M. and Cannon, J. (1989) *The Management of Construction Firms: Aspects of Theory* (London: Macmillan).

Hillebrandt, P.M. and Cannon, J. (1990) *The Modern Construction Firm* (London: Macmillan).

Hodgson, G. (2002) The Legal Nature of the Firm and the Myth of the Firm-Market Hybrid. *International Journal of the Economics of Business*, **9** (1), 37–60.

Hughes, T.P. (1983) *Networks of Power: Electrification in Western Society, 1880–1930* (Baltimore: Johns Hopkins University Press).

Hughes, T.P. (1989) *American Genesis: A Century of Invention and Technological Enthusiasm 1870–1970* (Chicago: University of Chicago Press).

Ive, G. (1990) Structures and Strategies: An Approach Towards International Comparison of Industrial Structures and Corporate Strategies in the Construction Industries of Advanced Capitalist Societies. *Habitat International*, **14** (2–3), 45–58.

Ive, G. (1994) A Theory of Ownership Types Applied to the Construction Majors. *Construction Management and Economics*, **12** (4), 349–364.

Ive, G. and Chang, C.Y. (2007) The Principle of Inconsistent Trinity in the Selection of Procurement Systems. *Construction Management and Economics*, **25** (7), 677–690.

Ive, G. and Gruneberg, S. (2000) *The Economics of the Modern Construction Sector* (London: Macmillan).

Kangari, R. and Sillars, D.N. (1997) Japanese Construction Alliances. *Journal of Construction Engineering and Management*, **123** (2), 146–152.

Lai, L.W.C. (2000) The Coasian Market-Firm Dichotomy and Subcontracting in the Construction Industry. *Construction Management & Economics*, **18** (3), 355–362.

Langford, D. and Male, S. (2001) *Strategic Management in Construction* (2nd edition) (Malden, MA: Blackwell Science).

Li, H., Arditi, D. and Wang, A. (2013) Factors That Affect Transaction Costs in Construction Projects. *Journal of Construction Engineering and Management*, **139** (1), 60–68.

Low, S.P. (1991) World Markets in Construction: II. A Country-by-Country Analysis. *Construction Management and Economics*, **9** (1), 73–78.
Low, S.P. and Lau, S.H. (2019) *Construction Quality and the Economy: A Study at the Firm Level* (Singapore: Springer Nature).
Machlup, F. (1967) Theories of the Firm: Marginalist, Behavioral, Managerial. *American Economic Review*, **57** (1), 1–33.
Male, S. (1991) Strategic Management in Construction: Conceptual Foundations. In: Male, S. and Stocks, R. (eds.) *Competitive advantage in Construction* (Oxford: Butterworth–Heinemann).
Male, S. (2003) Faster Building for Commerce: NEDO!988. In: Murray, M. and Langford, D. (eds.) *Construction Reports 1944–98* (Oxford: Blackwell).
Masten, S.E. (1991) A Legal Basis for the Firm. In: Williamson, O.E. and Winter, S.G. (eds.) *The Nature of the Firm: Origins, Evolution, and Development* (Oxford and New York: Oxford University Press).
McAfee, R.P. and McMillan, J. (1987) Auctions and Bidding. *Journal of Economic Literature*, **25**, 699–738.
McCloughan, P. (2004) Construction Sector Concentration: Evidence from Britain. *Construction Management and Economics*, **22**, 979–990.
Murray, A. and Kulakov, A. (2019) Make/Buy Decisions in International Construction Firms. In: Gruneberg, S. (ed.) *Global Construction Data* (London: Taylor & Francis).
Nelson, R. and Winter, S. (1982) *An Evolutionary Theory of Economic Change* (Cambridge, MA: Harvard University Press).
Norwood, S. and Mansfield, N.R. (1999) Joint Venture Issues Concerning European and Asian Construction Markets of the 1990s. *International Journal of Project Management*, **17** (2), 89–93.
OECD (2008) *Competition in the Construction Industry*. www.oecd.org/daf/competition/cartels/41765075.pdf.
Ofori, G. (2003) Frameworks for Analysing International Construction. *Construction Management & Economics*, **21** (4), 379–391.
Ofori, G. (1990) *The Construction Industry: Aspects of Its Economics and Management* (Singapore: Singapore University Press).
Richardson, G.B. (1972) The Organisation of Industry. *Economic Journal*, **82**, 883–896.
Rodrick, D. (2016) *Economics Rules: The Rights and Wrongs of the Dismal Science* (New York: W.W. Norton).
Runeson, G. (2000) *Building Economics* (Geelong: Deakin University Press).
Runeson, G. and de Valence, G. (2009) The New Construction Industry. In: Ruddock, L. (ed.) *Economics for the Modern Built Environment* (Oxford: Taylor and Francis).
Schmalensee, R. (1989) Industrial Economics: An overview. In: Oswald, A.J. (ed.) *Surveys in Economics* (Vol. 2) (Oxford: Blackwell).
Skitmore, M. (1991) The Construction Contract Bidder Homogeneity Assumption: An Empirical Test. *Construction Management & Economics*, **9** (5), 403–429.
Stinchcombe, A.L. (1959) Bureaucratic and Craft Administration of Production: A Comparative Study. *Administrative Science Quarterly*, **4**, 168–187.
Syverson, C. (2008) Markets: Ready-Mixed Concrete. *Journal of Economic Perspectives*, **22** (1), 217–233.
Syverson, C. (2019) Macroeconomics and Market Power: Context, Implications, and Open Questions. *Journal of Economic Perspectives*, **33** (3), 23–43.
Tirole, J. (1988) *The Theory of Industry Organization* (Cambridge, MA: MIT Press).

UN Statistical Division (2008) *Standard Industrial Classification of All Economic Activities* (New York: United Nations).

US Census Bureau (2012) *Enterprise Statistics*. www.census.gov/econ/esp/2012/esp2012.html.

Vickery, W. (1961) Counterspeculation, Auctions and Competitive Sealed Tenders. *Journal of Finance*, **16**, 8–37.

Walker, A. and Wing, C.K. (1999) The Relationship Between Construction Project Management Theory and Transaction Cost Economics. *Engineering, Construction and Architectural Management*, **6** (2), 166–176.

WEF/BCG (2016) *Shaping the Future of Construction: A Breakthrough in Mindset and Technology* (Geneva: World Economic Forum and the Boston Consulting Group).

Williamson, O.E. (1975) *Markets and Hierarchies: Analysis and Antitrust Implications* (New York: The Free Press).

Williamson, O.E. (2000) The New Institutional Economics: Taking Stock, Looking Ahead. *Journal of Economics Literature*, **38** (3), 595–613.

Winch, G. (1989) The Construction Firm and the Construction Project: A Transaction Cost Approach. *Construction Management & Economics*, **7**, 331–345.

Winch, G. (2006) Towards a Theory of Construction as Production by Projects. *Building Research & Information*, **34** (2), 154–163.

Editorial Comment

It's a big mess, frankly, and it is a problem that has been coming for a number of years (Colley, cited in de Freytas-Tamura 2018).

The big mess was the collapse of Carillion, the UK's second largest construction firm with debts well in excess of USD 1 billion. The failure of any construction firm, whether it is a giant like Carillion or a small cottage builder, has serious, often devastating effects on individuals who lose their jobs and other businesses who are left with unpaid accounts. Clients face expensive delays and the major headache of having to find another contractor who is willing to take over a project that has been partly completed by someone else with the potential of uncovering and having to deal with latent defects. Construction is a risky business at the best of times, with typically low profit margins; inevitably clients will pay a premium as potential replacement contractors factor the increased risk into their bids.

Carillion's failure had far-reaching effects as it employed thousands of people in the UK and abroad and had major linkages to many other businesses as well as to the UK government. In Colley's view the problem had been 'coming for a number of years', so why was action not taken sooner to correct the trajectory that the company was on before it reached the point of no return?

11 Using financial concepts to understand failing construction contractors

Alex Murray and Alexandr Kulakov

Introduction

Looking at the past to predict the future supports the operation of our economies, everything from forecasting crop yields to calculating insurance premiums. As firms grow and look to grow ever larger, they typically make recourse to external finance, the terms for which will be highly dependent on how the firm has performed in the past. This will involve investigation of the firm's success in winning work, controlling cost and ultimately how profitable it has been to date. The more historic evidence a firm has on its performance, the more likely it is to convince providers of credit to extend their beliefs[1] that good performance will continue, and that they can share in those fortunes. A firm's financial accounts form a central pillar in the information set used to judge their future prospects.

This chapter will look at the plight of a small set of large UK construction contractors who failed in recent years. Before presenting analysis on key concepts from firm accounts, the chapter will discuss the definition of firm failure and consider what the literature provides on insolvency prediction modelling (IPM). Following this, we will consider the particular characteristics of construction contracting firms to assess the specific risks their business models create in contrast to other types of firms. Discussion covers the related insolvency risks of subcontracting strategies, particularly for large construction firms who subcontract much of their overall gross output. This chapter does not present an IPM or combine multiple financial concepts to predict failure. Rather its contributions are in providing what many IPM studies lack – deeper exploration of the economic insight of financial ratios for the firm they represent.

Consideration of Carillion's demise will serve as a benchmark for a number of other UK contracting firms. The analysis will present a range of financial concepts in elapsed time relative to the point of failure as a basis for an explanation and discussion around their insight and predictive value. This chapter is not intended to provide an improved basis for IPM of firms generally, or contractors specifically. Rather, it provides an avenue into understanding how firm accounts can help identify alternative business models as well as a critique on the reliability and power of accounting data in predicting failure.

DOI: 10.1201/9781003102403-11

The business model of large modern tier 1 contractors can be characterised by the subcontracting of the vast majority of construction works, procured and delivered via internal expertise in tendering, and program and cash flow management. Such strategies are the source of high turnover growth, low margins and use of financial structures that displace long-term debt and equity finance with short-term trade credit, provided by the supply chains. *Ceteris paribus*, low margins factor into reduced profitability, yet all other things are not equal – business strategies have trade-offs, including between commercial, production and financial aspects. This chapter explores the business models of modern tier 1 contracting firms and explains why higher levels of subcontracting correlate with low margins, while still maintaining competitive profitability (return on capital invested).

The conclusions will consider the relatively low ability of even the most powerful ratios to predict insolvency, suggesting that the economic characteristics of construction operations make such business models less susceptible to risk-based forecasting of financial failure because of their higher rate of turnover to capital.

Classification of UK construction firms

The UK construction industry is categorised using three distinct two-digit Standard Industrial Classification (SIC) codes (ONS 2016). The inclusion of 'Development of building projects' within the 41 (building) code leads to some misclassification of firm activity, particularly in respect of developers (higher margin firms) returning their activity as contractors (lower margin firms). The set of firms in SIC 42 (civil engineering) also tends to include firms that meet these characteristics based on accounting reporting. Firms under the 43 'Specialised construction activities' SIC code supply the diverse range of more specific processes required for construction such as 43.21 Electrical installation and 43.92 Roofing activities.

Developers (SIC 41.1 – development of buildings) include clients that commission buildings that they will own and subsequently use or sell (Barkham 1997). Developers can be characterised by having ownership of land and so require high long-term leverage to finance this purchase and development phases (Ive and Gruneberg 2000). It is important to remember that not all purchasers of contractors' output are themselves within the SIC 41.1 classification. Many firms in other industries demand construction services such as Retail (SIC 47), Manufacture of electrical equipment (SIC 27) and Water collection, treatment and supply (SIC 36).

Within construction, there are both issues of misclassification of firms' activity besides unresolved challenges for the classification system as a whole. One of the challenges for the SIC system lies in its inability to reflect the diversity of firms within construction and the economy as a whole. While firms can indicate primary and secondary codes to reflect the different industries in which they operate, often, common means of production span different industries. A pertinent example includes a large tier 1 contractor that operates in both building and civil engineering. While the technology for undertaking these different forms of works is sufficiently distinct to deserve separate industrial classifications, the firm capabilities required to tender for and manage the delivery of works through sub-contracting of packages within these alternative markets are much more widely shared. Such capabilities and skills are arguably worthy of their own distinct industrial classification. It should be noted that some large tier 1 contractors also undertake developer investments as part of a diversified business models. In such cases, the use of either contractor or developer codes fails to represent the diversity of business operations. The main source of information by which we classify the activity of firms within the construction industry, namely SIC codes, are poor indicators of the underlying economic characteristics of contracting firms. They provide little information on the propensity to subcontract and resulting position on the 'make-buy' continuum as a differentiating feature of alternative business models. This 'make-buy' position of firms reflects the extent to which they source their gross output from other firms. This is partly a function of the transaction cost environment in which they operate (Williamson 1979).

The largest main contractors grow the turnover of their business through taking on contracts, the delivery of which they then predominantly subcontract to other firms. Such business models may be described as *outstensive* (Murray and Kulakov 2020). This contrasts with extensive and intensive forms of firm growth. Extensive growth results from increases in inputs used to deliver more output, while intensive growth results from more efficient use of inputs to increase output.

Outstensive growth is achieved by firms increasing the purchase of intermediate output from other firms in the supply chain. As such, outstensive growth is characterised by an increased detachment from decisions regarding the use of inputs in production processes and greater attention to commercial operations. The adoption of outstensive business models moves the firm towards the buy end of the 'make/buy' continuum involving more commercial, rather than productive, forms of value adding activity.

Contractor business models

To understand the business model of large contractors, as with any other firm, one needs to consider the performance of the sector in relation to the wider economy. Construction output is highly dependent on the wider economy. As much of the output of construction might be considered derived demand (it facilitates the consumption of other goods and services), output for the sector is more volatile than underlying changes in aggregate output, albeit with a lag. This lag reflects the relatively long production cycle for buildings and civil engineering works compared to other tangible assets. As figure 11.1 illustrates, there are large swings in construction output over time, particularly at the sub-sector level. The reduction in construction output following the Global Finance Crisis is much greater than in the wider economy.

In 2009 while gross domestic product (GDP) reduced by 4.2 percent, the construction industry witnessed an 18.8 percent reduction (ONS 2021). This business cycle backdrop sets the context for how growing contractors make strategic decisions about how to remain in business beyond the next downturn. This long-term

Figure 11.1 GDP and construction industry output change: 1997–2020 [Chained Volume Measure, Seasonally Adjusted] (ONS 2021)

volatility in construction demand drives some contractors to grow their capacity to take on more work through subcontracting business models. One major benefit of subcontracting is foregoing the need to maintain the internal capacity to deliver construction output. This reduces the need to finance working capital by making use of suppliers' capital to finance delivery of works. Maintaining capacity contributes to higher overheads (fixed costs), which in periods of reducing demand can lead to negative profitability and ultimately failure.

This use of outstensive business models (those that grow from purchasing more intermediate output of other firms to deliver gross output) by large and growing contractors goes hand in hand with another defining characteristic of contracting firms, their higher reliance than other firms in the wider economy on trade credit from other firms (Ive and Murray 2013). While large tier 1 contractors are not alone in their uses and abuses of trade credit strategies, some of these cash farming techniques have clearly been a core commercial feature of modern leading contractors, including and especially Carillion. By paying suppliers over longer periods following delivery of output than tier 1 contractors get paid by clients, they displace the need to use either equity or debt capital to maintain solvency. Construction contractors are unable to secure high levels of long-term debt financing given limited collateral (Metz and Cantor 2006) and volatile demand for their output.

Longer payment terms with suppliers can very effectively improve the amount of profit made by higher tier contractors, not on turnover, but rather on the capital employed (debt plus equity). The ability to generate profit on invested capital is the key concern of investors (shareholders and debt providers). This is not so from the perspective of suppliers of trade credit, whose profitability is in vertical competition with the purchasers of their output, the higher tier firm, to whom they extend trade credit. One of the trade-offs for larger firms becoming so reliant on trade credit is the risk of paying higher prices, because lower tiers are having to fund the additional finance costs of extending trade credit and so may increase prices to compensate for this. While firms who use high levels of trade credit do tend to have lower margins, there is evidence of 'free lunches' to be had (BIS 2013). The lower tier 1 profit margin resulting from suppliers increased prices (compensation for waiting longer for payment), has a smaller effect on overall profitability than the benefits of generating higher turnover on capital. A large part of this higher profitability stems from a form of economic rent on the monopoly status of tier 1 contractors on large projects. This limits the extent to which lower tier suppliers can fully 'price in' and charge higher tier contractors the costs incurred for extending trade credit by accepting longer payment terms. Smaller firms in lower tiers of supply chains anecdotally are less diverse in their client base, leading to reliance on a few, or even a single, source of demand from higher tiers.

When trade credit is used to fuel the turnover growth of a firm, the economic consequences of failure are very different in their effect on labour and sector capacity than they are when other forms of finance are used, such as developers' reliance on long-term liabilities (debt). Development is more capital intensive and so draws more heavily on the resources of the financial sector to buy land and finance the construction of buildings. Conversely, tier 1 contractors are heavily dependent on both the financial capital and productive capacity (physical capital and labour) of other construction firms (lower tiers) to deliver their gross output. As such, when a large contractor fails, the impact is more disruptive within the industry than if a developer fails, whose main sources of finance are from outside the construction industry with financial institutions. In this sense, the balance sheet of a firm provides insight not just on the firm itself via the assets it owns, but also how it relates to other firms in the wider economy through its liabilities – both sides of the balance sheet can inform the characterisation of firm business models.

At this point, it is worth noting the presence of the following characteristic tendencies of large tier 1 construction contractors. Firstly, over-optimism about revenues arising from often-used percentage-of-completion methods often applied to forecast revenues on construction projects. Although such methods provide means of recognising revenues earlier, due to the risky nature of the projects, contractors can overestimate. Further, if optimism creeps into areas where there is accounting discretion, this may lead to challenges to sustaining liquidity when actual payments are realised (Tserng *et al.* 2012). Secondly, tier 1 contractors also exhibit low levels of fixed assets in comparison to firms which possess land, such as developers, or other durable assets like property, plant and equipment (including plant hire firms).

The success of a construction firm ultimately depends on the success of the projects within its portfolio of works. To be more solvent, a suitable proportion of a firm's cash flow should be employed in operations with a reduced cash flow in investment (Arditi *et al.* 2000; Chen 2012). This is because the cash flow characteristics of contractors, where payments are received from clients only for completed work usually on a monthly basis. A cash flow plan for operations is essential to avoid shortages of cash from unforeseen delays to progress in project completion and related client payments. This may lead to firms having a negative cash flow, creating considerable risk for the survival of the firm (Kale and Arditi 1999). This is especially the case when high turnover growth is achieved through high subcontracting business models.

Business failure

Business failure can be defined in a number of ways, including:

> *situations in which a company cannot pay lenders, preferred stockholders, suppliers, etc. or a bill is overdrawn, or the company is bankrupt according to law.*
>
> (Ahn *et al.* 2000: 65)

Such situations are often referred to as financial failure, insolvency or bankruptcy,[2] often resulting in losses to creditors and shareholders, together with substantial social and economic costs (Dimitras *et al.* 1996). Directors of a firm can make a decision to cease all business activities, which can be considered a voluntary closure. Technically, in both scenarios, insolvency and voluntary closure, firms could be considered as failures. However, we focus here on instances of involuntary insolvency via reference to Gazette data. Gazette notices provide information about the legal processes that brings about the liquidation of company assets and include indicators of the type of insolvency, such as whether it was voluntary or involuntary.

Figure 11.2 illustrates the age of construction firms relative to the proportion that fail (dark grey section) (Kulakov 2017). As is evident, the vast majority of firms younger than two years fail and the typical firm fails to reach five years of

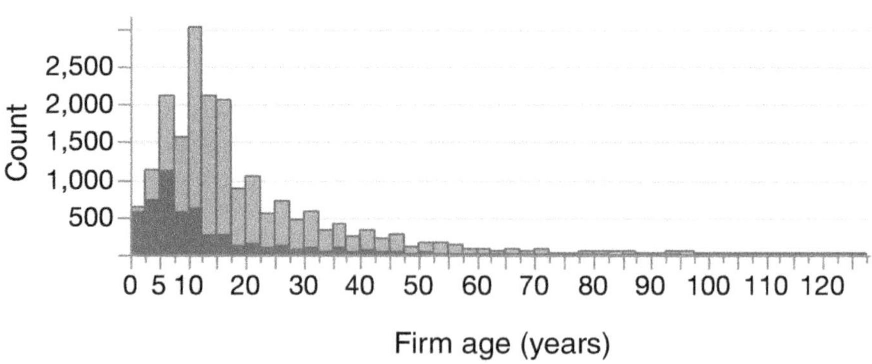

Figure 11.2 Distribution of construction industry firms by age (SIC 41, 42, 43.2008–2017 data) (Kulakov 2017)

operation before ceasing trading. It is only beyond ten years of trading where failure risk materially reduces.

Figure 11.3 illustrates the annual rates of change (%) in turnover for the firms considered here (see sampling method that follows) leading up to their failure over the prior 5-year period, finishing with the Last Available Year (LAY) accounts. The light grey shaded density area is a smoothed illustration of the distribution of values on the vertical axis for each period. The width of the area indicates the frequency of values whereby the wider the area, the more instances of firms with such values there are in that part of the range. The thickest dark line represents Carillion. The thick grey lines represent the larger firms (turnover of GBP 6 million[3] in at least one of the years before failure) and the fainter lines are the rest. Even when such firms are large, the turnover of construction firms approaching failure is volatile. Surges and contractions of over 10 percent are not uncommon, which against margins of less than 5 percent (see Figure 11.4), paints a backdrop where consistent cost control and securing of internal capacity or external supply to deliver gross output are everyday challenges. It is against this backdrop that we consider the sustainability of contractor business models.

Insolvency prediction

Developing IPM involves using financial ratios to predict firm failure by observation across large samples of firms. The general approach is to compare the financial results of firms which have failed with relevant sets of firms that continue to

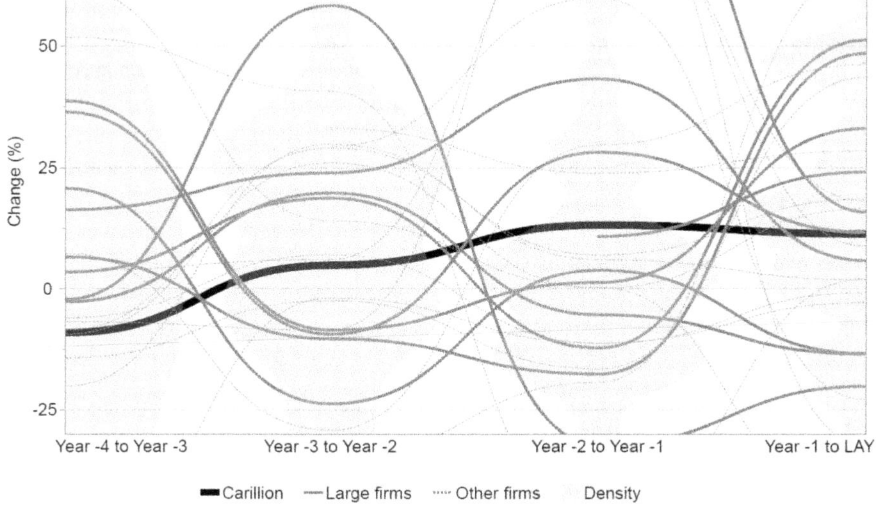

Figure 11.3 Rates of change in turnover approaching failure (Year -4 to LAY)

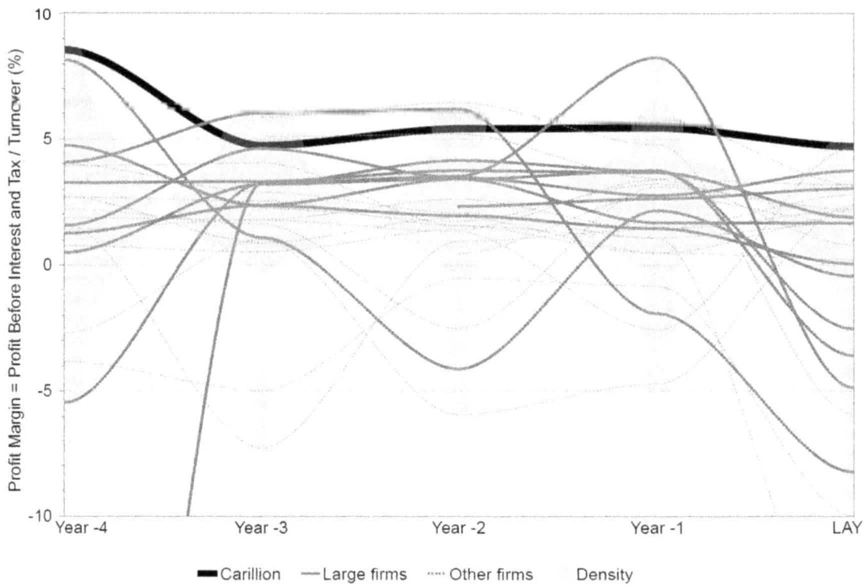

Figure 11.4 Time elapsed profit margin for failing contractors (Year -4 to LAY)

operate. In this sense, it is the analysis of correlations (regression analysis) with associated incremental additions of failure risk, that form the basis for discussion around underlying causal effects. Russell and Jaselskis (1992) suggest that all-sector models may be too general and therefore lack the desired predictive power for a specific industry, for example, construction contractors.

A review by Alaka *et al.* (2016a) of construction business failure prediction studies suggests efforts have coalesced around building high-performing IPMs through siloed studies of specific industries. The pioneer of developing such models was William H. Beaver, an accountant who in 1966 investigated how financial ratios can be used to predict failure of firms (Beaver 1966). Shortly after, in 1968, Edward Altman developed his study based upon Beaver's work (Beaver 1966) using a multivariate data analysis, specifically a multiple discriminant analysis. Multiple discriminant analysis 'is a statistical technique used to classify an observation into one of several groupings dependent upon the observation's individual characteristics. It is used primarily to classify and/or make predictions in problems where the dependent variable appears in dichotomous (discrete binary) form, e.g. male or female, bankrupt or non-bankrupt' (Altman 1968: 591).

Altman's study 'proved to be very successful and has more or less become a template for most . . . prediction models . . ., as most subsequent studies have used a similar approach' (Alaka *et al.* 2016a: 808). Altman developed the Z-Score model, which applies predetermined multipliers to a small number of key financial ratios to generate an insolvency risk indicator. Many studies have developed powerful and insightful models that can credibly claim to identify statistically significant predictors of firm failure, such as the payables-to-receivables ratio for contractors (Kulakov 2017).

The application of IPM to construction contracting requires, like any other industrial context, an appreciation of their particular economic characteristics. These include the extensive use of trade credit and/or supply chain finance strategies by firms in higher tiers and low margins compared to firms in the wider economy, resulting from high competition as well as faster rates of capital turnover. The existence of varying levels of subcontracting amongst contractors is another key indicator of failure risk.

The established way of thinking about how firms conduct themselves is based on the prevailing technology[4] used in production processes. Industries vary in their structure, levels of competition and business models through diverse supply chains which define the levels of vertical and horizontal integration. The technological drivers of industry structure that influence prevailing business models also contribute to the factors that lead to failure. Actual failure risks vary for firms in different industries, even with similar-looking financial positions (Chava and Jarrow 2004). Failure risk predictors based on financial information should be adjusted for specific economic conditions and industries (Kapliński 2008). Managerial and firm-specific soft variables might be considered alongside traditional financial ratio models to enhance their predictive power (Severson *et al.* 1994; Russell and Zhai 1996), though they can be difficult to observe empirically for large samples of firms.

The majority of studies on failure prediction of firms aim to identify them as either a failure or non-failure. This presumes that failure prediction is an equal pay-off state (Tserng *et al.* 2012). In other words, the importance of predicting a failure is equal to that of predicting non-failure. Arguably, in the real world, Type I errors are substantially more costly than Type II errors, that is, failing to predict a failure is worse than failing to predict a non-failure. For example, awarding a contract to a failing firm will incur significantly greater costs than rejecting a non-failing firm. Furthermore, it may be useful for decision-makers to have the ability to rank contractors according to their default probabilities. Examples include selection by the client of the most stable construction contractor, or bank lenders may use this information as a basis for credit terms across their construction industry loan books.

Balcaen and Ooghe (2006) highlighted various challenges associated with applying statistical tools to predict business failure, proposing that application of artificial intelligence (AI) tools could resolve methodological challenges. Alaka *et al.* (2016a: 809) state 'AI tools have become gradually more popular, . . . [as] yet only little or no progress has been made'. More recent IPM studies have made use of artificial intelligence/machine learning techniques for the selection of predictors. This is especially useful with approaches, involving a large number of financial concepts, which assess potential to predict failure (Kulakov 2017). Lasso techniques, a machine learning statistical approach,[5] apply penalties to variables that are both highly correlated to other variables which are more powerful predictors, as well as penalising other variables where data availability is inconsistent through time.

The time dimension of failure is a particular source of insight from IPM approaches. The backdrop to any firm failure is the business cycle as well as other exogenous factors. Without regard to time dependent effects of failure (unique events such as the Global Financial Crisis), IPM approaches can lead to sub-optimal predictions (Balcaen and Ooghe 2006). IPM methods can lead to the use of a number of concepts across time as consistent predictors[6] (Kulakov 2017), contributing to the power of models between periods of varying economic environments and interactions with business models. For example, the quick ratio, measuring the value of liquid assets relative to current liabilities, is a consistent indicator of a firm's ability to meet its short-term financial obligations through time. As such, it is often relied upon by analysts regardless of the wider economic conditions.

If a model does not directly take account of macroeconomic factors (e.g. inflation, interest rates, and/or the phases of the business cycle) the issue of data instability should be considered (Mensah 1984). The differences in size and significance of predictive effects over time suggest that even when predictors are common between models at different points in time, exogenous factors, such as the competitive nature of the industry, sectoral shifts in the use of technology and the different phases of the business cycle influence which variables are most appropriate to predict failure at that specific point in time (Wood and Piesse 1987). This undermines the generalisability of IPM models over longer periods.

The consistency in the selection of some predictors between years suggests that the economic characteristics of firms which contribute to failure risk can be reliably detected by models (Chava and Jarrow 2004; Kapliński 2008). If predictors were inconsistent through time, this would suggest the underlying endogenous firm factors were weak relative to non-firm level wider exogenous drivers. The latter being more time dependent would increase the inconsistency in predictors through time.

Formation of sample

To establish a sample of UK construction firms, the Fame database (Bureau van Dijk 2020) was used. The search criteria in Table 11.1 were applied to reach a set of 77 candidate firms. Developers were identified via reference to high levels of land relative to total assets and removed. Gazette data was interrogated to remove firms which dissolved voluntarily to focus on compulsory insolvencies.

The final sample of 27 firms constitutes a tiny proportion of the firms that failed in the years 2017–19, but importantly is made up of generally much larger firms. The application in step 4 (in Table 11.1) of a minimum GBP 4 million trade creditors (monies owed to the supply chain) results in a sample made up of firms that will tend to be more 'buy' than 'make' in nature. That is, they will tend to subcontract more of their gross output, and in so doing, increase the value of trade

Table 11.1 Search steps applied in Fame and subsequent sample cleaning steps

Search step	*Condition*	*Step result*
1. Status	Active (in administration), Active (receivership), Dissolved, Inactive (no precision). In liquidation	7,683,385
2. Year of last available accounts	2016, 2017, 2018, 2019, 2020	5,097,748
3. UK SIC (2007)	41 – Construction of buildings, 42 – Civil engineering, 43 – Specialised construction activities	1,079,676
4. Trade Creditors	Minimum of GBP 4 million in the last available year	18,554
5. Turnover	Minimum of GBP 50 million in the last available year	18,358
6. Remove any subsidiary of another company from the current set		512
7. Boolean sampling[7]	6 from (1 and 3 and 6 and (4 or 5))	77
8. Manual review	Remove developers and other non-contractor firms	55
9. Manual review	Remove voluntary insolvency through reference to Gazette data	31
10. Manual review	Remove individual firms for specific reasons[8]	27

credit on their balance sheets. While this can help us understand the factors at play in their demise, their economic structure will be very different from the wider population of contractors and, as such, it is not a representative sample of the circa 120,000 failures in the construction industry between 2017 and 2019 (Table 11.2 shows approximately 40,000 firm failures in the year preceding the period of study). This itself leads to obvious limitations in the generalisability of IPM results when based on the data of published accounts. The near entirety of construction firms to fail over our period do not meet the criteria to report sufficient financial data for analysis. This is the case across other industries as well. The availability of financial information is subject to a 'buy and large' bias (Murray and Kulakov 2020). This bias refers to the greater availability of financials for firms which have achieved outstensive growth through buy strategies, and thus meet the thresholds for reporting more complete accounts applying to larger firms under the UK Companies Act 2006.

Looking at net value-added (NVA) in our sample of 27 firms relative to industry level gross value added (GVA)[9] suggests about a 1 percent coverage – that is, our small number of firms (representing less than 100th of 1 percent of active firms in the wider construction industry; Table 11.2) produced about 1 percent of the industry's output in the years leading up to their failure. Carillion represents around 70 percent of the sample's NVA. Given Carillion's significance, it is illustrated in the analyses via use of distinct formatting. The results are presented in elapsed time to failure, that is concepts are in annual periods relative to the LAY accounts on the right, with preceding years to the left.

Table 11.2 Firm demography – construction and the wider business economy

Year	*All UK businesses (economy)*					*UK construction industry*									
	Active	*Birth*		*Death*		*Active*	*Birth*			*Death*			*Gross value added*		
	Count (000's)	*Count (000's)*	*Rate*	*Count (000's)*	*Rate*	*Count (000's)*	*Count (000's)*	*Rate (industry)*	*Rate (economy)*	*Count (000's)*	*Rate (industry)*	*Rate (economy)*	GBP *billions*	Δ	*% of economy*
2010	2,351	235	10.0%	249	10.6%	325.5	27.4	8.4%	11.7%	39.1	12.0%	15.7%	80.6	1.33%	5.6%
2011	2,343	261	11.1%	230	9.8%	314.9	29.8	9.5%	11.4%	34.6	11.0%	15.0%	84.1	4.40%	5.7%
2012	2,373	270	11.4%	252	10.6%	308.0	29.3	9.5%	10.9%	36.2	11.8%	14.4%	87.9	4.41%	5.7%
2013	2,449	346	14.1%	237	9.7%	308.9	38.3	12.4%	11.1%	32.4	10.5%	13.7%	92.6	5.41%	5.8%
2014	2,551	350	13.7%	246	9.6%	317.4	41.0	12.9%	11.7%	31.3	9.9%	12.7%	98.9	6.84%	6.0%
2015	2,699	383	14.2%	282	10.4%	335.0	48.8	14.6%	12.7%	31.1	9.3%	11.0%	106.2	7.33%	6.2%
2016	2,834	414	14.6%	281	9.9%	358.0	52.0	14.5%	12.6%	40.0	11.2%	14.2%	108.3	1.97%	6.1%
2017*	2,926	382	13.1%	362	12.4%	378.0	53.0	14.0%	13.9%	39.0	10.3%	10.8%	112.2	3.65%	6.1%
2018*	2,940	381	13.0%	336	11.4%	389.0	50.0	12.9%	13.1%	39.0	10.0%	11.6%	116.0	3.34%	6.1%

Data Sources: ONS, 2021

* – provisional

Table 11.2 provides further information on the construction industry's contribution to the wider economy as well as firm births and deaths. Correlating construction industry deaths' share of those in the wider economy with the industry's share of aggregate output reveals a strong negative correlation of −0.85 (highly significant) – as the industry grows its share in the economy, the proportion of failing firms reduces. While this data covers a period of construction industry growth, this relationship will hold when recession hits and construction industry share of output contracts leading to higher rates of construction firm failure than the wider economy.

Analysis

In the following sections, a number of concepts from financial statement analysis research are reviewed. These concepts were chosen on the basis of both their common relevance to IPM modelling generally, and to contracting business models specifically, based on prior research by the authors (Kulakov 2017).

The co-efficient of profitability – margin

The profit margin is calculated by dividing profit before interest and tax by the turnover of the firm. Low or negative levels of profit on turnover for contractors often results from aggressive bidding in response to tenders, as well as the passing down of construction risks to lower tiers, which can later lead to delay or cost overruns. The higher the level of profit on turnover, the more solvent a firm is (Alaka *et al.* 2016b). Profitability is different to the firm's profit margin on turnover. Profitability indicates how much profit is made per unit of a meaningful stock concept from the balance sheet (assets and liabilities). Turnover, being a concept from the profit and loss (income) statement is a flow concept. Pure profit margin is the key co-efficient of profitability, deciding whether it is positive or negative, but other factors within the business model need to be considered to measure profitability. To get to the first profitability concept of Return on Total Assets (ROTA), profit margin, is multiplied by the asset turnover ratio (turnover per unit of assets). One step further, with the addition of current gearing, (assets per unit capital employed, we get to the most frequently used profitability measure of Return on Capital Employed (ROCE). Since, none of turnover, assets or liabilities can be negative values, only profit margin can change the direction of profitability.

Profitability indicators are understandably often significant predictors in a range of IPMs. Presenting our sample of profit margins in elapsed years relative to failure, from Figure 11.4 it is seen that most firms actually report positive margins in their LAY accounts, consistent with prior years. Only one third of the sample report negative profit margins in the LAY. Low or negative profitability is of course a significant factor adding to insolvency risk. The more profitable a firm is, the more likely it is to cover its financial liabilities and therefore remain solvent (Alaka *et al.* 2016b). This particular finding creates a challenge for IPM methods. Reflecting on the earlier results, the co-efficient of additional insolvency risk from lower margins will have to find an effect somewhere between the minority that witnessed quite extreme negative margins, and the normal failing firm that maintained positive margins. This co-efficient would predict neither set of firms particularly well.

A more sophisticated IPM model might, instead of comparing financial concepts between just samples of failed and non-failed firms, compare sub-samples of failed firms based on clustering around the values of key dependent variables. Such a model would look to estimate separate coefficients for the incremental risk of lower margins based on whether the margin itself is below 0, or whether its volatility from year to year has been higher than a benchmark of all surviving firms. Such approaches may increase the power of IPMs – the key challenge from achieving this sophistication is creating meaningful criteria by which to sub-divide samples. It is unlikely 0 percent margin would be the optimal defining value to improve predictive power.

The presence of only a small proportion of firms in the sample with negative profitability in the LAY is somewhat surprising, especially given the cost risks that larger contracting firms attempt to pass on through high levels of subcontracting. It is the emergence of these risks that are often the cause of failure. While many of the firms examined here do not see a deterioration in profitability before failure, it is likely that if these firms had produced another set of accounts, even interim, their margins would deteriorate. While cash is king, such are the nature and scale of cost risks in contracting, that combined with low levels of capitalisation, what might be seen as survivable negative margins in some industries, can exhaust the balance sheet of contractors within a single period. The promptness of failure in contracting means annual reporting is often too infrequent to reveal the deterioration in margins and generally profitability.

Carillion's global revenue for 2016 amounted to GBP 5.2 billion, a third of which was UK public sector contracts (NAO 2018). The losses on three UK PFI contracts only accounted for 27 percent of the GBP 845 million provisions made by Carillion on 10th July 2017. As such these cannot be considered the primary reason for Carillion's failure. By the time of their second profit warning in September 2017, Carillion had made a GBP 1.15 billion loss in the first six months of 2017 against a half year turnover of GBP 2.25 billion leaving net assets (equivalent to equity value) of GBP –405 million (see Figure 9 in NAO 2018). Based on net assets in the previous period of GBP 730 million, the firms would have needed to increase equity capital by over 50 percent going into 2017 to stand a chance of absorbing such losses and continuing to operate. The losses that take any firm into insolvency are only one side of the coin. Carillion's persistent under-capitalisation from years of intensive cash farming strategies exposed their balance sheet to excessive risks given the scale of revenues being generated. Balfour Beatty Group PLC posted a −4.2 percent margin in 2014 (GBP –304 million profit on GBP 7.26 billion turnover) and managed to return from the brink. It has yet to increase its margin above 2 percent, averaging 1.8 percent over 2017–2019 and a return on capital of just 5.7 percent over the same period – this is some way from the average 25 percent return on capital seen in firms in the construction industry and wider economy between 2005 and 2011, or even that typically seen in larger firms of around 15 percent (BIS 2013).

Public procurement and use of financial statements

The Cabinet Office in the UK published its first *Outsourcing Playbook* in February 2019 (Cabinet Office 2019). This suite of documents provides 'Central Government Guidance on Service Delivery, including Outsourcing, Insourcing, Mixed Economy Sourcing and Contracting'. Chapter 3 of the main document provides guidance for public procuring authorities on specifically 'make/buy' decisions, recommending the use of delivery model assessment to decide between in-house, outsourced or hybrid delivery approaches. In the context of construction procurement, the in-house option is somewhat limited given the lack of public works (in-house) contractors. This option is more pertinent to the delivery of Facilities Management services in public estates, where in-house provision is a credible option. While, within the *Outsourcing Playbook*, there is a specific Market Management Guidance Note, broad monitoring of financials across the industry is not common practice. More recently the *Construction Playbook* (2020) highlights the role of digitisation of payment in improving transparency of payments down the supply chain.

The *Outsourcing Playbook* also encourages the use of potential suppliers' financials as indicators of their delivery capability and failure risk. The guidance note on Assessing and Monitoring the Economic and Finance Standing of Suppliers uses a number of financial ratios as indicators of supplier risk, suggesting that, for construction firms in particular, a margin below four percent would indicate that a firm bidding for an important or critical procurement is deemed high risk. In turn, that would mean the near entirety of our sample of firms would fail to pass this test based on their margins. While that might be appropriate for these set of firms known to go on to fail, there are few large contracting firms that would meet this criterion. Carillion manages to meet this threshold (i.e. four to seven percent is deemed medium risk) but this could be by virtue of its development activity profits as a diversified corporate. While assessments will look at the financials for specific subsidiaries bidding for a contract, as well as requirements to look into key subcontractors used by bidders for public contracts, these exercises require particular expertise. As such, there is a high risk of inferring misleading information from raw published accounts, without concerns for the 'make/buy' nature of the supply chain. For example, there is a risk in assessing the margins of multi-tier delivery models versus more vertically integrated suppliers to government – without reliable benchmarks, the lower margins of a subcontracted model, because profit is split between tiers, could lead to sub-optimal procurement outcomes.

Liquidity – current ratio

The current ratio is calculated by dividing the value of current assets by current liabilities and provides an indicator as to a firms' ability to meet its short-term liabilities without liquidating longer term fixed assets (Ng *et al.* 2011). A value below one suggests that a firm is approaching insolvency unless it is able to access additional cash (a form of current asset) in the short-term. The current ratio is often the first choice as an indicator of short-term solvency. Although liquidity factors might be poor early warning signs (Altman 1984), they are effective for near-immediate and immediate predictions of failure. Furthermore, a relatively high liquidity is required for construction firms as cash availability is vital for project execution (Alaka *et al.* 2016b).

Figure 11.5 illustrates the current ratio values for our sample of failing contractors. As is apparent, values remain above one for all but a small number of firms in the path to failure. The relative stability of this ratio (compared to other ratios) attests to its poor predictive power. Looking at the LAY, a similar number of the large firms witness increases as do those that see decreases. An IPM would struggle to find a clear signal from these results given the low correlation between commonly failing firms. Carillion's results hover just above one for the years approaching its insolvency.

While the current ratio is widely available for many firms, its reliance on balance sheet concepts, which are in effect a snapshot in time, meaning its insight is less than if one had more up to date information on actual cash positions. As firms approach their reporting period, they can change their normal practices to make their financial statements look more favourable. This is sometime referred to as window dressing and involves improving the values of key concepts, like the current ratio, for the period of time at which annual accounts are drawn. Practices and strategies are diverse and complex – the relevant point to be made here is that firms do this as they know their accounts provide external parties signals as to the potential risks and future prospects. The incentive to undertake such practices is greatest for publicly listed firms, whose accounts are a key source of information and heavily influence share prices.

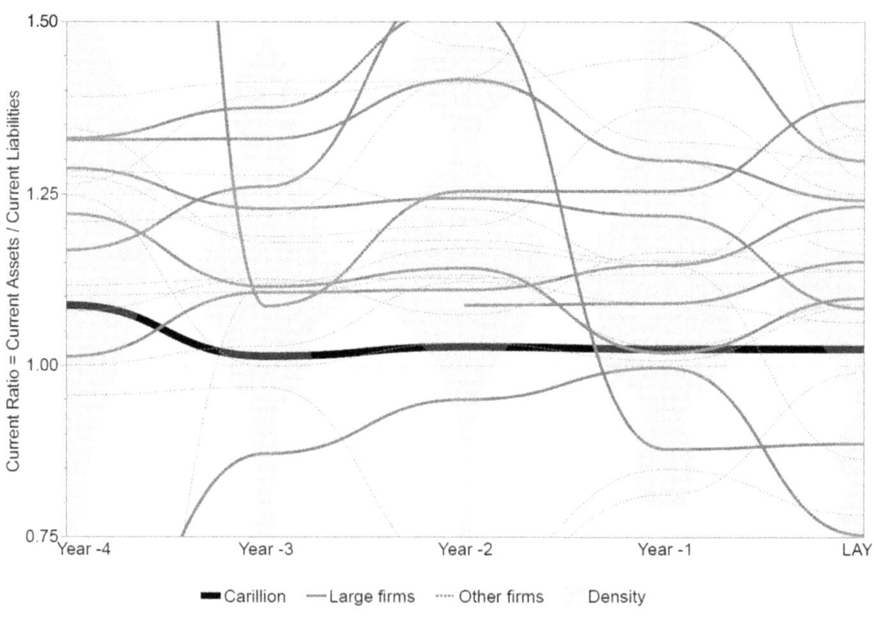

Figure 11.5 Time elapsed current ratio for failing contractors (Year -4 to LAY)

Carillion had over 310 companies within the UK PLC corporate group and employed over 40,000 people on the day of its compulsory liquidation on 15th January 2018. Compulsory liquidation is the most drastic procedure in UK insolvency law. Two months after Carillion's insolvency the rating agency S&P produced a technical accounting analysis of the firm's balance sheet looking back five years (S&P 2018). This analysis raises questions about the reliability of published accounts of even this behemoth listed company. It suggests that accounting discretion, particularly around how current liabilities are represented, can skew key financial measures that creditors might use to assess insolvency risk. The inadequacy of monitoring and audit of such a large company to deliver transparency does not bode well for what might be hidden in smaller firms' accounts, naturally subject to less scrutiny. Returning to Carillion's current ratio value, had they fully accounted for their current liabilities, they may have shown values below one and indicated higher risk of failure. This example makes clear that the quality of information available from published accounts is fundamentally linked to the use of accounting discretion in how to represent a firm's financial position.

Interest cover ratio

The interest cover ratio shows the value of profit made as a proportion of interest paid on debts. Other things being equal, the higher its profits, the more able a firm is to maintain its debt position by covering interest payments. For contractors, this measure is more volatile than for other firms due to two reasons: firstly, the varying performance of contractors leading to periods of unstable and negative profits leading to volatility in the numerator, and secondly, low levels of long-term debt generally compared to other companies (BIS 2013) leading to lower interest payments (lower denominator). Low levels of debt partly result from banks' high aversion to lending to contractors. There is anecdotal evidence this aversion has materially increased in the UK following the failure of Carillion, and with the significant woes of both Kier and Interserve.

The changing direction of profits to losses renders this ratio meaningless in the sense that a negative value does not mean interest was not paid, rather it was paid via additional finance (debt or equity) instead of from a surplus in that year, or an alternative agreement was reached, such as rolling up interest into the loan principal. Its high variability can hamper its use as a predictor of failure, although it is often chosen in IPM models where firms are more consistently profitable. It remains an important risk indicator in business models other than those used by contractors, especially in firms with much higher reliance on long-term leverage such as developers. Analysis of this concept for the set of firms discussed here revealed very volatile values which are of limited value for visual presentation.

Receivables to payables

The receivables-to-payables ratio measures how much trade credit a firm receives from its suppliers relative to how much trade credit it extends to its clients. This may be used as an identifier of the firm's business model with regard to cash flow management by reference to their net position. When a firm's payables (including both trade credit and accruals) are greater than their receivables (trade debt), then a firm is a net recipient of trade credit. In other words, overall, they extend less trade credit to their clients than they receive from their suppliers. The results in Figure 6 suggest, as might be expected, these larger firms record lower levels of receivables relative to payables, suggesting they are more 'buy' in nature and sub-contract relatively more of their gross output than other construction firms. It is this flow of work to other firms on which they accumulate stocks of payables. The density areas show a tendency towards failure to compress to lower values, consistent with the expectation that less assets (trade debtors) relative to liabilities (trade creditors) increases the risk of failure.

S&P (2018) suggests looking at trade debtors relative to wider measurements of current liabilities. The need to consider accruals in a firm's true exposure to business-to-business finance is one example of the importance of broadening the set of balance sheet liabilities considered to accurately reflect a firm's financial position (DeFond and Jiambalvo 1994). This requirement to broaden standardised measures to account for particular firms' financials reiterates the subjective nature of accounting practices, as well as the requirement for deep scrutiny and technical expertise in forensic accounting and IPM modelling. Such efforts are difficult to apply sensitively across industries and the alternative business models within them. Varying business models create unique biases in the way that firm

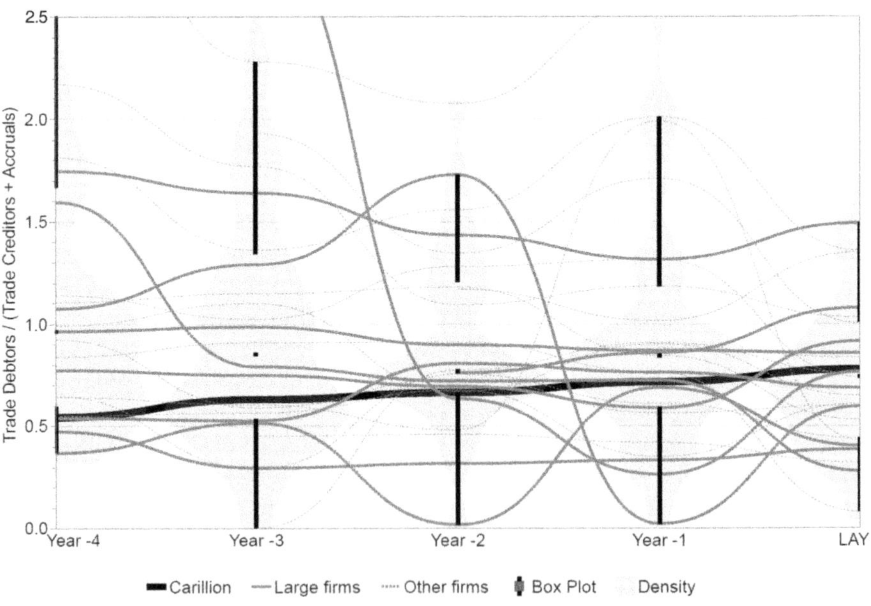

Figure 11.6 Time elapsed receivables-to-payables (incl. accruals) ratio for failing contractors (Year -4 to LAY)

level accounts represent the economic characteristics and failure risks of the firm. These are not accurately attributable by use of SIC codes.

Prior research (Kulakov 2017) suggests an increase in the ratio of receivables to payables would reduce failure risk, as working capital (current assets – current liabilities) increases suggesting an improved ability to finance firm operations (Kenley 2003). The effect size is small compared to other predictors. Some annual results for specifically civil engineering contractors however suggest effects that are at odds with this general finding. Their ratios for days receivables lead to reduced insolvency risk. This suggests that civil engineering contractors who extend credit for longer (i.e. allow longer periods for being paid) are less likely to fail. Such a result may be explained by either, the possibility that increased receivables indicate better terms of contract, such as lower unit costs aiding profit margins, or arguably, and more likely, that an increase in receivables will be passed further down the supply chain.

Equity multiplier – total assets to equity

The equity multiplier ratio is measured by dividing total assets by equity (also known as shareholders' funds). Since it is assets that drive the need for finance, this ratio reveals how much a firm is using other forms of finance beside equity to finance assets. If a firm had no debt (also known as long-term liabilities) or current liabilities, its assets would equal its equity. This is rarely the case and firms can vastly increase their asset base by use of debt, such as that used by developers to finance the purchase of land and subsequent building, or by contractors, who typically use more current liabilities (trade credit) to supplement equity in financing their assets.

As use of external finance comes with risk, it is not surprising that increases in the equity multiplier promote failure risk (Arditi *et al.* 2000) as an indicator of gearing. Figure 11.7 shows that some of the firms see distinct increases in this ratio in the LAY, including Carillion. However, it should be noted that in some cases, higher equity multipliers can result in lower chances of insolvency (Arditi *et al.* 2000). Debt is a cheaper form of finance and can aid the accumulation of net profits as retained earnings, providing a larger buffer if risks do emerge. Despite this, higher use of debt tends to increase failure risk. If cash flows reduce due to economic downturns or inability to obtain sufficient workload to utilise the assets being financed, this can lead to default on repayment.

Overtrading risks – capital turnover ratio

The final ratio to consider is the capital turnover ratio, calculated by dividing a firm's turnover by its capital employed. This ratio is a good indicator as to whether a firm may be overtrading by taking on more work than its capital base is able to finance working capital or absorb losses. Contractors record much higher levels of turnover to capital than other firms, turning over their capital at twice the rate of firms in the wider economy (BIS 2013). Averages for UK contractors between 2005 and 2011 show their turnover is around 3.7 times their capital compared

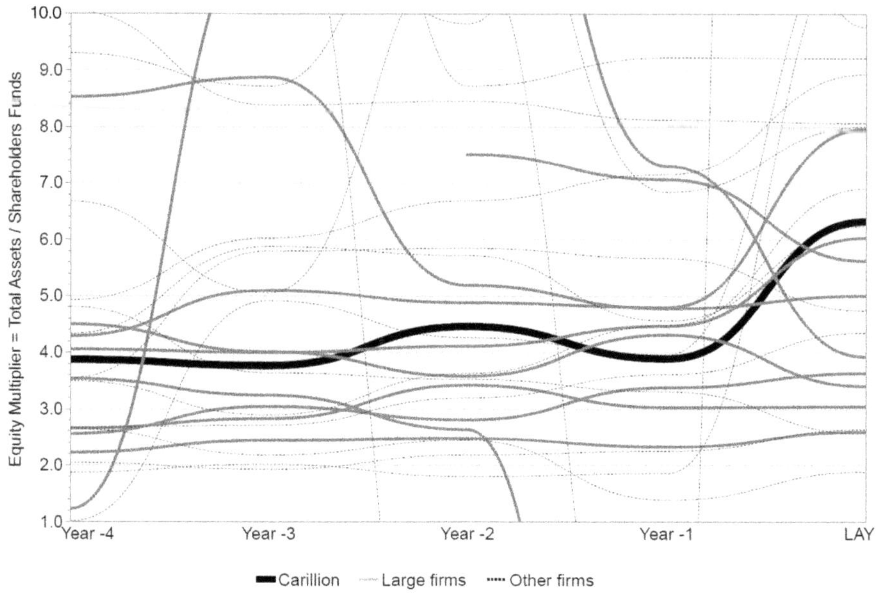

Figure 11.7 Time elapsed equity multiplier for failing contractors (Year -4 to LAY)

to a rate of 1.5 for firms in the wider economy. This will reduce IPMs' predictive power for contractors, as the speed of insolvency arising from a single rotation of the capital base is higher. Other types of firms that have slower rates of turnover to capital, such as manufacturing or plant hire firms, approach insolvency over longer periods of time as it will take more successive periods of losses to exhaust their capital.

Multiplying the profit margin by the capital turnover ratio produces return on capital employed (ROCE). A 4 percent profit margin and a capital turnover ratio of 4 would produce a ROCE of 16 percent, a normal result for large construction contractors between 2005 and 2011 (see Table 1B in BIS 2013). Margins and capital turnovers are negatively correlated. Margins tend to decrease with increased subcontracting, as the value added on production is passed down to lower tiers. Capital turnovers will tend to increase with subcontracting, as both the need to invest in productive capacity (assets) and the associated need for additional capital (debt or equity) reduces through use of trade credit strategies.

Outstensive 'buy' business models serve to both grow turnover via the subcontracting of output and reduce capital base via use of trade credit. This provides IPM approaches a greater challenge in predicting the insolvency of such business models, as financial accounts are less frequent relative to the rate at which such firms place their capital at risk through trading and turnover generation. Further, stocks of trade credit have accumulated via high rates of turnover growth, and which can only be settled with continued operation; even a relatively small loss on turnover is sufficient to exhaust the balance sheet and lead to failure. When heightened turnover is achieved without a matching growth in the capital base of a firm, that most essential element of financial sustainability, it should not be surprising that low or negative margins resulting from poor management of construction delivery increases the risk of financial failure. Outstensive 'buy' business models often increase the potential for negative margins through over reliance on risk transfer down supply chains.

The firms in Figure 11.8 approaching failure report higher capital turnover ratios than wider industry benchmarks, consistent with their higher risk of failure. This aligns with results in previous studies with weighted capital turnover ratios for non-failing contractors of 3.7 (see Table 1B in BIS 2013). While there is no consistent increase in the ratio across the sample going into the final year, some of the larger firms do show increases above their preceding year values. As with margins aforementioned, an IPM would struggle to find a clear signal from these results because the comparator sample of non-failed firms may demonstrate similar changes in capital turnovers without corresponding failure. Carillion's value is much lower than other firms, in part because of the higher capital base required for financing their development activities and the purchase of other contracting firms in the preceding years. The stability in their results from year -4 to year -2, followed by a 13 percent increase in the ratio into t-1 provided a balance sheet more prone to failure from losses on turnover. This contributed to their existing undercapitalisation. Combined with losses on such inflated turnover, this resulted in the collapse of this leading tier 1 construction contractor.

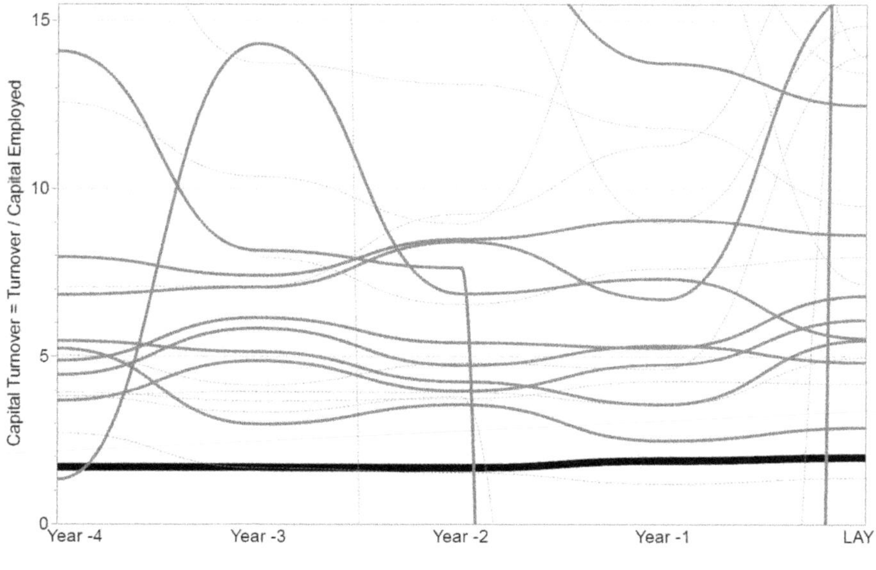

Figure 11.8 Capital turnover ratio for failing contractors (Year-4 to LAY).

Conclusion

This review of the financial ratios of failing contractors demonstrates the insight they can provide on both insolvency and the wider business models of contracting firms. Powerful IPM models draw on multiple predictors to capture the various causalities leading to failure. Results suggest the lack of timeliness of financial data, such as margins, limit their ability to signal the realisation of construction cost risks before failure. Poor management of construction cost risk and resulting negative margins, no doubt, contribute to failure, but the firm is often no longer around to report their losses. Consideration of accounting discretion was seen as an important subjective driver of financial ratio values, particularly the recognition of payables as a key factor in contractors' financing. Finally, a review of capital turnover ratios reminds us that contractor failure can happen faster than firms from other sectors, as a result of inflated levels of turnover relative to their capital base. While the risks of losses from project cost and time overruns are often the cause of firm failure, the undercapitalisation of contractors makes them more prone to insolvency than other types of firms.

The relative stability between years of a number of the financial ratios in Carillion's accounts, particularly current ratio (Figure 11.5), raises a contradiction in the business model they adopted. On the one hand, the larger a firm is, the more scrutiny and monitoring there is of their accounts, especially when they are publicly listed as Carillion was. On the other, Carillion had, as any firm, incentives to present information in ways that painted a better picture of their financial position than was the case. Window dressing of results via creative accounting practices is not confined to presentation of financial information. It can be effective in maintaining creditors' faith in the short term. In the longer term, keen observation of accounting information often reveals signals of misleading practices. One such example is the use of other creditors for the accounting of reverse factoring facilities (S&P 2018). This allowed Carillion to move trade credit (monies owed to supplier firms) to 'other creditors', through use of bank facilities providing prompter payment for works due to firms in the supply chain finance scheme (also known as the 'reverse factoring scheme' – Prime Minister's Office 2012). This leads us to a potentially universal challenge for the prediction of failure using IPM methods, or indeed the use of accounts for reliable insight on a firm's performance and risk. The data from which IPMs generate information should not be considered a purely objective and precise measure of firms, individually or collectively. Accounting discretion can undermine the impartiality of financial data.

Like any firm strategy, outstensive growth business models that rely on the purchase of intermediate output of other firms have their trade-offs. This prevailing tier 1 model is considered by many to have had its day. Since Carillion, some creditors, including banks, have further reduced their appetite for exposure and investment in construction industry firms. This has implications for the prospective growth of offsite construction processes. These will require much greater external financial investment into the industry. If such business models emerge, they will pose new challenges for IPM approaches designed to predict typical contractor failure. Applying insights from investigation of manufacturing firms could no doubt be valuable in such modelling.

Recommendations

One area for advancement in IPM modelling includes alternative classification of industry through a business models lens based on their economic characteristics, for example their 'make/buy' position. SIC codes are not sensitive enough indicators in this regard, often referring more to the form of output produced rather than the technology or processes used in production. 'Make/buy' perspectives allow an alternative approach from which to generate new classification of firms based on cost and financial structures, and as such can draw on the financial accounts of firms as a basis. Such alternative classifications might prove a more effective basis for grouping firms in IPMs instead of SIC codes.

Cash flow statements provide additional valuable information for predicting insolvency. However, they are only available for the largest firms due to reporting requirements. With advancements in data science approaches and increased accessibility of financial information, there is scope for development of automated cash flow statement imputation. This would involve using balance sheet and income statement data for firms to reverse engineer the cash positions of firms. Applied systematically across whole industries, this would materially increase the dataset for IPMs to draw upon.

With regard to the frequency for firms' reporting of financials, the arbitrary use of yearly periods belies the reality of the varying levels of risk taken by firms over this period. Firms which turn over their capital much more quickly by definition expose their balance sheets to more opportunity for failure from loss-making activities. As such, a more sensitive, if not practical, periodic basis for reporting of financials might include the requirement for firms to report accounts in periods related to the rate at which they expose themselves to failure risks. For example, a requirement for firms to report accounts every time capital is turned over, say, twice, this would lead to a typical contractor providing financials twice a year (given turnover capital ratios of 3.5–4.5 are usual).

Government's wider use of supplier financials is an encouraging sign for improving public procurement practices. However, like any other analyst using financial indicators, they are constrained by the common challenges discussed earlier. One aspect the government guidance neglects through deeper application of 'make/buy' concepts, includes the extension of this framework to consider where suppliers to public authorities themselves are on the 'make/buy' continuum. Given that the delivery model of a tier 1 supplier to public authorities influences the financial characteristics of the firm, guidance would be more relevant if it encouraged outsourcing procuring authorities to consider their suppliers' 'make/buy' characteristics to inform government's commercial strategy. As indicators approach zero, such as margins for contractors, greater precision is needed in their measurement and estimation. Whether a 2, 3, or 4 percent profit margin is deemed high risk has significant implications for the number of firms considered to have poor economic and financial standing. Government guidance often applies a market lens focusing on the demand which public procurement generates. This is in contrast to an industrial perspective by considering the supply context from the perspective of the firms they procure from. A shift towards the latter may help public procurement achieve their strategic aims in anticipating how firms individually, as well as the industry as a whole, will adapt to changes in policy.

Notes

1 The term *credit* originates from the Latin term *credo*, meaning 'I believe'.
2 There are important differences between insolvency and bankruptcy. Insolvency is the point at which a firm is unable to pay liabilities when due because of insufficient access to cash, and as such is a financial conception. Such entities may still have illiquid assets that, in time, can be sold to pay creditors who can continue to seek settlement. Bankruptcy, is typically a later stage and is a legal process that determines an entity is no longer able to settle any outstanding debts, having either liquidated, or had ceased, all assets to meet liabilities. Once bankruptcy is reached, creditors are no longer able to seek settlement.
3 Note that all monetary amounts in this chapter are in GBP (£).
4 Technology, coming from the Greek word *techne*, meaning making or doing. The application of this terms to industries goes beyond the conventional concept of physical machinery or tools, but also includes techniques, skills, knowledge. In the context of contractors, this may include expertise around tendering or programme management practices, knowledge of specific regulations and financial management of suppliers.
5 LASSO is an abbreviation for 'least absolute shrinkage and selection operator'. It is a form of regression analysis that assist in both variable selection and model optimisation.
6 Liquid assets to total assets; cash flow return on total assets; total assets turnover; receivables to payables; long-term liabilities to capital employed; return on total assets.
7 Each search step is applied independently of one another. This means the results from each can have larger samples than alternative stages. The Boolean sampling step is the essential stage at which firms that meet the desired characteristics are aggregated.
8 Four firms were removed: a Special Purpose Vehicle for a Private Finance Initiative contract (Severn River Crossing PLC) was removed as its liquidation was always anticipated on completion of the contract; LL Realisations Limited and MFL Realisations 2017 Limited were removed as they were subsidiaries of Lakesmere (a firm within the sample); one other firm was removed as it was confirmed to be in-house works contractors for a large retail corporate.
9 The difference between net value added (remuneration + profits) and gross value added is depreciation of capital. This tends to be low in construction generally, as a labour-intensive activity, and particularly low for some contractors that rely on the use of leased plant when undertaking capital-intensive works in-house.

References and further reading

Abidali, A.F. and Harris, F. (1995) A Methodology for Predicting Company Failure in the Construction Industry. *Construction Management and Economics*, **13** (3), 189–196.

Ahn, B.S., Cho, S.S. and Kim, C.Y. (2000) The Integrated Methodology of Rough Set Theory and Artificial Neural Network for Business Failure Prediction. *Expert Systems with Applications*, **18** (2), 65–74.

Alaka, H.A., Oyedele, L.O., Owolabi, H.A., Ajayi, S.O., Bilal, M. and Akinade, O.O. (2016a) Methodological Approach of Construction Business Failure Prediction Studies: A Review. *Construction Management and Economics*, **34** (11), 808–842.

Alaka, H.A., Oyedele, L.O., Owolabi, H.A., Oyedele, A.A., Akinade, O.O., Bilal, M. and Ajayi, S.O. (2016b) Critical Factors for Insolvency Prediction: Towards a Theoretical Model for the Construction Industry. *International Journal of Construction Management*, **17** (1), 25–49.

Altman, E.I. (1968) Financial Ratios, Discriminant Analysis and the Prediction of Corporate Bankruptcy. *The Journal of Finance*, **23** (4), 589–609.

Altman, E.I. (1984) The Success of Business Failure Prediction Models. *Journal of Banking & Finance*, **8** (2), 171–198.

Altman, E.I., Haldeman, R.G. and Narayanan, P. (1977) ZETA Analysis A New Model to Identify Bankruptcy Risk of Corporations. *Journal of Banking & Finance*, **1** (1), 29–54.

Arditi, D., Koksal, A. and Kale, S. (2000) Business Failures in the Construction Industry. *Engineering Construction and Architectural Management*, **7** (2), 120–132.

Balcaen, S. and Ooghe, H. (2006) 35 Years of Studies on Business Failure: An Overview of the Classic Statistical Methodologies and Their Related Problems. *The British Accounting Review*, **38** (1), 63–93.

Ball, M., Farshchi, M. and Grilli, M. (2000) Competition and the Persistence of Profits in the UK Construction Industry. *Construction Management and Economics*, **18** (7), 733–745.

Barkham, R. (1997) The Financial Structure and Ethos of Property Companies: An Empirical Analysis. *Construction Management and Economics*, **5** (15), 441–456.

Beaver, W.H. (1966) Financial Ratios as Predictors of Failure. *Journal of Accounting Research*, **4**, 71–111.

Bureau van Dijk (2020) *Fame*. www.bvdinfo.com/en-gb/our-products/data/national/fame.

Cabinet Office (2019) *The Outsourcing Playbook: Central Government Guidance on Service Delivery, Including Outsourcing, Insourcing, Mixed Economy Sourcing and Contracting* (London: Cabinet Office). https://assets.publishing.service.gov.uk/government/uploads/system/uploads/attachment_data/file/816633/Outsourcing_Playbook.pdf.

Cabinet Office (2020) *The Construction Playbook: Government Guidance on Sourcing and Contracting Public Works Projects and Programmes* (London: Cabinet Office). https://assets.publishing.service.gov.uk/government/uploads/system/uploads/attachment_data/file/941536/The_Construction_Playbook.pdf.

Chava, S. and Jarrow, R.A. (2004) Bankruptcy Prediction With Industry Effects. *Review of Finance*, **8**, 537–569.

Chen, J.-H. (2012) Developing SFNN Models to Predict Financial Distress of Construction Companies. *Expert Systems with Applications*, **39** (1), 823–827.

Companies Act (2006) www.legislation.gov.uk/ukpga/2006/46.

DeFond, M. and Jiambalvo, J. (1994) Debt Covenant Violation and Manipulation of Accruals. *Journal of Accounting and Economics*, **17** (1–2), 145–176.

de Freytas-Tamura, K. (2018) Collapse of U.K. Construction Giant Rattles the Government. *New York Times*, January 15. www.nytimes.com/2018/01/15/world/europe/carillion-bankruptcy-outsourcing-britain.html

Dimitras, A.I., Zanakis, S.H. and Zopounidis, C. (1996) A Survey of Business Failures With an Emphasis on Prediction Methods and Industrial Applications. *European Journal of Operational Research*, **90** (3), 487–513.

Gruneberg, S.L. and Ive, G. (2000) *The Economics of the Modern Construction Firm* (Basingstoke: Macmillan).

Ive, G. and Murray, A. (2013) *Trade Credit in the UK Construction Industry: An Empirical Analysis of Construction Contractor Financial Positioning and Performance*. BIS Research Paper No. 118 (London: Department for Business, Innovation and Skills).

Ive, G.J. and Gruneberg, S.L. (2000) *The Economics of the Modern Construction Sector* (Basingstoke: Palgrave Macmillan).

Kale, S. and Arditi, D. (1999) Age-Dependent Business Failures in the US Construction Industry. *Construction Management and Economics*, **17** (4), 493–503.

Kapliński, O. (2008) Usefulness and Credibility of Scoring Methods in Construction Industry. *Journal of Civil Engineering and Management*, **14** (1), 21–28.

Kenley, R. (2003) *Financing Construction: Cash Flows and Cash Farming* (London: Routledge).

Kulakov, A. (2017) *UK Construction Industry Business Models: Empirical Failure Prediction of Developers and Contractors With Use of Financial Accounts*. BSc Dissertation (London: University College). bit.ly/2kxl3KU.

Mensah, Y. (1984) An Examination of the Stationarity of Multivariate Bankruptcy Prediction Models: A Methodological Study. *Journal of Accounting Research*, **22** (1), 380–395.

Metz, A. and Cantor, R. (2006) *The Distribution of Common Financial Ratios by Rating and Industry for North American Non-Financial Corporations*. Moody's Special Comment. www.moodys.com/sites/products/defaultresearch/2005700000436062.pdf.

Murray, A. and Kulakov, A. (2020) Make/Buy Decisions in International Construction Industry Firms. In: Gruneberg, S.L. (ed.) *Global Construction Data* (1st edition, pp. 149–166) (Abingdon: Routledge).

ONS (2016) *UKSIC2007* (London: Office of National Statistics). www.ons.gov.uk/methodology/classificationsandstandards/ukstandardindustrialclassificationofeconomicactivities/uksic2007.

ONS (2021) *Output in the Construction Industry* (London: Office of National Statistics). www.ons.gov.uk/businessindustryandtrade/constructionindustry.

NAO (2018) *Investigation Into the Government's Handling of the Collapse of Carillion*. National Audit Office. www.nao.org.uk/wp-content/uploads/2018/06/Investigation-into-the-governments-handling-of-the-collapse-of-Carillion.pdf.

Ng, S.T., Wong, J.M.W. and Zhang, J. (2011) Applying Z-Score Model to Distinguish Insolvent Construction Companies in China. *Habitat International*, **35** (4), 599–607.

Prime Minister's Office (2012) *Prime Minister announces Supply Chain Finance Scheme*. Gov.UK. https://bit.ly/2THPYpo.

Ross, S., Westerfield, R. and Jaffe, J. (2005) *Corporate Finance* (7th edition) (New York: McGraw-Hill).

Russell, J. and Jaselskis, E. (1992) Predicting construction contractor failure prior to contract award. *Journal of Construction Engineering and Management*, **118** (4), 791–811.

Russell, J. and Zhai, H. (1996) Predicting Contractor Failure Using Stochastic Dynamics of Economic and Financial Variables. *Journal of Construction Engineering and Management*, **122** (2), 183–191.

Severson, G., Russell, J. and Jaselskis, E. (1994) Predicting Contract Surety Bond Claims Using Contractor Financial Data. *Journal of Construction Engineering and Management*, **120** (2), 405–420.

S&P (2018) *Carillion's Demise: What's at Stake? S&P Global Market Intelligence*. www.capitaliq.com/CIQDotNet/CreditResearch/RenderArticle.aspx?articleId=2011452&SctArtId=450735&from=CM. Accessed 21 April 2020.

Tserng, H.P., Liao, H.-H., Jaselskis, E., Tsai, L.K. and Chen, P.-C. (2012) Predicting Construction Contractor Default with Barrier Option Model. *Journal of Construction Engineering and Management*, **138** (5), 621–630.

Williamson, O. (1979) Transaction-Cost Economics: The Governance of Contractual Relations. *Journal of Law and Economics*, **22**, 233–261.

Wood, D. and Piesse, J. (1987) The Information Value of MDA Based Financial Indicators. *Journal of Business Finance & Accounting*, **14** (1), 27–38.

Editorial Comment

In 1946 cartoonist Chester Gould introduced Dick Tracy's famous Two-Way Wrist Radio, an invention that lived in the realm of science fiction 75 years ago but now appears rather old fashioned as smart watches and other digital devices driven by tiny computer chips are produced in the billions every year. The rapid development of digital technologies has driven profound changes in the way most, perhaps all, industries operate, from farmers who use satellite navigation systems to monitor livestock and control harvesters and other equipment, to micro businesses that can only survive now if they have electronic payment options for their customers, and it is the ever more powerful chips that lie at the heart of the latest industrial revolution.

The construction sector is often described as being slow to adopt change, and examples such as the continued practice of laying masonry units (bricks, blocks and similar) in a traditional manner that has, in some respects, changed little in centuries, suggest that there is some truth in this perception. While there have been plenty of attempts to design viable bricklaying machines none have been particularly successful due, at least in part, to our liking for bespoke building designs that generally require some degree of traditional, craft-based workmanship for them to be realised on site. There are also plenty of practical problems that limit the use of such machines, including job-specific issues such as sloping and / or uneven terrain, restricted site access and limited room to manoeuvre on site.

Even within such traditional practices as bricklaying there have been significant advances; for example, laser levelling devices have largely replaced the old string line, and even bricks themselves have been re-imagined with the advent of products such as polymer-modified mortar that allow thinner mortar joints and faster laying. The basic process, however, still relies mostly on the units being laid, by hand, one by one, and many of the changes have been small tweaks to existing methods rather than any sort of radical re-engineering of the construction process.

A parallel can be drawn between, for example, remotely controlled excavators and remotely controlled airborne drones that deliver ordnance to targets many thousands of kilometres from where their 'pilots' are sitting. In neither case is the device acting by itself, and in the case of the drone, the decision of whether to fire or not is still up to the controller. It may not be long, however, before the drones become autonomous, as developments in artificial intelligence (AI) are likely to eventually make the creation of 'thinking' machines a reality, at which time the decision to fire on a target may be taken by the machine rather than a human controller. If that is possible then the advent of autonomous construction robots will surely follow, and we are likely to see radical shifts in the way buildings are built.

The development of AI is just one of the possibilities discussed in the following chapter as the author puts forward some suggestions on where the construction may be headed in years to come.

12 The race to the future for the construction sector

Roger Flanagan

An international perspective

Annual global construction of work put in place in 2020 is estimated to be worth around USD 9 trillion, exclusive of land, local taxes and construction professional services. Estimates of annual output rely upon national statistics; amounts may be distorted by output from the informal sector[1]/black economy which is unreported in many countries. Lack of consistent available data on construction output is a major challenge, particularly for developing countries. Annual global construction output for the sector represents around 10 percent of global gross national product, demonstrating the vital importance of the construction sector to the economy in every country. There has been year-on-year growth in construction output since 2010, but just like the financial crisis in 2008–2009 that disrupted the markets, COVID-19 is impacting and influencing change in the markets.

Structure of the construction sector

The terms 'construction industry' and 'sector' are often used interchangeably; that is wrong, the sector comprises a much wider grouping than the industry.[2]

The construction sector is fragmented and complex, with four principal groups:

1. Knowledge intensive professional service (KIPS) providers, the architectural, civil, structural, mechanical, electrical and plumbing engineers, as well as cost, project and programme management consultants.
2. Construction contractors and specialty trade contractors.
3. Materials, plant, equipment and component manufacturers and suppliers, including merchants.
4. Ancillary service providers, such as IT developers, financiers, insurers, transport, security, logistics, accountants, legal advisers and regulators, all of whom may work across several sectors.

They are all dependent upon the market for construction works but have different net profit margins, see Figure 12.1; the return on capital employed and perceptions of risk are also an influence. Net profit margins of 1 to 4 percent for

DOI: 10.1201/9781003102403-12

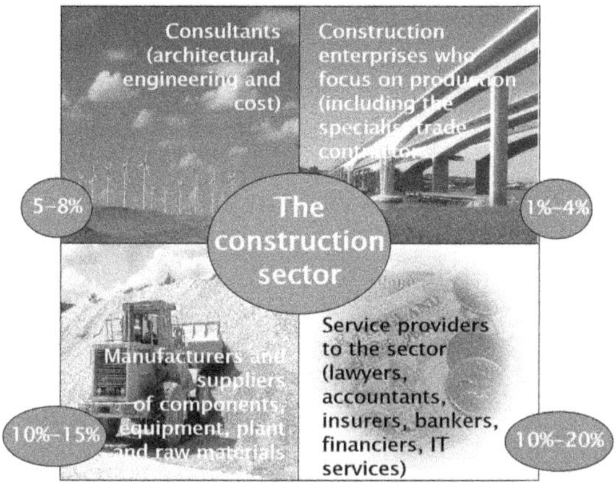

Figure 12.1 The four principal groups in the sector and their net profit margins

construction enterprises are not conducive to building resilient companies. There are drivers and issues that influence the market, and this chapter focuses upon the construction market, which ultimately impacts all the players.

Construction projects are location-specific, and economies of scale are not easily achieved, with the need for local labour and local materials to keep down costs. Planning and regulatory codes and standards are local and are designed to meet local conditions and requirements. Some standards for materials, workmanship and design are international, such as ISO standards.

The perfect storm

International construction is being transformed by the influence of the fourth industrial revolution (FIR), with COVID-19 changing the way the industry works. There are changes to market making and project creation, project procurement, architectural and engineering design, new governance requirements and more regulations, and more off-site site production.

Four drivers are coming together to create the perfect storm. Firstly, the FIR, with digitalisation and new technology transforming and disrupting all industry sectors. Secondly, the growing importance of combatting climate change and achieving zero carbon. Thirdly, the tension between internationalisation and localisation, influenced by geopolitical factors. Fourthly, pandemics, with COVID-19 causing economic and social disruption and re-alignment of production. Construction sector enterprises must respond because they are such an important part of helping countries return to economic growth and prosperity. There is a need for enterprises to rebuild their balance sheets and investment programmes and to re-think their business models.

The old world of international construction was about enterprises using models that worked in their home country, believing that they must be superior and would work overseas. The separation of architectural and engineering design from site production is such an example. The new world is about flexibility, integration, adaptation, modernisation, localisation, governance, transparency and conformance to new standards. Western enterprises no longer believe that home country technology or systems are superior. They source globally and act locally.

The design, build and throwaway society is no longer appropriate, with the move towards a circular economy. A circular economy is an industrial system that is restorative or regenerative by intention and design. It replaces the end-of-life concept with restoration, shifts towards the use of renewable energy, eliminates the use of toxic chemicals, which impair reuse and return to the biosphere, and aims to eliminate waste through the superior design of materials, products, systems and business models.

Change is fast moving and not respectful of geographic boundaries. Charles Darwin wrote, 'It is not the strongest of the species that survives, nor the most intelligent that survives. It is the most adaptable to change.'[3] That statement is still true in 2021.

The nature of the construction business requires local production when working in any country, with few exceptions. Site production involves the technical, financial and management capacity, the production process and the assembly of an integrated supply chain. Financial capacity is the glue that bonds the process together. Scale is an important part of international success, and micro, small and medium-sized enterprises find it difficult to go overseas. Chinese enterprises are an example of how scale has provided the ability to exploit overseas opportunities.

The 'new kids on the block' have arrived, many of whom were the Asian tigers in the 1990s. Turkish materials and component producers, and construction firms have exploited the new opportunities by using their low-cost advantage in developing countries. The 'new kids' have been quick to adopt new technologies and to build new business models that better reflect the international nature of the construction sector. Their flexibility and innovativeness have given them opportunities to compete against more established companies. However, the increasing complexity of construction and the need to adopt collaborative funding opportunities (especially for infrastructure) has led to the large players being able to create competitive advantage through project creation.

The aim to be global but act local is still prevalent in many large companies in the sector. The way they achieve this differs across countries/regions and business types. For example, construction professional service firms, whose 'currency' is their skill base, competencies and intellectual capital, use different growth strategies to contractors, primarily because they are more risk averse. However, the sector is witnessing international growth being a popular policy. Mergers and acquisitions have played a major role with the acquisition of companies established in a particular country/region with a proven track record.

Being global but acting on a local basis relies upon trading freedoms and operating on a level playing field; but the field is skewed by increasing protectionism. Localisation policies are not a new phenomenon and are prevalent in some countries, including Argentina, Brazil, China, India, Indonesia, Russia, Saudi Arabia

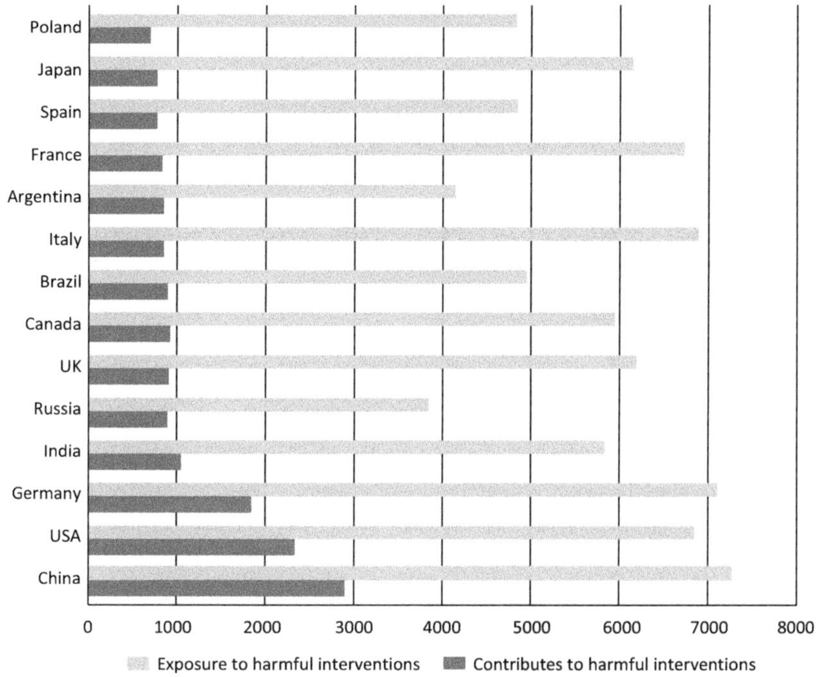

Figure 12.2 Exposure/contribution to harmful interventions

and the USA (Deringer *et al.* 2018). Figure 12.2 shows the number of interventions by the countries ranking highest (for numbers of interventions). Interventions are trade-related policies that are import restrictive and export related.

The interventions were across all sectors. Some interventions are liberalising. For example, China benefited from 2001 of these in the latest Global Trade Alert (2021) figures.

> *[Localisation] policies are imposed by governments that require firms to use domestically manufactured goods or domestically supplied services in order to operate in an economy.*
> (OECD 2016: 4)

Localisation

Whilst the aim may be to improve/protect local business, in an economy with inefficient firms, a high degree of required local content thwarts competition. Increasing localisation/protectionism is changing the face of global competitiveness. China is engaging in 'import substitution' through a 'buy Chinese' policy. Licensing agreements discriminate against foreign companies and Chinese companies benefit from cheap capital through a state-directed financial system (Springford 2020). India has announced its intention to seek self-reliance. It is prohibiting global tenders for some government procurements to help local businesses sell to government. The intention is to build local brands and make them world class. It's not about looking inwards or being isolationist; it's about a confident India that contributes to the globe.

Localisation is one of the tensions between internationalisation and protectionism. Globalisation has changed the landscape of international construction with more international design teams, the emergence of Chinese construction companies as a global force, greater migration of labour, and many 'new kids on the block' from a variety of countries seeking to win construction projects. Scale and influence have been important with big is best, big meaning greater competence. However, higher revenue has not always been converted to bigger profits.

Fourth industrial revolution and construction

The FIR, unlike its predecessors, has not merely built upon the innovations and progress of the third industrial revolution. Instead, it presents a distinct change because of its speed, scope and impact on systems. The FIR comprises a fusion of technologies from different disciplines with their edges getting increasingly blurred, blending the physical, the digital and the biological worlds. The FIR is disruptive by bringing rapid change, which is exponential, rather than linear as in previous revolutions. Production and management systems and governance need to adjust to reflect the changes the FIR brings.

The four industrial revolutions followed a shallow curve at their inception, gathering pace, but nothing like the speed of the third and fourth industrial revolutions; see Figure 12.3.

The technologies shown at the start of the fourth industrial revolution are in no particular order, and each of them has been increasing exponentially but at different

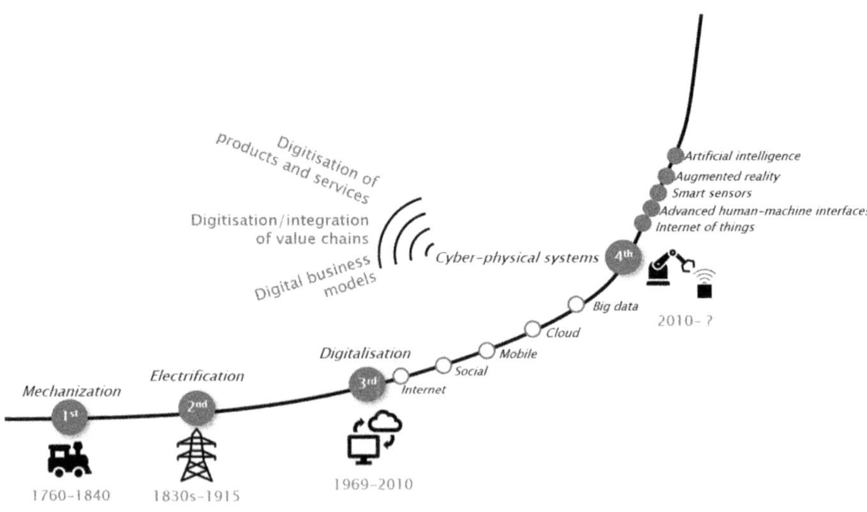

Figure 12.3 Moving towards the fourth industrial revolution

rates. The take-up of these technologies has varied from sector to sector. The common factor is the information overload that accompanies this rate of innovation.

Digitalisation is a core part of FIR, but it has experienced a slower pick-up in the construction sector, compared with other sectors, although BIM, off-site fabrication, sensors and automated equipment are evidence of change. Construction still relies heavily on manual labour and technology that has not dramatically improved over the last few decades. Digital innovations in the industry could see costs cut and productivity increased but its fragmentation is a considerable barrier. There is a 'long tail' of small- and medium-sized enterprises (SMEs) in most countries' construction sectors which have low profit margins and limited cash flow.

As technologies become cheaper this can change. Smartphones have led the way towards greater accessibility to information and communication technologies, with wireless technologies no longer leaving a construction site a disconnected island.

Facing the future

The six World Economic Forum (WEF) working groups, part of the *Shaping the Future of Construction* project, developed three infrastructure and urban development (IUD) scenarios: *Factories Run the World, The Green Reboot* and *Building in a Virtual World* – see Figure 12.4.

The WEF is calling for action for a more circular economy with a shift from 'grey' to 'green'. Increasing life cycle thinking in the construction sector is a major step towards this goal. The WEF states that growth should not be the sole aim of the global economy but, instead, a greater distribution of wealth. These changes will be enabled by greater innovations in terms of energy, food, safety and shelter. The threat of climate change requires innovation in self-healing systems as well as being smart and robust. Affordability is key if innovations are to get a widespread take-up. Human-centred approaches are also important with the increase of cyber-physical systems in the fourth industrial revolution.

> *Art, culture, music, a sense of place and a feeling of belonging will increasingly be seen as critical to the physical and mental happiness and well-being of the global*

Factories run the world
In an era where people are immersed in virtual reality in all aspects of life, intelligent systems and robots run the construction industry

The green reboot
A corporate-dominated society uses prefabrication and modularization to create cost-efficient structures

Building in a virtual world
A world, with increasing conflicts over scarce resources and climate change, rebuilds using environmentally-friendly construction methods and sustainable materials

Figure 12.4 The World Economic Forum's future scenarios

> *population, thus fostering a culture of enjoyment and a personal creative responsibility for socially stable and vibrant communities.*
>
> (WEF 2018: 5)

Change – the drivers, issues, disruptors, enablers and actions

An approach to make sense of the future is to use drivers, issues, disruptors, enablers and actions required.

Drivers

Figure 12.5 shows the drivers of global construction as a system of cogs, a dynamic process that faces constant and rapid change. The eight drivers selected have a major impact on global construction. They are not self-sustaining but are interdependent and interconnected. All are impacted by the growing complexity, risk and speed of change within the sector.

The drivers lead to issues around the eight cogs that must be confronted by governments, policy makers and companies.

The environment is an example of interconnectivity; the issues that arise are closely connected with both tackling climate change and embracing new technologies. The new technologies can help to find solutions for climate change, but they involve social responsibility with actions to be taken by individuals to reduce their carbon footprint.

Drivers are difficult to control, whilst the issues must be addressed. They manifest themselves on both a global and local scale and are influenced by the growing complexity, speed of change and attitudes that exist today.

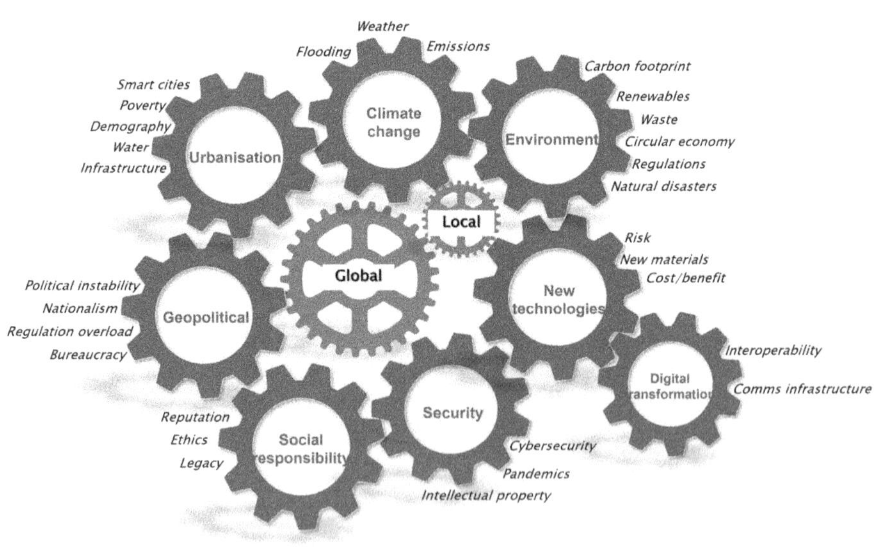

Figure 12.5 Drivers and issues impacting global construction

Complexity

Projects have become more complex with new technologies, increased demands of clients and requirements for sustainability. Complexity affects the business environment as well as the workforce. It puts pressure on the ability to manage change in both processes and technologies.

Speed of change

The speed of change is apparent in technologies, the competitiveness landscape and the regulatory system (both local and national). Firms in the construction sector are having to be nimbler and more flexible to meet the pressures of rapid change. The choice is to be a leader or a follower. Lessons can be learned from the approach used for innovation discussed later.

Globalisation

Globalisation has brought about stronger connections. Pandemics are more prevalent as social and business mobility increases. Some regulations have moved beyond domestic borders to become supranational. Business risks are different from just a decade ago. For example, cybersecurity is a more prevalent issue. Compliance risks have changed with the increase in regulations.

Issues

Issues may be local, national or firm-specific, and their level of impact can change over time. The urbanisation driver is an example. Figure 12.6 shows some of the issues derived from urbanisation.

By 2050, 6.5 billion people will live in urban centres – two-thirds of the projected world population. This movement of the population has huge impact on cities' resources as well as their environment. There is a growing need for *smart cities* that make better use of smart technologies. This requires three things: 1) the development of a technology base for connected networks and devices, 2) smart applications and the ability for data analysis and 3) applications and usage, which means educating users and behavioural changes. The pressure on water increases with rapid urbanisation. Both a safe and accessible *water* supply and the

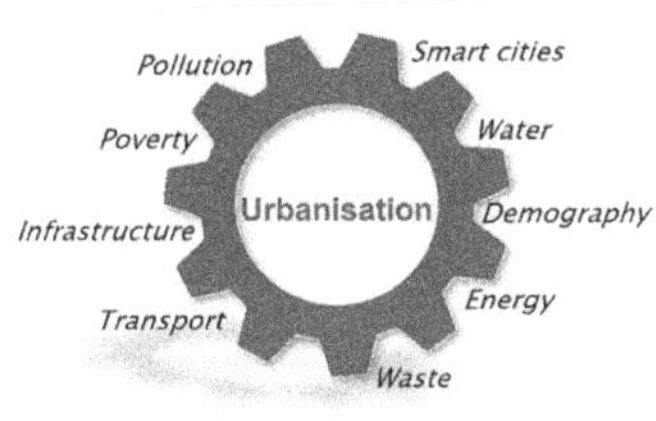

Figure 12.6 Urbanisation drivers and issues

management of wastewater including sewerage helps to maintain the sustainability of a city and the health of its inhabitants.

Demographics are changing with increasing single-person households and, in many countries, an ageing population. Demography will impact the types of construction needed and the ability to future-proof buildings. *Energy* requirements in a city can be huge and diverse. Whatever the size of either the city or the energy requirements, there are three essential obligations that need to be met: a reliable and secure supply; long-term affordability; and the reduction of greenhouse gas emissions associated with energy supply. Constructing new buildings or refurbishing/retrofitting existing ones provides an opportunity to achieve better energy management, reducing usage, costs and emissions.

Reduction and management of *waste* has many benefits: reduction, recycling and re-use (the 3 Rs), which are part of the zero-waste hierarchy; see Figure 12.7. Zero-waste policies are becoming increasingly popular and have a significant impact on the construction sector, especially as it is responsible for a large percentage of the world's waste.

The volume of construction waste generated worldwide every year will nearly double to 2.2 billion tons by 2025 (Transparency Market Research 2018).

The need for sustainable *transport* systems within cities increases as the population grows. This provides an opportunity for construction companies for both the infrastructure and the associated facilities. Construction work may be affected if a city lacks a good transport system, which may cause severe congestion, adding time and money to a project.

The availability and quality of *infrastructure*, whether it is for transport, communication or utilities, is an issue for construction. Although the lack, or inefficiency,

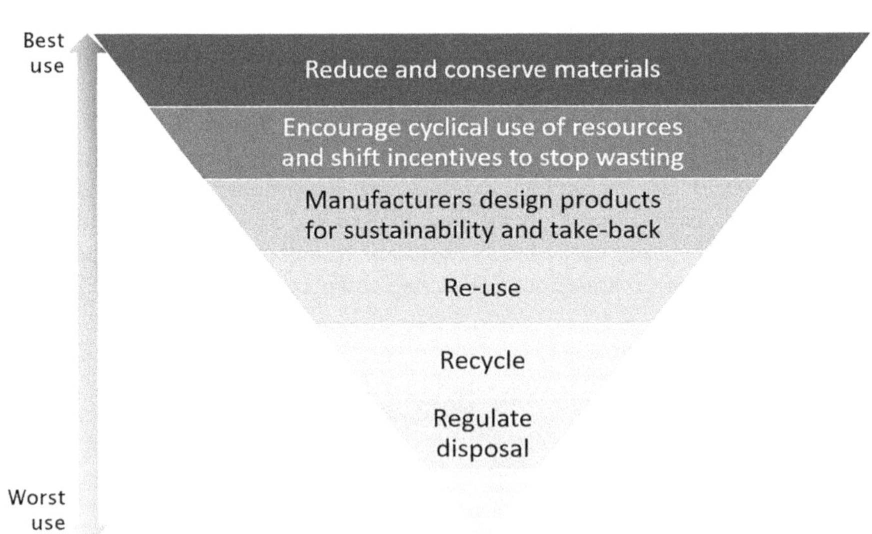

Figure 12.7 The zero-waste hierarchy

of any infrastructure will be a challenge, it can present an opportunity for construction firms.

Where infrastructure development/growth has not kept pace with rapid urbanisation, *poverty*/inequalities can/are likely to exist. International agencies focusing on the eradication of poverty are often the source of funding for the development world – an opportunity for construction.

Air, ground, water and noise *pollution* are all impacted and created by construction. Environmental legislation (local, national and international) needs to be understood and complied with. Pollution is an important issue in terms of site operations and building system management.

Disruptors, enablers and actions

Disruptors can negatively impact or exacerbate issues; they must be minimised/managed to avoid significant effects on projects and on the business. Disruptors can be anything from severe weather to a workforce strike to civil unrest. *Enablers* are the ways in which the disruptors can be prevented/ameliorated. These need to be enhanced and the disruptors reduced through formulated company-specific *actions*.

Disruptors can be on a local, national, or global scale. Global-scale disruptors are pandemics, such as the recent COVID-19 pandemic, and the global financial crisis. Events on this sort of scale are difficult to combat. Finding enablers that would reduce the disruption is difficult. However, on a more local scale, the disruption caused by bad weather can be foreseen (unless it is a catastrophic event) and steps taken by, say, providing site protection or planning construction at a different time of the year. Workforce disputes may be avoided by better communication and/or employee empowerment.

There are market disruptors, such as the influence of Chinese companies in the international construction market. China's domestic construction market is the largest single market in the world, yet there are few foreign firms who have entered the market.

The concept of drivers, issues and disruptors is important with the huge changes the fourth industrial revolution is bringing.

The emergence of China as a major influence in construction

Large Chinese enterprises have followed a strategy for international growth over the past 20 years based upon the competitive advantage of low price, access to project finance, leveraging of China's geopolitical influence overseas and market making/project creation abilities. Price is always a critical factor in any project. Building upon the low-cost Chinese supply chain to go overseas with materials, plant, equipment and human resources proved a good base for winning overseas work. A mobile and flexible Chinese workforce was prepared to work overseas in difficult conditions, but that is changing as the living standards improve and wages rise in China.

Chinese companies have the advantage of a largely protected domestic market in China. Foreign construction companies are prohibited by law from participating in all public tenders in the construction sector, whilst so-called foreign companies registered as 'wholly-foreign owned enterprises' are only allowed to take part in tenders that are financed by non-Chinese authorities, i.e. foreign investors and multilateral institutions.

Geopolitical influence is leading to conflict between internationalisation and localisation in many overseas markets, with demands for capacity building for local industries with employment opportunities, social responsibility, stringent local contract requirements and local standards.

Chinese enterprises need to create a new operational and business model going forward, built upon the success of the past and recognising the challenges of the future. The propensity to use Chinese inputs for project delivery on international projects will change with digitalisation, technology and integration of the design, with off-site and on-site production playing a bigger role. It makes economic sense to source locally instead of artificially sourcing inputs from overseas; that principle works in developed countries, but not in developing countries where the quality standards can be poor and there are limited skills availability.

Chinese enterprises have used project finance at attractive loan rates to engage in project creation/market making. China has a large overseas development aid programme in developing countries, with development assistance provided in the form of infrastructure projects given as gifts, grants, interest free and concessional loans, disaster relief and other forms of technical assistance. Work undertaken on the projects is generally undertaken by Chinese enterprises. For example, the Belt and Road Initiative (BRI) has created opportunities overseas, alongside projects funded by the multilateral banks, such as the World Bank/IFC, and the regional development banks.

The BRI (also known as the Silk Road) has opened new opportunities for Chinese investment overseas. The BRI links China with south-east Asia, south Asia, central Asia and Europe by land as well as by a sea route connecting China's coastal regions – see Figure 12.8.

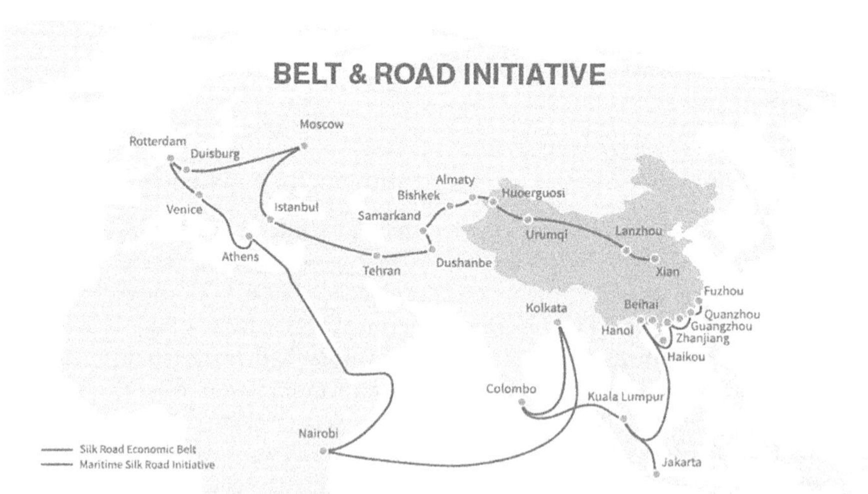

Figure 12.8 The map of the Belt and Road initiative (McKinsey 2016)

Competitors in the international market, often with a longer history of working overseas, have watched, and learned from China's international strategy. They are seeking new ways to compete based upon operational excellence by investing in market making. This means creating projects through PPP and other ways of creating opportunities. Competitors are demanding a level playing field,[4] where enterprises do not gain an unfair advantage by receiving aid or subsidies from their home country.

Chinese construction companies are growing organically in developing and other countries; they are also growing through acquisition, as was the case in Australia and the USA. Competition in the global market from Chinese construction companies changed the strategy adopted by many western construction enterprises. China Communications Construction Company (CCCC) acquired the Australian firm John Holland in 2015, but failed to acquire the Canadian construction company, Aecon, in 2017, following intervention by the Canadian government. The number of Chinese construction companies has increased with the top five in ENR's[5] 2020 Top 250 Global Contractors (based on construction revenue generated domestically and overseas) being Chinese.

An inhibitor to international growth is the lack of acceptance of Chinese standards and codes of practice as being to the highest international standards for materials, plant and equipment in the international market. Whilst China is a member of the International Standards Organisation (ISO), standards are very important in gaining acceptance of Chinese manufactured goods to be incorporated in overseas projects. Standards can be at the national level, local level, industry level and enterprise level. Having Chinese-manufactured steel used in a project in the USA means the material must meet the highest international standard, and conform to the local standards.

Another inhibitor is the development of KIPS for independent design and consultancy. China's fourteenth five-year plan (2021–2025) emphasises the development of professional services as a priority, which means more Chinese KIPS firms will enter the international market. An enabling factor is that many large Chinese design institutes are owned by large state-owned construction enterprises, which provides the opportunity to deliver integrated design and construct projects.

How companies respond to the future

Leader, fast follower, pack member, late mover and laggard

Firms in the construction sector are having to be nimbler and more flexible to meet the pressures of the rapid change shown in the cogs (Figure 12.5). The choice is to be a leader, fast follower, pack member, late mover and laggard; Figure 12.9 shows the innovation hierarchy, with the different types of enterprise behaviour. The concept is relevant to international construction enterprises from the perspective of how projects are procured and delivered. KIPS providers can leverage a technology, such as 5D digital design with BIM, through the integration of design with site production and into use. Construction enterprises can leverage advantage through the integration of an off-site production manufacturing technology

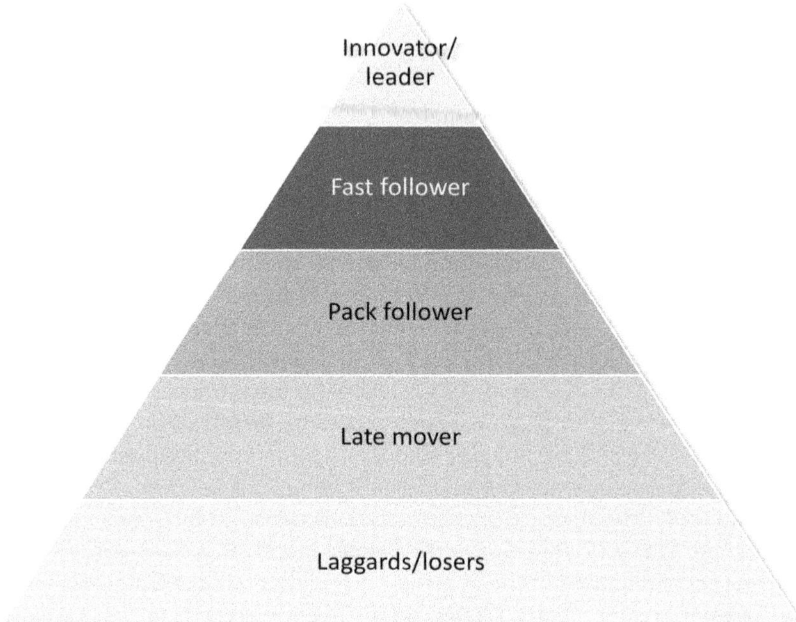

Figure 12.9 The innovation hierarchy model

through site installation, such as in the development of de-salination equipment, or photovoltaic panels through the design, manufacture and installation process. Construction equipment manufacturers exploit technology to produce intelligent and more efficient site equipment that embodies artificial intelligence.

The innovator is the leader pioneering the approach, the trailblazer, and being first to market hoping to achieve the largest market share and best profit potential. The fast followers replicate and duplicate. Once success is clear, the pack followers enter the market. The pack follower and late mover can be left behind, by relying on proven tried and tested approaches; many traditional KIPS firms fit this mould, believing professional services are different. The laggards struggle to catch up, which means being left behind.

The construction sector has reached a tipping point in 2021. The business model based upon price competition is hard to sustain. The traditional triangle of cost, time and quality has changed, and it is now supplemented by the importance of health and safety, sustainability, zero carbon emissions, green issues, ethical and social responsibility, diversity and exploiting digital transformation. Most importantly, project creation has become an important part of winning work. The knowledge of high-speed rail infrastructure in China and Japan has involved bringing together the train manufacturers with the infrastructure design and delivery teams to produce an integrated approach.

Globalisation has brought about stronger connections. Pandemics are more likely as social and business mobility increases. Some regulations have moved beyond domestic borders to become supranational. Business risks are different; cybersecurity, for example, is a more common issue. Compliance risks have changed with the increase in regulations and the dangers of blacklisting. The innovators, sometimes from outside the construction sector, are seeing the opportunities, exemplified by the way Google is applying artificial intelligence to systems.

Stand out, break out, watch out and stall out

Another way of looking at the way companies seek to grow in the international market is the approach often used by digital organisations. They look at countries adopting digitalisation as being stand out, break out, watch out and stall out – see Figure 12.10, and this can be adapted to suit enterprises in construction.

Stand out

Stand out enterprises show high levels of adaptation, and flexibility, especially in innovation and digital development. They remain on an upward trajectory, and they fit with the notion of 'sustaining internationalisation' shown in Figure 12.10. Stand out enterprises in other industry sectors, for example, Uber Technologies Inc. (taxis and ridesharing), Tesla Inc. (automotive) and Ryanair DAC (air transport) are all examples of international enterprises who developed a new business delivery model; they were innovators and, in some cases, fast followers. In construction, many enterprises have looked at moving up the value chain to higher margin activities, and many have become more specialised and focused.

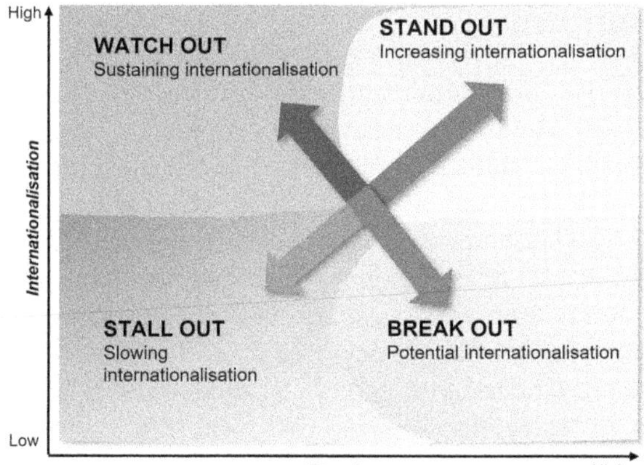

Figure 12.10 Internationalisation and enterprises

Traditional construction has low barriers to entry, it can be highly mechanised and carries high risk. Building projects fit this category, where local enterprises know the local market and can compete on price because they understand the local conditions and local suppliers. High risk is endemic, and companies must ensure that they leverage their competitive advantage. Competition is fierce and low price dominates. By engaging in large infrastructure projects, the barriers to entry into the market are restricted because of the need for specialist competencies and a strong balance sheet to sustain high risks. Many of the large state-owned Chinese enterprises such as CCCC, China State Construction Engineering Company (CSCEC) and China Railways Construction Corporation Ltd (CRCC) are Standing Out; all their businesses are at the leading edge. They are devising new delivery models with focus on their competitive advantages.

Break out

Break out enterprises have the potential to develop strong growth through innovative ideas, or by diversification, either laterally or horizontally, through the supply chain. They are moving upward and are poised to become Stand Out companies in the future. They fit the 'increasing internationalisation' shown in Figure 12.10. Break out companies are emerging in the KIPS enterprises, where many mergers and acquisitions resulted in KIPS design and engineering enterprises becoming very large with 50,000-plus staff on payroll. They believe that larger KIPS enterprises can engage on megaprojects more easily, and they have economies of scale, and invest in technology to create competitive advantage. They have also diversified into the defence, automotive and aviation sectors by providing high value design and engineering services.

In construction, Turkish construction enterprises are winning more overseas work whilst diversifying into investment and manufacturing.

Watch out

Watch out enterprises face significant opportunities, and they fit the 'potential internationalisation' category. A new way of looking at companies is needed, other than profitability and revenue growth. Some overcome limitations with clever innovations and stopgap measures, while others seem to be stuck. Scale is part of the process. The sector is a mix of micro, small, medium and large enterprises. Micro and small enterprises have high rates of failure, hence the desire for growth as a survival principle. Watch out companies face many challenges. Some spectacular company failures occurred recently, caused by a mixture of bad luck, poor management decisions, poor focus and a weak balance sheet. Arabtec in the UAE was a construction business that went into liquidation in late 2020. They grew very fast from 2010 in the Middle East construction market and won many high-risk projects in a market renowned for volatility and intense competition. They suffered losses and went into administration. Carillion plc in the UK, the largest UK construction contractor by revenue, diversified into providing industrial services to hospitals. They won too many public-private partnership projects, and worked on large loss-making projects in the Gulf region. Leadership was poor, financial controls were weak, and they went into liquidation with liabilities of USD 8 billion.

Stall out

Stall out enterprises have achieved a high level of evolution but are losing momentum and risk falling behind; unless they can adapt quickly to the new world of construction, they have slowing internationalisation. Many western construction enterprises fall into this category. Enterprises from the UK, New Zealand, South Africa, Brazil and Malaysia are examples of large construction enterprises who have scaled back their international activities from the international construction market; because of increasing competition they are focusing on domestic and more regional work, which is regarded as having lower risk and uncertainty. Stall out companies include some of the traditional construction enterprises from Brazil, Germany, Italy and the UK, who continued to believe that competing using the traditional model of price competition would secure growth. It failed because of local competition, and enterprises from break out countries, such as Turkey and Vietnam creating new competition. Stall out can occur for several reasons, such as the Lava Jato[6] scandal in Brazil. In 2016, Odebrecht, the Brazilian-based engineering and construction group, signed what has been described as the world's largest leniency deal with the US and Swiss authorities; it confessed to corruption and paid USD 3.5 billion in fines. Dozens of companies acknowledged paying bribes to politicians and officials in exchange for contracts with the state oil company. Odebrecht executives confessed to paying bribes in exchange for contracts in Brazil, and other parts of the world, including Argentina, Colombia, Ecuador, Peru and Venezuela. The company president was sentenced to 19 years' imprisonment. The government has introduced draconian legislation to ensure Brazil is not embroiled in corruption scandals in the future. Whereas companies in Brazil used to boast about their profits, they are now more likely to emphasise how they stick to compliance rules.

Infrastructure and urban development industry scenarios

Tomorrow will not be like today; the current business models are not fit for meeting the challenges of global megatrends with climate change, resource depletion and rapid urbanisation.

The changes will bring disruption to the sector which needs to be addressed by players across the construction value chain. Improving the skill base will be as important as increasing the levels of collaboration and integration. Advanced technologies are key but only if they can operate within a system that allows their take-up, growth and success.

Conclusion

There will be winners and losers in the race to the future. The old business model used by the construction sector for governance, regulation, procurement, design and site production evolved over the past decades; it is robust and stable. Change is now faster moving than ever before, and disruption is a fact of life. The new world is about flexibility, integration, adaptation, modernisation, resilience, localisation, governance, transparency and conformance to new standards. Digital transformation is an enabler, driving the fourth industrial revolution. The construction sector will change; it is resilient, and the companies within the sector will change. Projects will be built faster, safer, and to a higher quality. Localisation will increase with more demands for local content and jobs. Geopolitics will play a bigger role as the pandemic has shown. Government to government agreements will create work, but will be tied and closed to outside competition.

Modern methods of construction will evolve, and productivity will improve. New players will disrupt those companies who believe that tomorrow will be like today.

Notes

1 Definition (OECD 2008): The informal sector typically operates at a low level of organisation, with little or no division between labour and capital as factors of production and on a small scale. Labour relations, where they exist, are based mostly on casual employment, kinship or personal and social relations rather than contractual arrangements with formal guarantees. Taxation, registration, insurances, and employment rights are issues. The construction sector uses a high number of enterprises in the informal sector in developing countries.

2 Although they may seem the same, the terms industry and sector have slightly different meanings. Industry refers to a much more specific group of companies or businesses, while the term sector describes a large segment of the economy.

3 Charles Darwin wrote on 'The Origin of Species by Means of Natural Selection, or the Preservation of Favoured Races in the Struggle for Life', published in 1859.

4 A level playing field is a trade-policy term for a set of common rules and standards that prevent businesses in one country gaining a competitive advantage over those operating in other countries. It is about fair and open competition goods in services and capital. Donald Tusk, President of the European Council, underlined the fact that 'our aim is to focus on achieving a balanced relation, which ensures fair competition and equal market access.' The rebalancing of the economic relationship remains a top objective and priority for the European Union.

5 Engineering News Record (ENR) is a US weekly news publication that annually publishes the top international firms in construction.

6 Operation Car Wash (Portuguese: Operação Lava Jato) is an ongoing criminal investigation by the Federal Police of Brazil, Curitiba Branch.

References and further reading

Brautigam, D. (2020) China, the World Bank, and African Debt: A War of Words. *The Diplomat.* https://thediplomat.com/2020/08/china-the-world-bank-and-african-debt-a-war-of-words/ Accessed 18 January 2021.

Deringer, H., Erixon, F., Lamprecht, P. and Marel, E.V.D. (2018) *The Economic Impact of Local Content Requirements: A Case Study of Heavy Vehicles*. ECIPE Occasional Paper 1/2018. European Centre for International Political Economy. https://ecipe.org/publications/the-economic-impact-of-local-content-requirements/.

Global Trade Alert (2021) *Global Dynamics*. www.globaltradealert.org/global_dynamics/area_all. Accessed 20 January 2021.

Horn, S., Rienhart, C.M. and Trebesch, C. (2020) *How Much Money Does the World Owe China?* https://hbr.org/2020/02/how-much-money-does-the-world-owe-china. Accessed 18 January 2021.

Kirkpatrick, G. (2009) The Corporate Governance Lessons from the Financial Crisis. In: *Financial Market Trends* (Paris: OECD). www.oecd.org/finance/financial-markets/42229620.pdf.

Liu, M., Su, C., Wang, F. and Huang, L. (2020) Chinese Cross-Border M&As in the 'One Belt One Road' Countries: The Impact of Confucius Institutes. *China Economic Review*, **61**, June.

McKinsey (2016) *'One Belt and One Road': Connecting China and the World*. McKinsey & Company. www.globalinfrastructureinitiative.com/article/one-belt-and-one-road-connecting-china-and-world.

OECD (2008) *Glossary of Statistical Terms* (Paris: OECD).

OECD (2016) *The Economic Impact of Local Content Requirements*. Trade Policy Note, February. www.oecd.org/trade/topics/local-content-requirements/. Accessed 20 January 2021.

Springford, J. (2020) *Building 'European Champions': An Effective Way to Fend Off Chinese Competition?* (Centre for European Reform). www.cer.eu/in-the-press/building-european-champions-effective-way-fend-chinese-competition. Accessed 20 January 2021.

Transparency Market Research (2018) *Construction Waste Market - Global Industry Analysis, Size, Share, Growth, Trends, and Forecast 2017–2025*. https://www.transparencymarketresearch.com/construction-waste-market.html

WEF (2018) *Infrastructure and Urban Development: Industry Vision 2050*. World Economic Forum. www3.weforum.org/docs/IU_Industry_Vision_report_2018.pdf.

Index

Note: Page numbers in *italics* indicate a figure and page numbers in **bold** indicate a table on the corresponding page.

For Product Safety Concerns and Information please contact our EU
representative GPSR@taylorandfrancis.com
Taylor & Francis Verlag GmbH, Kaufingerstraße 24, 80331 München, Germany

www.ingramcontent.com/pod-product-compliance
Ingram Content Group UK Ltd.
Pitfield, Milton Keynes, MK11 3LW, UK
UKHW021724240926
475702UK00030BA/2681
9780367608910